ISAAC

Series on Analysis, Applications and Computation – Vol. 12

Metric Space Topology

Examples, Exercises and Solutions

Series on Analysis, Applications and Computation

ISSN: 1793-4702

Published

Vol. 12: *Metric Space Topology: Examples, Exercises and Solutions*
by W-S Cheung

Vol. 11: *Hardy Operators on Euclidean Spaces and Related Topics*
by S Lu, Z Fu, F Zhao & S Shi

Vol. 10: *Fractional Differential Equations and Inclusions: Classical and Advanced Topics*
by S Abbas, M Benchohra, J E Lazreg, J J Nieto & Y Zhou

Vol. 9: *Nonlinear Waves: A Geometrical Approach*
by P Popivanov & A Slavova

Vol. 8: *The Linearised Dam-Break Problem*
by D J Needham, S McGovern & J A Leach

Vol. 7: *The "Golden" Non-Euclidean Geometry: Hilbert's Fourth Problem, "Golden" Dynamical Systems, and the Fine-Structure Constant*
by A Stakhov & S Aranson, Assisted by S Olsen

Vol. 6: *An Introduction to Pseudo-Differential Operators Third Edition*
by M W Wong

Vol. 5: *Asymptotic Behavior of Generalized Functions*
by S Pilipović, B Stanković & J Vindas

More information on this series can be found at http://www.worldscientific.com/series/saac

ISAAC

Series on Analysis, Applications and Computation – Vol. 12

Metric Space Topology

Examples, Exercises and Solutions

Wing-Sum Cheung

The University of Hong Kong, Hong Kong

World Scientific

NEW JERSEY • LONDON • SINGAPORE • BEIJING • SHANGHAI • HONG KONG • TAIPEI • CHENNAI • TOKYO

Published by

World Scientific Publishing Co. Pte. Ltd.
5 Toh Tuck Link, Singapore 596224
USA office: 27 Warren Street, Suite 401-402, Hackensack, NJ 07601
UK office: 57 Shelton Street, Covent Garden, London WC2H 9HE

Library of Congress Cataloging-in-Publication Data
Names: Cheung, Wing-Sum (Mathematician), author.
Title: Metric space topology : examples, exercises and solutions /
Wing-Sum Cheung, the University of Hong Kong, Hong Kong.
Description: New Jersey : World Scientific, [2024] | Series: Series on analysis, applications
and computation, 1793-4702 ; Vol. 12 | Includes bibliographical references and index.
Identifiers: LCCN 2023023176 | ISBN 9789811266973 (hardcover) |
ISBN 9789811266980 (ebook for institutions) | ISBN 9789811266997 (ebook for individuals)
Subjects: LCSH: Metric spaces--Problems, exercises, etc.
Classification: LCC QA611.28 .C44 2024 | DDC 514/.325--dc23/eng20230919
LC record available at https://lccn.loc.gov/2023023176

British Library Cataloguing-in-Publication Data
A catalogue record for this book is available from the British Library.

First published 2023 (hardcover)
Reprinted 2025 (in paperback edition)
ISBN 9789819819232 (pbk)

For any available supplementary material, please visit
https://www.worldscientific.com/worldscibooks/10.1142/13160#t=suppl

Desk Editors: Nimal Koliyat/Lai Fun Kwong

Typeset by Stallion Press
Email: enquiries@stallionpress.com

To my parents
in heaven

Preface

This is a volume on the *Topology of Metric Spaces*, which is a subject the author has been teaching in the last couple of decades. It can serve as a textbook or a reference on the subject concerned, mainly for advanced level undergraduates concentrating on Mathematics, Physics, Economics and Finance, etc. A common serious drawback of most existing references/textbooks at this level is the lack of worked examples/exercises with detailed solutions. So the readers are frequently frustrated for not being able to solve the exercises or not being able to tell whether their solutions are valid. To overcome this drawback, it is the purpose of the present volume to provide plenty of worked examples and exercises, including True-or-False type questions and open-ended questions, with detailed solutions. True-or-False type questions and open-ended questions are particularly effective in helping the readers to get a birds eye view of the subject and to master the materials. They are also instrumental in nurturing the mathematical insight and the mathematical maturity of the readers.

The emphasis of most standard textbooks or reference books on the subject is the abstract and theoretical aspect, but such an approach may not be readily digestible by most readers. In order to master a new concept, it is most effective to first keep in mind a few simple, concrete and familiar examples, and then when dealing with a problem in the general setting, consider first the problem in such simple concrete settings. In many situations, the treatment of a problem in a simple setting could serve as a clue to tackle the problem in the general setting. On the other hand, pictorization or visualization of abstract problems into simple pictures is very often instrumental to the rigorous treatment of the problems. However, this is rarely provided in most standard references. In view of this, in this volume, within each section in each chapter, blended with the concise and rigorous treatments on the materials and before tackling problems in the more abstract settings, there will be a number of concrete examples/exercises, supplemented with simple pictorizations as appropriate, to help the readers master the concepts.

The learning outcomes of the volume include:

- Demonstrate knowledge and understanding of the basic features of mathematical analysis and point set topology (e.g., able to identify objects that are topologically equivalent).
- Apply knowledge and skills acquired in mathematical analysis to analyze and handle novel situations in a critical way (e.g., able to determine whether a specific function is uniformly continuous).
- Think creatively and laterally to generate innovative examples and solutions to non-standard problems (e.g., able to construct counterexamples to inaccurate mathematical statements).
- Acquire sufficient background for further studies in Functional Analysis, Real Analysis, Complex Analysis, Differential and Algebraic Geometry, Probability Theory, Mathematical Physics, Engineering, Economics and Finance, etc.

Throughout the 30+ years of teaching experience of the author, the most frequently asked question is "What is the use of the subject?" To be absolutely frank, many branches of pure mathematics do not find any direct application in the real world whatsoever. In fact, except for a small portion of the students who would go on for further studies and take mathematics research as their careers, most theorems or results they have learnt in their mathematics courses will not be seen or used for the rest of their lives. So it is perfectly fine for them to forget all the mathematics contents they have learnt at school or in college. But one thing they should never forget is the study process of and the way of thinking in mathematics, that is, the mathematics training they have undergone. It is not the mathematical contents they have learnt that are important but the mathematics training they received. Proper mathematics training nurtures logical arguments, analytical power, critical thinking, and rational thinking. With these attributes, one could excel in any career one may go into. So the target of this volume is, through numerous worked examples and exercises, to guide the readers argue logically, think critically, analyze various possible scenarios, and to nurture a mathematical and logical mind that is able to tackle novel and ill-posed problems they may encounter in the future. With that, the so-called life long self-learning would no longer be a burden but instead, would turn into a basic instinct.

The volume is perfect as a textbook or a reference for a one-semester course on Introduction to Metric Space Topology. It is the intention of the author to make it concise instead of stuffing it with unessential materials, as the latter may distract the readers from the natural flow of the development of the subject. Less is more. For advanced undergraduate level students, mastering the central theme of a subject would be much more beneficial than dabbling in a wider range of materials shallowly and prematurely.

Finally, the author would like to express his gratitude to the many students who have taken his course on the subject. Their enthusiastic comments and feedback on the teaching materials have provided strong incentive for his determination to write up this book. The professional assistance rendered by the staff of World Scientific, especially Lai Fun Kwong, is also gratefully acknowledged.

A Note on the Convention

There are two different conventions for the notations of "subset" and "proper subset". In this volume, we will be exclusively adopting the following convention:

$$\begin{aligned} A \text{ is an arbitrary subset of } X: &\quad A \subset X \\ A \text{ is a proper subset of } X: &\quad A \subsetneqq X \end{aligned}$$

About the Author

Wing-Sum Cheung was a full Professor of the Department of Mathematics of the University of Hong Kong before he retired in 2022. He is currently the Director of Undergraduate Admissions of the Faculty of Science and an Honorary Professor of the Department of Mathematics of the University of Hong Kong. He holds a BSc (with 1st class honors) from the Chinese University of Hong Kong, an MA and a PhD from Harvard University, USA. He has served as Head of Department of Mathematics and Associate Dean of the Faculty of Science of the University of Hong Kong, Vice-President of the Hong Kong Mathematical Society, Council Member of the Southeast Asian Mathematical Society, Council Member of the Hong Kong Institute of Science, Leader of the Hong Kong International Olympiad Team, and Honorary Consultant of the Ministry of Education, Youth and Sports of the Government of Cambodia. He has published over 200 journal articles, conference proceedings and book chapters, in which over 140 appeared in ISI journals. He has been named a top 1% highly cited researcher in the world by Clarivate Analytics' Essential Science Indicator for 6 times in the last decade. He is on the editorial board of a number of international mathematical journals including *Abstract*

and Applied Analysis, Asian European Journal of Mathematics, Far East Journal of Mathematical Sciences, Journal of Inequalities and Applications, etc. His current research interests include Differential Geometry, Exterior Differential Systems, Calculus of Variations, Analytic Inequalities, and Differential Equations.

Contents

Chapter 1

Metric Spaces

In this chapter, the basic concept of metric spaces will be introduced. Naively, they are simply nonempty sets equipped with a structure called *metric*. For the less matured students, at the beginning, this structure may appear to be a bit abstract and difficult to master. But in practice, this seemingly new concept is nothing more than a tiny little abstractization of the familiar space $\mathbb{R}^n$ and so all one needs to do is that whenever one needs to work on a problem in an abstract metric space, one first looks at the problem on $\mathbb{R}^n$, then one would be able to see the clue of how to proceed in the general case. In fact, in general, the most effective way to master a new concept in any branch of mathematics is to keep in mind a couple of typical concrete examples and think of these examples all the time. It is just that easy.

1.1 Definitions and Examples

Definition 1.1.1. *Let X be a nonempty set. A metric on X is a real-valued function*

$$d : X \times X \to \mathbb{R}$$

satisfying

(M1) $d(x, y) \geq 0$ *and* $d(x, y) = 0$ *if and only if* $x = y$,

(M2) (*symmetry*) $d(x, y) = d(y, x)$,

(M3) (*triangle inequality*) $d(x, y) \leq d(x, z) + d(z, y)$

for all $x, y, z \in X$. Given $x, y \in X$, $d(x, y)$ is also known as the distance between x and y with respect to d. The pair (X, d) is called a metric space and elements in X are referred to as points in X. For the sake of convenience, in case there is a clearly defined metric d on X, we shall simply call X a metric space.

Example 1.1.2.

(i) Let $X := \{a, b, c\}$, consider d_1, d_2, and $d_3 : X \times X \to \mathbb{R}$ given by

d_1	a	b	c
a	0	2	3
b	2	0	4
c	3	4	0

;

d_2	a	b	c
a	0	1	2
b	1	0	3
c	2	3	0

;

d_3	a	b	c
a	0	1	2
b	1	0	4
c	2	4	0

.

Then d_1 and d_2 are well-defined metrics on X but d_3 is not (why?)

(ii) Let X be any nonempty set and $d(x, y) := 1 - \delta_{xy}$ for any x, $y \in X$, where δ_{xy} is the *Kronecker delta function* defined by

$$\delta_{xy} := \begin{cases} 0 & \text{if } x \neq y \\ 1 & \text{if } x = y . \end{cases}$$

Then d is a well-defined metric called the *discrete metric* and (X, d) is called a *discrete metric space*. Notice that in a discrete metric space, all distinct points have the same distance 1, no matter how "far" or "close" they are from each other.

(iii) Let $X := \mathbb{R}^n$ and $d_e(x, y) := \|x - y\|$ for any x, $y \in \mathbb{R}^n$, where as usual, $\| \ \|$ stands for the Euclidean norm of $\mathbb{R}^n$, that is, $\|z\| := \sqrt{\sum_{i=1}^{n} z_i^2}$ for any $z = (z_1, \ldots, z_n) \in \mathbb{R}^n$. Hence

$$d_e(x, y) = \sqrt{\sum_{i=1}^{n} (x_i - y_i)^2}$$

for any $x = (x_1, \ldots, x_n)$, $y = (y_1, \ldots, y_n) \in \mathbb{R}^n$. Then d_e is a well-defined metric called the *Euclidean metric* on $\mathbb{R}^n$ and $(\mathbb{R}^n, d_e)$ is called a *Euclidean metric space*. In general, when we refer to the "Euclidean space $\mathbb{R}^n$", we mean the Euclidean metric space $(\mathbb{R}^n, d_e)$. In the sequel, unless it is specifically mentioned, all $\mathbb{R}^n$'s will be equipped with the Euclidean metric d_e and so sometimes we will be sloppy and write d for d_e and $\mathbb{R}^n$ for $(\mathbb{R}^n, d_e)$ in case no confusion may arise.

(iv) Let $X := \mathbb{C}$ and $d_{\mathbb{C}}(x, y) := |x - y|$ for any x, $y \in \mathbb{C}$, where, as usual, $|z| :=$ the modulus of $z \in \mathbb{C}$. Then $d_{\mathbb{C}}$ is a well-defined metric, and we will see that $(\mathbb{C}, d_{\mathbb{C}})$ is "equivalent" to the Euclidean space $(\mathbb{R}^2, d_e)$. In fact, it is easy to see that under the usual identification $C \cong \mathbb{R}^2$, any x, $y \in \mathbb{C}$ can be considered as points in $\mathbb{R}^2$, and under this identification, $d_{\mathbb{C}}(x, y) = d_e(x, y)$.

(v) Let $X := \mathbb{R}^2$ and $\tilde{d}(x, y) := \sqrt{100(x_1 - y_1)^2 + (x_2 - y_2)^2}$ for any $x = (x_1, x_2)$ and $y = (y_1, y_2) \in \mathbb{R}^2$. Then $\tilde{d}$ is a well-defined metric on X. Note that as real-valued functions of $X \times X$ into $\mathbb{R}$, $\tilde{d} \neq d_e$.

(vi) Let $X := S^1 \subset \mathbb{R}^2$ and $d(x, y) :=$ the arc length of the smaller arc in S^1 joining x and $y \in X$. Then d is a well-defined metric on X. Note that $d \neq d_e$ as real-valued functions of $X \times X$ into $\mathbb{R}$. This is a typical example of "intrinsic metric" that can be defined on a metric space. It is called intrinsic as the distance between two points can be measured without referring to the ambient space $\mathbb{R}^2$. In fact, imagine you are an ant walking on a rubber band. Then for you, the "distance" between two points A, B on the rubber band is $d(A, B)$ instead of $d_e(A, B)$, because the latter is the Euclidean distance between the two points A, B which is attained by the straight line segment joining the two points, but for you, the ant, unless you can fly, you cannot travel from A to B along the straight line segment joining the two points.

(vii) Let $X := S^2 \subset \mathbb{R}^3$ and x, $y \in X$. Observe that if $x \neq \pm y$, then there is a unique "great circle" on X passing through x, y. Here, a great circle means a planar circle on S^2 with radius 1. Define $d(x, y) :=$ the arc length of the smaller arc on the unique great circle in S^2 joining x and y. If $x = y$, we define $d(x, y) := 0$. Finally, if $x = -y$, then they are antipodal points on S^2 and so they are joined by infinitely many great circles. In this case we define $d(x, y) := \pi =$ half of the arc length of *any* such great circles. Then, similar to Example (vi), d is a well-defined intrinsic metric on X.

(viii) Let $X := \mathbb{R}^n$ and $d_S : X \times X \to \mathbb{R}$ be defined by

$$d_S(x, y) := \sum_{i=1}^{n} |x_i - y_i|$$

for any $x = (x_1, \dots, x_n)$, $y = (y_1, \dots, y_n) \in \mathbb{R}^n$. Then d_S is a well-defined metric on $\mathbb{R}^n$. Note that $d_S \neq d_e$.

(ix) Let $X := \mathbb{R}^n$ and $d_\infty : X \times X \to \mathbb{R}$ be defined by

$$d_\infty(x, y) := \max\{|x_i - y_i| : i = 1, \dots, n\}$$

for any $x = (x_1, \dots, x_n)$, $y = (y_1, \dots, y_n) \in \mathbb{R}^n$. Then d_∞ is a well-defined metric on $\mathbb{R}^n$. Note that $d_e \neq d_\infty \neq d_S$.

(x) Let $X := \mathbb{R}^n$. For any $k \in \mathbb{N}$, let $d_k : X \times X \to \mathbb{R}$ be defined by

$$d_k(x, y) := \left(\sum_{i=1}^{n} |x_i - y_i|^k \right)^{1/k}$$

for any $x = (x_1, \dots, x_n)$, $y = (y_1, \dots, y_n) \in \mathbb{R}^n$. Then d_k is a well-defined metric on $\mathbb{R}^n$. Note that $d_k \neq d_e$, d_S, nor d_∞.

(xi) Let $X := C[a, b]$. Clearly, X is a real vector space under pointwise operations. For any f, $g \in X$, define

$$d_1(f, g) := \int_a^b |f - g|(x)dx \ ,$$

where, as usual,

$$|f - g|(x) := |f(x) - g(x)| \quad \text{for any } x \in [a, b] \ .$$

Then d_1 is a well-defined metric and so (X, d_1) is a metric space. [See Exercise 1.1, Part B, Problem #4.]

(xii) Similar to Example (xi) above, let $X := C[a,b]$. For any f, $g \in X$, define

$$d_2(f,g) := \left[\int_a^b |f-g|^2(x)dx \right]^{\frac{1}{2}} .$$

Then d_2 is a well-defined metric and (X, d_2) is a metric space. [See Exercise 1.1, Part B, Problem #4.]

(xiii) Again, let $X := C[a,b]$. For any $f, g \in X$, define

$$d_\infty(f,g) := \sup\{|f(x)-g(x)| : \ x \in [a,b]\} .$$

Then (X, d_∞) is a metric space. Clearly, d_∞ is different from d_p, $p = 1, 2$, and as we will see in the sequel (see, for example, Exercise 1.2, Part B, Problem #19 for details), (X, d_∞), (X, d_1), and (X, d_2) are very different metric spaces.

(xiv) Let ℓ^∞ be the space of all bounded sequences of real numbers. So an element in ℓ^∞ is of the form $x = \{x_n\}_{n\in\mathbb{N}}$ with $\sup\{|x_n| : n \in \mathbb{N}\} < \infty$. For any elements $x = \{x_n\}_{n\in\mathbb{N}}$ and $y = \{y_n\}_{n\in\mathbb{N}} \in X$, define

$$d(x,y) := \sup\{|x_n - y_n| : n \in \mathbb{N}\} .$$

Then it is elementary to verify that (ℓ^∞, d) is a metric space.

Remark. From Example 1.1.2, we see that on the same nonempty set, different metrics could be defined and they could in turn make the same underlying set into metric spaces with very different properties.

Definition 1.1.3. *Let (X,d) be a metric space, $x \in X$, and $A \subset X$ be a nonempty subset. The distance from x to A is defined as*

$$d(x,A) := \inf\big\{d(x,a) : a \in A\big\}$$

and the diameter of A is defined as

$$d(A) := \sup\{d(a_1, a_2) : a_1, a_2 \in A\}.$$

Note that $d(A)$ can be $+\infty$. For the sake of convenience, we extend the notion of diameter to an empty set by defining $d(\phi) := -\infty$ (see Exercise 1.1, Part B, Problem #1.) A is said to be bounded if $d(A) \neq \infty$. A mapping from any nonempty set into X is said to be bounded if its image is bounded.

Example 1.1.4. In $(\mathbb{R}^3, d)$, where as usual, d stands for the Euclidean metric d_e, we have

$$\begin{aligned} d\big((2,0,0), S^2\big) &= 1, \\ d(S^2) &= 2. \end{aligned}$$

Note that both values are *attained* in S^2. That is, there exist a point $p \in S^2$ (namely, $p = (1,0,0)$) such that $d\big((2,0,0), p\big) = d\big((2,0,0), S^2\big) = 1$ and a pair of points r, $s \in S^2$ (any pair of antipodal points in S^2 will do) such that $d(r,s) = d(S^2) = 2$.

Example 1.1.5. In $(\mathbb{R}, d)$, if $A := (0,1)$, we have

$$\begin{aligned} d(2, A) &= 1, \\ d(A) &= 1. \end{aligned}$$

Note that neither of these values is attained in A.

Definition 1.1.6. *Let (X, d) be a metric space, and A, B be nonempty subsets of X. The distance between A and B is*

$$d(A, B) := \inf\{d(a, b) : a \in A,\ b \in B\}.$$

Clearly, if $A \cap B \neq \phi$, then $d(A, B) = 0$. But the converse is in general not true.

Example 1.1.7. In $(\mathbb{R}, d)$, let $A := (0,1)$, $B := [3,4]$, and $C := (4,\infty)$. Then $d(A,B) = 2$ and $d(B,C) = 0$. Note, however, that neither of the infima is attained.

Example 1.1.8. A good source of metric spaces is the "inner-product spaces". In general, an *inner-product space* is an ordered pair $(V, \langle\ ,\ \rangle)$, where V is a vector space and $\langle\ ,\ \rangle$ is a function

$$\langle\ ,\ \rangle : V \times V \to \mathbb{R}$$

called the *inner-product* in V satisfying

(I1) $\langle x, x\rangle > 0$ for all $0 \neq x \in V$,

(I2) $\langle x, y\rangle = \langle y, x\rangle$ for all $x, y \in V$,

(I3) $\langle\ ,\ \rangle$ is linear in the first argument, that is,

$$\langle \alpha_1 x_1 + \alpha_2 x_2, y\rangle = \alpha_1\langle x_1, y\rangle + \alpha_2\langle x_2, y\rangle$$

for any x_1, x_2, $y \in V$ and any α_1, $\alpha_2 \in \mathbb{R}$.

Note that by (I2) and (I3), $\langle x, 0\rangle = \langle 0, x\rangle = 0$ for all $x \in V$ and $\langle\ ,\ \rangle$ is bilinear. If $(V, \langle\ ,\rangle)$ is an inner-product space, define

$$\|\quad\| : V \to \mathbb{R}$$

by

$$\|x\| = \langle x, x\rangle^{\frac{1}{2}} .$$

Then $\|\quad\|$ satisfies

(N1) $\|x\| \geq 0$ for all $x \in V$ and $\|x\| > 0$ if $0 \neq x \in V$,

(N2) $\|\alpha x\| = |\alpha|\,\|x\|$ for all $\alpha \in \mathbb{R}$, $x \in V$,

(N3) $\|x + y\| \leq \|x\| + \|y\|$ for all $x, y \in V$.

Note that by (N2), we have $\|0\| = 0$. In general, any vector space V together with a function $\|\quad\| : V \to \mathbb{R}$ satisfying (N1)–(N3) is called a *normed vector space* and the function $\|\quad\|$ is called the *norm* of V. Hence in particular, every inner-product space is automatically a normed vector space.

Now if $(V, \| \ \|)$ is a normed vector space, define

$$d : V \times V \to \mathbb{R}$$

by

$$d(x, y) = \|x - y\| \quad \text{for any } x, y \in V.$$

Then it is easily checked that d satisfies (M1) – (M3) and thus (V, d) is a metric space. In general, we have the relations

$$\begin{aligned}\{\text{inner product spaces}\} &\subsetneqq \{\text{normed vector spaces}\} \\ &\subsetneqq \{\text{metric spaces}\}.\end{aligned}$$

For instance, each $\mathbb{R}^n$ is naturally an inner product space and it gives rise to the usual metric space $\mathbb{R}^n$. Similarly, $C[a, b]$ is an inner product space with

$$\langle f, g\rangle := \int_a^b f(x)g(x)dx \quad \text{for any } f, g \in C[a, b]$$

and it gives rise to the metric space in Example 1.1.2(xii).

Definition 1.1.9. *Let (X, d) be a metric space and $Y \subset X$ be a nonempty subset. If we denote by $d|_Y$ the restriction of d to Y, then $d|_Y$ is a well-defined metric on Y and hence $(Y, d|_Y)$ is also a metric space which is known as a metric subspace of X. For the sake of simplicity, unless ambiguity may arise, we normally drop the restriction notation and write $d|_Y$ simply as d, and call it the relative metric induced by d on Y.*

Example 1.1.10.

(i) Let $Y := (0, \infty) \subset \mathbb{R}$. Then (Y, d) is a metric subspace of $\mathbb{R}$ with $d = d_e$. In particular, for any x, $y \in Y$, $d_Y(x, y) = d_e(x, y) = |x - y|$.

(ii) Let $Y := \mathbb{N} \subset \mathbb{R}$. Then (Y, d) is a metric subspace of $\mathbb{R}$ with $d = d_e$. In particular, for any m, $n \in Y$, $d_Y(m, n) = d_e(m, n) = |m - n|$.

(iii) Let $X := S^1 \subset \mathbb{R}^2$. Using the intrinsic metric d defined in Example 1.1.2 (vi), (X, d) is a well-defined metric space. On the other hand, when considered as a subset of $\mathbb{R}^2$, X inherits the Euclidean metric d_e and so (X, d_e) is also a metric space, or more precisely, a metric subspace of $\mathbb{R}$. Observe that (X, d) is different from (X, d_e). For example, if we denote by n the "north pole" and s the "south pole" of S^1, then $d(n, s) = \pi$, while $d_e(n, s) = 2$.

For the rest of this chapter, unless otherwise specified, X will always stand for a metric space, and S, T, etc., will stand for arbitrary subsets of X.

Exercise 1.1

Part A: True or False Questions

For each of the following statements, determine if it is true or false. If it is true, prove it. If it is false, give a counterexample or provide proper justification.

1. Let X be a nonempty set and $d : X \times X \to \mathbb{R}$ a function satisfying

   (i) $d(x, y) = 0 \iff x = y$,

   (ii) $d(x, y) \leq d(x, z) + d(z, y)$

   for all x, y, $z \in X$. Then d is a metric on X.

   Answer: False.

   Example: Let $X := \{a, b, c\}$, with $d : X \times X \to \mathbb{R}$ defined by

d	a	b	c
a	0	1	1
b	2	0	1
c	1	1	0

   ,

   Then d satisfies (i) and (ii) but it fails to satisfy (M2).

2. Let X be a nonempty set and $d : X \times X \to \mathbb{R}$ a function satisfying

   (i) $d(x, y) = 0 \iff x = y$,

   (ii) $d(x, y) = d(y, x)$,

   (iii) $d(x, y) \leq d(x, z) + d(z, y)$

   for all x, y, $z \in X$. Then d is a metric on X.

   Answer: True.

   Proof. (ii) and (iii) are exactly (M2) and (M3). For (M1), for any x, $y \in X$, by (i), (iii) and (ii), we have

   $$0 = d(x, x) \leq d(x, y) + d(y, x) = 2d(x, y).$$

   Thus (M1) is also satisfied.

3. $d(x, y) := \sqrt{|x - y|}$ is a metric on $\mathbb{R}$.

   Answer: True.

Proof. (M1) and (M2) are clearly satisfied. For any x, y, $z \in \mathbb{R}$,

$$\begin{aligned} d(x,y)^2 &= |x-y| \\ &\leq |x-z| + |z-y| \\ &= d(x,z)^2 + d(z,y)^2 \\ &\leq \big[d(x,z) + d(z,y)\big]^2, \end{aligned}$$

hence (M3) is also satisfied.

4. The function $d : \mathbb{R}^2 \times \mathbb{R}^2 \to \mathbb{R}$ defined by

$$d((x_1,y_1),(x_2,y_2)) := \left(\sqrt{|x_1 - x_2|} + \sqrt{|y_1 - y_2|}\right)^2,$$

for any $(x_1,y_1),(x_2,y_2) \in \mathbb{R}^2$, is a metric on $\mathbb{R}^2$.

Answer: False.
Example: Take $(x_1,y_1) = (1,0)$ and $(x_2,y_2) = (0,1)$. Then

$$\begin{aligned} &d\big((x_1,y_1),(x_2,y_2)\big) \\ &\quad = 4 > 2 = d\big((x_1,y_1),(0,0)\big) + d\big((0,0),(x_2,y_2)\big) \end{aligned}$$

and so the triangle inequality is violated.

5. Let $X := \mathbb{R}^n$. Then $d(x,y) := \min\big\{|x_i - y_i| : i = 1, \ldots, n\big\}$ for any $x = (x_1, \ldots, x_n)$ and $y = (y_1, \ldots, y_n) \in \mathbb{R}^n$ is a well-defined metric on $\mathbb{R}^n$.

Answer: False.
Example: Take $n = 2$, $x := (1,0)$, $y := (-1,0)$. Then $x \neq y$ but $d(x,y) = 0$.

6. If d_1 and d_2 are metrics on X, then $d := \min\{d_1, d_2\}$ is a metric on X.

Answer: False.
Example: Let $X := \{p,q,r\}$ and d_1, d_2 be defined by

d_1	p	q	r
p	0	10	1
q	10	0	10
r	1	10	0

,

d_2	p	q	r
p	0	10	10
q	10	0	1
r	10	1	0

.

It is easily shown that d_1 and d_2 are well-defined metrics on X. However,

$$d(p,q) = \min\{d_1(p,q), d_2(p,q)\} = \min\{10, 10\} = 10$$
$$d(p,r) = \min\{d_1(p,r), d_2(p,r)\} = \min\{1, 10\} = 1$$
$$d(r,q) = \min\{d_1(r,q), d_2(r,q)\} = \min\{10, 1\} = 1$$

and so

$$d(p,q) = 10 \not\leq 2 = 1 + 1 = d(p,r) + d(r,q) \ .$$

7. $d(x,S) = 0$ if and only if $x \in S$.

   Answer: False.
   Example: Let $X := \mathbb{R}$, $S := (0,1)$, then $0 \notin S$ but $d(0,S) = \inf\{d(0,s) : s \in S\} = \inf\{s : 0 < s < 1\} = 0$.

8. $d(S,T) = 0$ if and only if $S \cap T \neq \phi$.

   Answer: False.
   Example: Let $X := \mathbb{R}$, $S := (0,1)$, $T := (1,2)$. Then $S \cap T = \phi$ but $d(S,T) = \inf\{d(s,t) : s \in S,\ t \in T\} = 0$.

9. If $S \cap T \neq \phi$, then $d(S \cup T) \leq d(S) + d(T)$.

   Answer: True.
   Proof. By assumption, there exists $x \in S \cap T$. For any p, $q \in S \cup T$, we have 4 cases.
   Case 1: p, $q \in S$.
   In this case we have $d(p,q) \leq d(S) \leq d(S) + d(T)$.
   Case 2: p, $q \in T$.
   In this case we have $d(p,q) \leq d(T) \leq d(S) + d(T)$.

Case 3: $p \in S$, $q \in T$.

In this case we have

$d(p,q) \leq d(p,x) + d(x,q) \leq d(S) + d(T)$.

Case 4: $p \in T$, $q \in S$.

In this case we have

$d(p,q) \leq d(p,x) + d(x,q) \leq d(T) + d(S)$.

Combining, we always have $d(p,q) \leq d(S) + d(T)$ and so $d(S \cup T) \leq d(S) + d(T)$.

10. If $S \cap T = \phi$, then $d(S \cup T) \geq d(S) + d(T)$.

Answer: False.

Example: Let $X := \mathbb{R}$, $S := \mathbb{Q} \cap [0,1]$, and $T := (\mathbb{R} \setminus \mathbb{Q}) \cap [0,1]$. Then $S \cap T = \phi$, but $d(S \cup T) = d([0,1]) = 1 \ngeq 2 = 1 + 1 = d(S) + d(T)$.

11. $d(S \cup T) \leq d(S) + d(T) + d(S,T)$.

Answer: True.

Proof. For any p, $q \in S \cup T$, similar to the considerations in Exercise 1.1, Part A, #9, it suffices to consider the case where $p \in S$ and $q \in T$. By the definition of infimum, for any $\varepsilon > 0$, there exist $s \in S$, $t \in T$ such that

$$d(S,T) \leq d(s,t) < d(S,T) + \varepsilon\ .$$

Hence

$$d(p,q) \leq d(p,s) + d(s,t) + d(t,q) \leq d(S) + d(S,T) + d(T) + \varepsilon\ .$$

Since this is true for all $p \in S$ and $q \in T$, we have

$$d(S \cup T) \leq d(S) + d(T) + d(S,T) + \varepsilon\ .$$

Since $\varepsilon > 0$ is arbitrary, we conclude that $d(S \cup T) \leq d(S) + d(T) + d(S,T)$.

Part B: Problems

1. In Definition 1.1.3, we defined $d(\phi) := -\infty$. Why?

   Solution: Recall that we defined $d(S) := \inf\{d(x, y) : x, y \in S\}$ for any $\phi \neq S \subset X$. So if we want to the extend the domain of definition of diameter so as to include the empty set, in order to make things defined in a "consistent" or "continuous" manner, we should define $d(\phi) := \inf\{d(x, y) : x, y \in \phi\} = \inf \phi$ and so we must first know what $\inf \phi$ should be. Now observe that for any A, $B \subset \mathbb{R}$, if $A \subset B$, we must have $\inf A \leq \sup B$. As $\phi \subset B$ for all $B \subset \mathbb{R}$, we must have $\inf \phi \leq \sup B$ for all $B \subset \mathbb{R}$. Hence this leaves us no choice but to define $\inf \phi := -\infty$ and so $d(\phi)$ must also be defined as $-\infty$.

2. Let (X, d) be a metric space and let $M > 0$. Define $\tilde{d} : X \times X \to \mathbb{R}$ by

$$\tilde{d}(x, y) := \min(M, d(x, y))$$

   for any x, $y \in X$. Is $(X, \tilde{d})$ a metric space?

   Answer: Yes.

   Proof. We need to verify the conditions (M1)-(M3).

   (M1): Since $d(x, y) \geq 0$ and $M > 0$, we have $\tilde{d}(x, y) \geq 0$ for any $x, y \in X$. Furthermore, for any x, $y \in X$, if $x = y$, then $\tilde{d}(x, y) = \min(M, d(x, y)) = \min(M, 0) = 0$. Conversely, if $\tilde{d}(x, y) = 0$, then as $M > 0$, we must have $d(x, y) = 0$ and so $x = y$. Hence (M1) is satisfied.

   (M2): It is easily seen that $\tilde{d}(x, y)$ is symmetric, hence (M2) is satisfied.

   (M3): For any x. y, $z \in X$,

   Case 1: $\tilde{d}(x, z) \geq M$. In this case we have

$$\begin{aligned}\tilde{d}(x, y) &= \min(M, d(x, y)) \\ &\leq M \leq \tilde{d}(x, z) \leq \tilde{d}(x, z) + \tilde{d}(y, z) .\end{aligned}$$

Case 2: $\tilde{d}(y,z) \geq M$. Similar to Case 1, we have

$$\begin{aligned}\tilde{d}(x,y) &= \min(M, d(x,y)) \\ &\leq M \leq \tilde{d}(y,z) \leq \tilde{d}(x,z) + \tilde{d}(y,z) \ .\end{aligned}$$

Case 3: $\tilde{d}(x,z) < M$ and $\tilde{d}(y,z) < M$. In this case we have

$$\tilde{d}(x,z) = d(x,z) \quad \text{and} \quad \tilde{d}(y,z) = d(y,z)$$

and so

$$\begin{aligned}\tilde{d}(x,y) &= \min(M, d(x,y)) \leq d(x,y) \\ &\leq d(x,z) + d(y,z) = \tilde{d}(x,z) + \tilde{d}(y,z) \ .\end{aligned}$$

Combining, we see that (M3) also holds.

3. Determine whether the following functions are metrics on $\mathbb{R}$.
   (a) $d(x,y) := [\,|x-y|\,]$.
   Here, the *Gaussian symbol* $[\,\cdot\,] : \mathbb{R} \to \mathbb{Z}$ is defined as $[x] :=$ the largest integer that is less than or equal to x, for any $x \in \mathbb{R}$.
   (b) $d(x,y) := (x-y)^2$.
   (c) $d(x,y) := |f(x) - f(y)|$, where $f(x) := \frac{x}{1+|x|}$.
   (d) $d(x,y) := \left|\tan^{-1} x - \tan^{-1} y\right|$.

Answer:
   (a) No.
   Justification. Since $d\left(1, \frac{1}{2}\right) = \left[\,\left|1 - \frac{1}{2}\right|\,\right] = \left[\frac{1}{2}\right] = 0$ but $1 \neq \frac{1}{2}$, (M1) is violated.
   (b) No.
   Justification. Since
   $d(0,2) = (2-0)^2 = 4 \not\leq 2 = 1^2 + 1^2 = d(0,1) + d(1,2)$, (M3) is violated.
   (c) Yes.

Proof. It is clear that $d(x,y) \geq 0$ and $d(x,x) = 0$ for all x, $y \in \mathbb{R}$. Conversely, if $d(x,y) = 0$, then

$$\frac{x}{1+|x|} = f(x) = f(y) = \frac{y}{1+|y|} \; . \qquad (*)$$

Taking absolute values on both sides and simplify, we have $|x| = |y|$. Putting it back to (*), we have $x = y$ and so (M1) is satisfied. (M2) is obvious. (M3) follows immediately from the usual triangle inequality for absolute value:

$$\begin{aligned} d(x,y) &= |f(x) - f(y)| \\ &\leq |f(x) - f(z)| + |f(z) - f(y)| \\ &= d(x,z) + d(z,y) \; . \end{aligned}$$

(d) Yes.

Proof. (M1) and (M2) are obvious. (M3) follows immediately from the usual triangle inequality for absolute value.

4. Prove Example 1.1.2 (xi) and (xii), that is, on $C[a,b]$,

$$d_p(f,g) := \left[\int_a^b |f-g|^p(x)dx \right]^{\frac{1}{p}} , \quad p = 1, 2 \; ,$$

are well-defined metrics. How about

$$d(f,g) := \int_a^b (f-g)^2(x)dx \; ?$$

Solution: For $p = 1$, it is clear that $d_1(f,g) \geq 0$ for all f, $g \in C[a,b]$ and $d_1(f,g) = 0$ for $f = g$. Furthermore, if $d_1(f,g) = 0$, then

$$\int_a^b |f-g|(x)dx = 0 \; .$$

Since $|f-g| \geq 0$ and is continuous on $[a,b]$, this forces $f - g = 0$ on $[a,b]$ and so $f = g$. Hence (M1) is satisfied. Next, it is obvious that (M2) also holds. Finally, for (M3), let f, g, $h \in X$, we have

$$\begin{aligned} d_1(f,g) &= \int_0^1 \big|f(x) - g(x)\big| dx \\ &\le \int_0^1 \big|f(x) - h(x)\big| + \Big|h(x) - g(x)\Big| dx \\ &= \int_0^1 \big|f(x) - h(x)\Big| dx + \int_0^1 \big|h(x) - g(x)\big| dx \\ &= d(f,h) + d(h,g) \ . \end{aligned}$$

Hence (M3) also holds and so d_1 is a metric.

Next, for $p = 2$, it is clear that $d_2(f,g) \ge 0$ for all f, $g \in C[a,b]$ and $d_2(f,g) = 0$ for $f = g$. Furthermore, if $d_2(f,g) = 0$, then

$$\left[\int_a^b (f-g)^2(x)dx\right]^{\frac{1}{2}} = 0 \ .$$

By the continuity of $f - g$, this forces $f - g = 0$ on $[a,b]$ and so $f = g$. Hence (M1) is satisfied. Next, it is obvious that (M2) also holds. Finally, for (M3), let f, g, $h \in X$. By the elementary Cauchy–Schwarz inequality, we have

$$\begin{aligned} d_2^2(f,g) &= \int_0^1 (f(x) - g(x))^2 dx \\ &= \int_0^1 (f(x) - h(x) + h(x) - g(x))^2 dx \\ &= \int_0^1 (f(x) - h(x))^2 dx + \int_0^1 (h(x) - g(x))^2 dx \\ &\quad + 2\int_0^1 (f(x) - h(x))(h(x) - g(x)) dx \\ &\le \int_0^1 (f(x) - h(x))^2 dx + \int_0^1 (h(x) - g(x))^2 dx \\ &\quad + 2\sqrt{\int_0^1 (f(x) - h(x))^2 dx \int_0^1 (h(x) - g(x))^2 dx} \\ &= (d(f,h) + d(h,g))^2 \ . \end{aligned}$$

Hence (M3) also holds. So d_2 is a metric.

However, $d(f,g) := \int_a^b (f-g)^2(x)dx$ is not a metric, as triangle inequality (M3) is violated. For instance, take $f :\equiv 1$, $g :\equiv 2$, and $h :\equiv 3$ on $[0,1]$. Then

$$\begin{aligned} d(f,g) + d(g,h) &= \int_0^1 1^2\,dx + \int_0^1 1^2\,dx = 1+1 = 2 \\ &< 4 = \int_0^1 2^2\,dx = d(f,h)\ . \end{aligned}$$

5. Let $\{(X_n, d_n)\}_{n\in\mathbb{N}}$ be a sequence of metric spaces and $\{c_n\}_{n\in\mathbb{N}}$ a sequence of positive numbers. Fix $N \in \mathbb{N}$. On the product space $P_N := \prod_{n=1}^N X_n$, define

$$d'_N(x,y) := \sum_{n=1}^N c_n d_n(x_n, y_n)$$

for any $x = (x_1, \ldots, x_N)$, $y = (y_1, \ldots, y_N) \in P_N$.

(a) Prove that d'_N is a metric on P_N.

(b) Prove that $d''_N(x,y) := \sum_{n=1}^N \frac{c_n d_n(x_n,y_n)}{1 + d_n(x_n,y_n)}$ is a metric on P_N.

(c) Prove that $d'''_N(x,y) := \sup_{1\le n\le N} d_n(x_n,y_n)$ is also a metric on P_N.

Proof.

(a) We need to show (M1)-(M3).

(M1): For any x, $y \in P_N$ and every $n = 1, \ldots, N$, since d_n is a metric on X_n, all $d_n(x_n, y_n)$ are non-negative real numbers. Hence being a finite sum of non-negative real numbers,

$d'_N(x,y) \geq 0$. Furthermore, for any x, $y \in P_N$,

$$\begin{aligned} d'_N(x,y) = 0 &\iff \sum_{n=1}^{N} c_n d_n(x_n, y_n) = 0 \\ &\iff d_n(x_n, y_n) = 0 \quad \text{for all } n = 1, \ldots, N \\ &\iff x_n = y_n \quad \text{for all } n = 1, \ldots, N \\ &\iff x = y \; . \end{aligned}$$

Hence (M1) is satisfied.

(M2): Trivial.

(M3): For any x, y, $z \in P_N$, since each d_n satisfies the triangle inequality, we have

$$\begin{aligned} d'_N(x,z) + d'_N(z,y) &= \sum_{n=1}^{N} c_n \big(d_n(x_n, z_n) + d_n(z_n, y_n)\big) \\ &\geq \sum_{n=1}^{N} c_n d_n(x_n, y_n) \\ &= d'_N(x,y) \; . \end{aligned}$$

(b) (M1): Similar to the considerations in (a) above, $d''_N(x,y) \geq 0$ for any $x, y \in P_N$. Furthermore, for any x, $y \in P_N$,

$$\begin{aligned} d''_N(x,y) = 0 &\iff \sum_{n=1}^{N} \frac{c_n d_n(x_n, y_n)}{1 + d_n(x_n, y_n)} = 0 \\ &\iff d_n(x_n, y_n) = 0 \quad \text{for all } 1 \leq n \leq N \\ &\iff x_n = y_n \quad \text{for all } 1 \leq n \leq N \\ &\iff x = y. \end{aligned}$$

(M2): Trivial.

(M3): First of all, it is easily checked that for $0 \leq a \leq b$,

$\dfrac{a}{1+a} \le \dfrac{b}{1+b}$. Hence for all $x,\ y,\ z \in P_N$,

$$\begin{aligned}
d''_N(x,y) &= \sum_{n=1}^{N} \frac{c_n d_n(x_n,y_n)}{1+d_n(x_n,y_n)} \\
&\le \sum_{n=1}^{N} \frac{c_n d_n(x_n,z_n)+c_n d_n(z_n,y_n)}{1+d_n(x_n,z_n)+d_n(z_n,y_n)} \\
&= \sum_{n=1}^{N} \frac{c_n d_n(x_n,z_n)}{1+d_n(x_n,z_n)+d_n(z_n,y_n)} \\
&\qquad + \sum_{n=1}^{N} \frac{c_n d_n(z_n,y_n)}{1+d_n(x_n,z_n)+d_n(z_n,y_n)} \\
&\le \sum_{n=1}^{N} \frac{c_n d_n(x_n,z_n)}{1+d_n(x_n,z_n)} + \sum_{n=1}^{N} \frac{c_n d_n(z_n,y_n)}{1+d_n(z_n,y_n)} \\
&= d''_N(x,z)+d''_N(z,y).
\end{aligned}$$

(c) (M1): Again, it is easily seen that $d'''_N(x,y) \ge 0$ for any $x, y \in P_N$. Furthermore, for any $x,\ y \in P_N$,

$$\begin{aligned}
d'''_N(x,y) = 0 &\iff \sup_{1\le n\le N} d_n(x_n,y_n) = 0 \\
&\iff d_n(x_n,y_n) = 0 \quad \text{for all } 1 \le n \le N \\
&\iff x_n = y_n \quad \text{for all } 1 \le n \le N \\
&\iff x = y.
\end{aligned}$$

(M2): Trivial.

(M3): For all $x,\ y,\ z \in P_N$,

$$\begin{aligned}
d'''_N(x,y) &= \sup_{1\le n\le N} d_n(x_n,y_n) \\
&\le \sup_{1\le n\le N} \left(d_n(x_n,z_n)+d_n(z_n,y_n)\right) \\
&\le \sup_{1\le n\le N} d_n(x_n,z_n) + \sup_{1\le n\le N} d_n(z_n,y_n) \\
&= d'''_N(x,z)+d'''_N(z,y).
\end{aligned}$$

6. Let $d_k (k = 1, 2)$ and d_∞ be metrics on $\mathbb{R}^n$ defined by

$$d_k(x,y) := \left(\sum_{i=1}^{n} |x_i - y_i|^k\right)^{1/k}, \quad k = 1, 2,$$
$$d_\infty(x,y) := \max_{i=1,\dots,n} |x_i - y_i|,$$

for any $x = (x_1, \dots, x_n)$, $y = (y_1, \dots, y_n) \in \mathbb{R}^n$. Prove that

$$\frac{1}{n}\, d_1(x,y) \le \frac{1}{\sqrt{n}}\, d_2(x,y) \le d_\infty(x,y) \le d_2(x,y) \le d_1(x,y) .$$

Proof. By Cauchy–Schwarz inequality,

$$n \sum_{i=1}^{n} |x_i - y_i|^2 \ge \left(\sum_{i=1}^{n} |x_i - y_i|\right)^2 .$$

Taking the square root on both sides, we have

$$\sqrt{n}\, d_2(x,y) \ge d_1(x,y) .$$

This proves the first inequality. Next, we obviously have

$$\sum_{i=1}^{n} |x_i - y_i|^2 \le n \max_{i=1,\dots,n} |x_i - y_i|^2$$

from which we have the second inequality:

$$\frac{1}{\sqrt{n}}\, d_2(x,y) = \frac{1}{\sqrt{n}} \left(\sum_{i=1}^{n} |x_i - y_i|^2\right)^{1/2}$$
$$\le \max_{i=1,\dots,n} |x_i - y_i| = d_\infty(x,y) .$$

The third inequality $d_\infty \le d_2$ is obvious. Finally, it is easy to see that

$$\max_{i=1,\dots,n} |x_i - y_i|^2 \le \sum_{i=1}^{n} |x_i - y_i|^2 \le \left(\sum_{i=1}^{n} |x_i - y_i|\right)^2 ,$$

which is exactly the last inequality.

7. Let $X = C[0,1]$ be the set of continuous real-valued functions on $[0,1]$, and d_∞, $d_2 : X \times X \to \mathbb{R}$ be given by

$$d_\infty(f,g) := \sup_{x\in[0,1]} |f(x) - g(x)| \ ,$$

$$d_2(f,g) := \left(\int_0^1 (f(x) - g(x))^2 dx\right)^{1/2}$$

for all $f, g \in X$.

(a) Show that (X, d_∞) and (X, d_2) are metric spaces.

(b) Write $\mathbb{Q} \cap [0,1] = \{r_k : k \in \mathbb{N}\}$. Define

$$d(f,g) := \sum_{k=1}^{\infty} 2^{-k} \min\{|f(r_k) - g(r_k)|,\ 1\} \ .$$

Prove that (X, d) is a metric space.

Proof.

(a) (X, d_∞) is a metric space:
Clearly, $d_\infty(f,g) \geq 0$ and $d_\infty(f,f) = 0$ for all f, $g \in X$. Furthermore, if $d_\infty(f,g) = 0$, then for all $t \in [0,1]$,

$$0 \leq |f(t) - g(t)| \leq \sup_{x\in[0,1]} |f(x) - g(x)| = d_\infty(f,g) = 0 \ .$$

It follows that $f = g$, hence (M1) holds. Obviously, (M2) holds. For (M3), let f, g, $h \in X$. We have

$$\begin{aligned} d_\infty(f,g) &= \sup_{x\in[0,1]} |f(x) - g(x)| \\ &\leq \sup_{x\in[0,1]} \left(|f(x) - h(x)| + |h(x) - g(x)|\right) \\ &\leq \sup_{x\in[0,1]} |f(x) - h(x)| + \sup_{x\in[0,1]} |h(x) - g(x)| \\ &= d_\infty(f,h) + d_\infty(h,g). \end{aligned}$$

Hence (M3) also holds.

Next, for (X, d_2), (M1) and (M2) are clear. For (M3), let f, g, $h \in X$. By Cauchy–Schwarz inequality,

$$\begin{aligned}
d_2^2(f,g) &= \int_0^1 (f(x)-g(x))^2 dx \\
&= \int_0^1 (f(x)-h(x)+h(x)-g(x))^2 dx \\
&= \int_0^1 (f(x)-h(x))^2 dx + \int_0^1 (h(x)-g(x))^2 dx \\
&\quad + 2\int_0^1 (f(x)-h(x))(h(x)-g(x))dx \\
&\leq \int_0^1 (f(x)-h(x))^2 dx + \int_0^1 (h(x)-g(x))^2 dx \\
&\quad + 2\left[\int_0^1 (f(x)-h(x))^2 dx\right]^{\frac{1}{2}} \\
&\quad \times \left[\int_0^1 (h(x)-g(x))^2 dx\right]^{\frac{1}{2}} \\
&= [d(f,h)+d(h,g)]^2 .
\end{aligned}$$

Hence (M3) also holds.

(b) (M1): It is obvious that $d(f,g) \geq 0$ for all f, $g \in X$ and $d(f,f) = 0$. On the other hand, suppose $d(f,g) = 0$. If there exists $\ell \in \mathbb{N}$ such that $f(r_\ell) \neq g(r_\ell)$, then $d(f,g) \geq 2^{-\ell}|f(r_\ell) - g(r_\ell)| > 0$. Hence we must have $f(r_k) = g(r_k)$ for all $k \in \mathbb{N}$. That is, $f = g$ on $\{r_k : k \in \mathbb{N}\}$ which is dense in $[0,1]$. Since both f and g are continuous, this forces $f = g$ on $[0,1]$.

(M2): Trivial.

(M3): Observe first the following elementary inequality

$$\min\{a,1\} \leq \min\{b,1\} + \min\{c,1\} \tag{*}$$

for all a, b, $c \geq 0$ with $a \leq b + c$. Hence for any f, g, $h \in X$,

we have

$$\begin{aligned}
d(f,g) &= \sum_{k=1}^{\infty} 2^{-k} \min\{|f(r_k) - g(r_k)|,\ 1\} \\
&\le \sum_{k=1}^{\infty} 2^{-k} \left[\min\{|f(r_k) - h(r_k)|,\ 1\}\right. \\
&\qquad \left. + \min\{|h(r_k) - g(r_k)|,\ 1\}\right] \\
&= \sum_{k=1}^{\infty} 2^{-k} \min\{|f(r_k) - h(r_k)|,\ 1\} \\
&\qquad + \sum_{k=1}^{\infty} 2^{-k} \min\{|h(r_k) - g(r_k)|,\ 1\} \\
&= d(f,h) + d(g,h)\ .
\end{aligned}$$

Note that we have made used of the fact that as all terms in the series above are non-negative and so the order of the summation does not matter.

8. If d and d' are metrics on a nonempty set X, must $D := d \cdot d'$ also be a metric on X?

   Answer: False.

   Example: Take $X = \mathbb{R}$ and $d = d'$ to be the Euclidean metric. Then D fails to be a metric as it violates the triangle inequality (M3). In fact, for $x = 1$, $y = 2$, and $z = 3$, we have

$$D(x,z) = 2 \cdot 2 = 4 > 1 + 1 = D(x,y) + D(y,z)\ .$$

9. Determine whether each of the functions d below defines a metric on $\mathbb{R}^2$. In case d is a metric, describe $B(0,1)$. In case d is not a metric, determine which condition in the definition is violated.

   (a) $d(u,v) := \sqrt{(u-v)^T A(u-v)}$ for any $u,\ v \in \mathbb{R}^2$, where $A := \frac{1}{2}\begin{pmatrix} 7 & -5 \\ -5 & 7 \end{pmatrix}$.

(b) $d(u,v) := \sqrt{(u-v)^T A(u-v)}$ for any $u,\ v \in \mathbb{R}^2$, where $A := \frac{1}{2}\begin{pmatrix} -5 & 7 \\ 7 & -5 \end{pmatrix}$.

(c) $d(u,v) := \max\{d_1(u,v), d_2(u,v)\}$ for any $u, v \in \mathbb{R}^2$, where $d_1,\ d_2$ are metrics on $\mathbb{R}^2$.

(d) $d(u,v) := \min\{d_1(u,v), d_2(u,v)\}$ for any $u, v \in \mathbb{R}^2$, where $d_1,\ d_2$ are metrics on $\mathbb{R}^2$.

Answer:

(a) True, d is a metric.

Proof. Note that $A = \frac{1}{2}PDP^T$, with

$$D := \begin{pmatrix} 1 & 0 \\ 0 & 6 \end{pmatrix}$$

and

$$P := \begin{pmatrix} 1 & -1 \\ 1 & 1 \end{pmatrix}$$

is an invertible matrix. For any $u = \begin{pmatrix} u_1 \\ u_2 \end{pmatrix}$ and $v = \begin{pmatrix} v_1 \\ v_2 \end{pmatrix} \in \mathbb{R}^2$, define $s := u_1 - v_1,\ t := u_2 - v_2$. It follows that

$$\begin{aligned} d(u,v) &= \sqrt{\frac{1}{2}\begin{pmatrix} s+t \\ -s+t \end{pmatrix}^T \begin{pmatrix} 1 & 0 \\ 0 & 6 \end{pmatrix} \begin{pmatrix} s+t \\ -s+t \end{pmatrix}} \\ &= \sqrt{\frac{1}{2}[(s+t)^2 + 6(-s+t)^2]} \geq 0 \end{aligned}$$

with equality holds if and only if $s + t = 0$ and $s - t = 0$, or equivalently, if and only if $s = 0$ and $t = 0$. It follows that $d(u,v) = 0$ if and only if $u = v$. So (M1) holds. On the other hand, as

$$\begin{aligned} d(u,v) &= \sqrt{(u-v)^T A(u-v)} \\ &= \sqrt{(v-u)^T A(v-u)} = d(v,u)\ , \end{aligned}$$

(M2) also holds. Finally, before checking the triangle inequality, we first prove that for any $u, v \in \mathbb{R}^2$, we have

$$\left|u^T Av\right| \leq \sqrt{u^T Au}\,\sqrt{v^T Av}\,.$$

In fact, write $\lambda := \sqrt{u^T Au}$ and $\eta := \sqrt{v^T Av}$. Then

$$\begin{aligned}
0 &\leq d(\eta u, \lambda v)^2 \\
&= (\eta u - \lambda v)^T A\,(\eta u - \lambda v) \\
&= \eta^2 u^T Au - 2\lambda\,\eta\, u^T A\, v + \lambda^2\, v^T Av \\
&= 2\lambda^2\,\eta^2 - 2\lambda\,\eta\, u^T A\; v \\
&= 2\lambda\,\eta\,\left(\lambda\,\eta - u^T A\; v\right)
\end{aligned}$$

and hence $u^T Av \leq \lambda\,\eta$. Similarly, by considering $d(\eta\, u, -\lambda\, v)$, we have $-\lambda\,\eta \leq u^T Av$. Thus we have

$$\left|u^T Av\right| \leq \sqrt{u^T Au}\,\sqrt{v^T Av}$$

for any $u,\ v \in \mathbb{R}^2$.

Back to our problem. For any $u,\ v,\ w\ \in\ \mathbb{R}^2$, since A is symmetric, we have

$$\begin{aligned}
&d(u,v)^2 \\
&= (u-v)^T A(u-v) \\
&= \left[(u-w)+(w-v)\right]^T A\left[(u-w)+(w-v)\right] \\
&= (u-w)^T A(u-w) + 2(w-v)^T A(u-w) \\
&\qquad + (w-v)^T A(w-v) \\
&\leq d(u,w)^2 + d(w,v)^2 \\
&\qquad + 2\sqrt{(w-v)^T A(w-v)}\sqrt{(u-w)^T A(u-w)} \\
&= \left(d(u,w) + d(w,v)\right)^2 .
\end{aligned}$$

The triangle inequality follows. Hence d is a metric.

(b) False, d is not a metric.

Justification. In fact, for $u = \begin{pmatrix} 1 \\ 0 \end{pmatrix}$ and $v = \begin{pmatrix} 0 \\ 0 \end{pmatrix}$, we have

$$(u - v)^T A(u - v) = \frac{-5}{2} < 0$$

and $d(u, v)$ is not defined and in particular, d is not a metric.

(c) Yes, d is a metric.

Proof. The conditions $d(u, v) \geq 0$, $d(u, u) = 0$ and $d(u, v) = d(v, u)$ follow directly from that d_1 and d_2 are metrics. On the other hand, if $\max\{d_1(u, v), d_2(u, v)\} = d(u, v) = 0$, we must have $d_1(u, v) = d_2(u, v) = 0$ and so $u = v$. Finally, for any u, v, $w \in \mathbb{R}^2$, we have either $d(u, v) = d_1(u, v)$ or $d(u, v) = d_2(u, v)$. Without loss of generality, assume that $d(u, v) = d_1(u, v)$. Then as $d_1 \leq d$, we have

$$\begin{aligned} d(u, v) = d_1(u, v) &\leq d_1(u, w) + d_1(w, v) \\ &\leq d(u, w) + d(w, v) \ . \end{aligned}$$

Hence (M1)–(M3) are satisfied and so d is a metric.

(d) No, d is not a metric.

Justification. For instance, take

$$d_1(u, v) := \sqrt{\frac{1}{100}(u_1 - v_1)^2 + (u_2 - v_2)^2}$$

and

$$d_2(u, v) := \sqrt{(u_1 - v_1)^2 + \frac{1}{100}(u_2 - v_2)^2} \ .$$

It is elementary to check that d_1 and d_2 are well-defined metrics on $\mathbb{R}^2$. However, for $u := \begin{pmatrix} 1 \\ 0 \end{pmatrix}$, $v := \begin{pmatrix} 0 \\ 1 \end{pmatrix}$, and $w := \begin{pmatrix} 0 \\ 0 \end{pmatrix}$, we have

$$d(u, v) = \sqrt{\frac{101}{100}} > \frac{1}{5} = d(u, w) + d(w, v) \ ,$$

which violates (M3).

10. Let X be the space of all sequences of real numbers. So an element in X is of the form $x = \{x_n\}_{n\in\mathbb{N}}$. For any elements $x = \{x_n\}_{n\in\mathbb{N}}$ and $y = \{y_n\}_{n\in\mathbb{N}} \in X$, define

$$d(x,y) := \sum_{n\in\mathbb{N}} \frac{1}{2^n}\frac{|x_n - y_n|}{1+|x_n - y_n|}.$$

Show that d is a metric on X.

Proof. By Comparison Test, it is elementary to see that the defining series of d is convergent and so d a well-defined real-valued function on $X \times X$. Also, it is evident that (M1) and (M2) are satisfied. For (M3), we first observe that if $0 \le a \le b$, then we must have $\dfrac{a}{1+a} \le \dfrac{b}{1+b}$. Hence for any x, y, $z \in X$, we have

$$\begin{aligned}
\frac{|x_n - y_n|}{1+|x_n - y_n|} &\le \frac{|x_n - z_n| + |z_n - y_n|}{1+|x_n - z_n| + |z_n - y_n|} \\
&= \frac{|x_n - z_n|}{1+|x_n - z_n| + |z_n - y_n|} \\
&\quad + \frac{|z_n - y_n|}{1+|x_n - z_n| + |z_n - y_n|} \\
&\le \frac{|x_n - z_n|}{1+|x_n - z_n|} + \frac{|z_n - y_n|}{1+|z_n - y_n|},
\end{aligned}$$

hence

$$\begin{aligned}
d(x,y) &= \sum_{n\in\mathbb{N}} \frac{1}{2^n}\frac{|x_n - y_n|}{1+|x_n - y_n|} \\
&\le \sum_{n\in\mathbb{N}} \frac{1}{2^n}\frac{|x_n - z_n|}{1+|x_n - z_n|} + \sum_{n\in\mathbb{N}} \frac{1}{2^n}\frac{|z_n - y_n|}{1+|z_n - y_n|} \\
&= d(x,z) + d(z,y) .
\end{aligned}$$

11. Let $\{(X_n, d_n) : n \in \mathbb{N}\}$ be a countable collection of metric spaces and let $X := \prod_{n=1}^{\infty} X_n$. For any $x = (x_n : n \in \mathbb{N})$ and

$y = (y_n : n \in \mathbb{N}) \in X$, define

$$d(x,y) := \sum_{n=1}^{\infty} \frac{d_n(x_n, y_n)}{2^n (1 + d_n(x_n, y_n))} .$$

Show that d is a metric on X.

Proof. Note first that the series $\sum_{n=1}^{\infty} \frac{1}{2^n}$ is convergent. For any $n \in \mathbb{N}$ and any x_n, $y_n \in X_n$, we have

$$0 \leq \frac{d_n(x_n, y_n)}{1 + d_n(x_n, y_n)} \leq 1$$

and so by comparison test, the series

$$\sum_{n=1}^{\infty} \frac{d_n(x_n, y_n)}{2^n(1 + d_n(x_n, y_n))}$$

is also convergent. Hence d is a well-defined function on $X \times X$ and hence it remains to check conditions (M1) to (M3).

(M1): Since all d_n's are metrics, it is clear that $d(x,y) \geq 0$ for all x, $y \in X$ and $d(x,y) = 0$ if $x = y$. Furthermore, if $d(x,y) = 0$, we must have $d_n(x_n, y_n) = 0$ for all $n \in \mathbb{N}$ and so $x_n = y_n$ for all $n \in \mathbb{N}$, that is, $x = y$. Hence (M1).

(M2): It is trivial as all d_n's are metrics.

(M3): By elementary considerations, the function $f(t) := \dfrac{t}{1+t}$ is monotonically increasing for $t \geq 0$. In particular, for any $n \in \mathbb{N}$ and any x_n, y_n, $z_n \in X_n$, by triangle inequality, we have

$$d_n(x_n, z_n) \leq d_n(x_n, y_n) + d_n(y_n, z_n)$$

and so

$$f\big(d_n(x_n, z_n)\big) \leq f\big(d_n(x_n, y_n) + d_n(y_n, z_n)\big) ,$$

that is,

$$\begin{aligned}\frac{d_n(x_n, z_n)}{1 + d_n(x_n, z_n)} &\le \frac{d_n(x_n, y_n) + d_n(y_n, z_n)}{1 + (d_n(x_n, y_n) + d_n(y_n, z_n))} \\ &= \frac{d_n(x_n, y_n)}{1 + d_n(x_n, y_n) + d_n(y_n, z_n)} \\ &\quad + \frac{d_n(y_n, z_n)}{1 + d_n(x_n, y_n) + d_n(y_n, z_n)} \\ &\le \frac{d_n(x_n, y_n)}{1 + d_n(x_n, y_n)} + \frac{d_n(y_n, z_n)}{1 + d_n(y_n, z_n)} .\end{aligned}$$

It follows that

$$\begin{aligned}&d(x, z) \\ &= \sum_{n=1}^{\infty} \frac{d_n(x_n, z_n)}{2^n(1 + d_n(x_n, z_n))} \\ &\le \sum_{n=1}^{\infty} \frac{d_n(x_n, y_n)}{2^n(1 + d_n(x_n, y_n))} + \sum_{n=1}^{\infty} \frac{d_n(y_n, z_n)}{2^n(1 + d_n(y_n, z_n))} \\ &= d(x, y) + d(y, z) .\end{aligned}$$

This checks the condition (M3). Hence d is a metric on X.

12. Consider the following two metrics on $\mathbb{R}^2$:

$$d_1(x, y) := \max_{1 \le i \le 2} |x_i - y_i|$$

$$d_2(x, y) := \sum_{i=1}^{2} |x_i - y_i|$$

for any $x = (x_1, x_2)$, $y = (y_1, y_2) \in \mathbb{R}^2$. Denote the unit ball in $(\mathbb{R}^2, d_i)$ by $B_i(0, 1)$, $= 1, 2$, respectively.

(a) Show that $B_2(0, 1) \subset B_1(0, 1)$.

(b) Does there exist a metric d_3 on $\mathbb{R}^2$ so that $B_3(0, 1) \not\subset B_2(0, 1)$ and $B_2(0, 1) \not\subset B_3(0, 1)$?

(c) Recall that any straight line in the **Euclidean space** $(\mathbb{R}^2, d)$ can be described by

$$L := \{a + tv : t \in \mathbb{R}\} ,$$

where $a, v \in \mathbb{R}^2$. Let $b \in \mathbb{R}^2 \setminus L$. Then there exists a unique $t_0 \in \mathbb{R}$ such that

$$d(b, L) = d(b, a + t_0 v) .$$

Is this still valid in $(\mathbb{R}^2, d_1)$ or $(\mathbb{R}^2, d_2)$?

(d) Let $\mathbb{H}$ be the "upper half plane"

$$\mathbb{H} := \{(x, y) : y > 0\} \subset \mathbb{R}^2 .$$

(i) Show that $\mathbb{H}$ is unbounded in $(\mathbb{R}^2, d_1)$.

(ii) Is there any metric d_4 on $\mathbb{R}^2$ such that $\mathbb{H}$ is bounded in $(\mathbb{R}^2, d_4)$?

Solution:

(a) For any $(a_1, a_2) \in B_2(0, 1)$, we have

$$|a_1| + |a_2| = d_2\big((a_1, a_2), (0, 0)\big) < 1 .$$

It follows that $|a_1|$, $|a_2| < 1$. Hence $d_1\big((a_1, a_2), (0, 0)\big) = \max\{|a_1|, |a_2|\} < 1$ and so $(a_1, a_2) \in B_1(0, 1)$.

(b) Yes. For example, define d_3 by

$$d_3(x, y) := \frac{4}{3}\, d_1(x, y) .$$

It is obvious that d_3 is also a metric, and

$$\begin{aligned} B_3(0, 1) &= \{(x_1, x_2) : \frac{4}{3} \max\{|x_1|, |x_2|\} < 1\} \\ &= \{(x_1, x_2) : |x_1| < \frac{3}{4},\ |x_2| < \frac{3}{4}\} . \end{aligned}$$

Note that

$$\left(\frac{7}{8}, 0\right) \in B_2(0,1) \setminus B_3(0,1)$$

and

$$\left(\frac{5}{8}, \frac{5}{8}\right) \in B_3(0,1) \setminus B_2(0,1) .$$

(c) It is not valid in either metric space.
Let $L_1 := \{(1,t) : t \in \mathbb{R}\}$, $L_2 := \{(t, 1-t) : t \in \mathbb{R}\}$, and $b := 0 \in \mathbb{R}^2$.
Then in $(\mathbb{R}^2, d_1)$,

$$d_1(b, L_1) = 1 = d_1\left((0,0),(1,t)\right)$$

for any $t \in \mathbb{R}$ satisfying $|t| \leq 1$, while in $(\mathbb{R}^2, d_2)$,

$$d_2(b, L_2) = 1 = d_2((0,0),(u,v))$$

for any u, $v \in \mathbb{R}$ satisfying $u \geq 0$, $v \geq 0$, $u + v = 1$.

(d) (i) For any $N \in \mathbb{N}$,

$$d_1((-N,1),(N,1)) = 2N \geq N .$$

Thus $\mathbb{H} \subset (\mathbb{R}^2, d_1)$ is unbounded.

(ii) Yes, such d_4 exists.
Define $d_4(x,y) := \dfrac{d_1(x,y)}{1 + d_1(x,y)}$ for any x, $y \in \mathbb{R}^2$.
It is easy to verify that d is a well-defined metric on $\mathbb{R}$ (or it follows directly from Exercise 1.1, Part B, Problem #11). For any x, $y \in \mathbb{H}$, we have $d_4(x,y) \leq 1$. Thus $\mathbb{H}$ is bounded in $(\mathbb{R}^2, d_4)$.

13. For any $1 \leq p < \infty$, let ℓ^p be the space of sequences $x = \{x_n\}_{n\in\mathbb{N}}$ of real numbers with $\sum\limits_{n\in\mathbb{N}} |x_n|^p < \infty$. For any elements $x = \{x_n\}_{n\in\mathbb{N}}$ and $y = \{y_n\}_{n\in\mathbb{N}} \in \ell^p$, define

$$d(x,y) := \left[\sum_{n\in\mathbb{N}} |x_n - y_n|^p\right]^{1/p} .$$

Show that d is a metric on ℓ^p by establishing the following:

(a) For any α, $\beta \geq 0$ and $0 < \lambda < 1$, we have

$$\alpha^\lambda \, \beta^{1-\lambda} \leq \lambda \, \alpha + (1 - \lambda) \, \beta \, . \tag{1}$$

(b) Let p, $q \geq 0$ be real numbers satisfying $\frac{1}{p} + \frac{1}{q} = 1$ (p, q are said to be conjugate pairs). Clearly this forces $p > 1$ and $q > 1$. For any $x = \{x_n\}_{n \in \mathbb{N}} \in \ell_p$ and $y = \{y_n\}_{n \in \mathbb{N}} \in \ell_q$, write $z_n := x_n \, y_n$, $n \in \mathbb{N}$. Then $z := \{z_n\}_{n \in \mathbb{N}} \in \ell_1$ and the following *Hölder's Inequality* holds:

$$\sum_{n \in \mathbb{N}} |x_n y_n| \leq \left[\sum_{n \in \mathbb{N}} |x_n|^p \right]^{1/p} \left[\sum_{n \in \mathbb{N}} |y_n|^q \right]^{1/q} . \tag{2}$$

(c) For any $p \geq 1$ and any elements $x = \{x_n\}_{n \in \mathbb{N}}$ and $y = \{y_n\}_{n \in \mathbb{N}} \in \ell^p$, the following *Minkowski Inequality* holds:

$$\left[\sum_{n \in \mathbb{N}} \left| x_n + y_n \right|^p \right]^{1/p} \leq \left[\sum_{n \in \mathbb{N}} \left| x_n \right|^p \right]^{1/p} + \left[\sum_{n \in \mathbb{N}} \left| y_n \right|^p \right]^{1/p} . \tag{3}$$

Proof.

(a) Define $\varphi : [0, \infty) \to \mathbb{R}$ by

$$\varphi(t) := (1 - \lambda) + \lambda \, t - t^\lambda \qquad t \in [0, \infty) \, .$$

It is clear that φ is continuous on $[0, \infty)$ and differentiable in $(0, \infty)$, with $\varphi'(t) = \lambda(1 - t^{\lambda - 1})$ and so

$$\varphi'(t) > 0 \quad \text{for } t > 1 \, ,$$
$$\varphi'(t) < 0 \quad \text{for } t < 1 \, .$$

Hence for all $t \neq 1$, $\varphi(t) > \varphi(1) = 0$ and so

$$(1 - \lambda) + \lambda \, t \geq t^\lambda \, , \tag{4}$$

with equality if and only if $t = 1$.

Back to the proof of inequality (1). If $\beta = 0$, the inequality is trivial. If $\beta > 0$, it follows immediately from (4) by letting $t = \frac{\alpha}{\beta}$.

(b) Putting

$$\alpha = \frac{|x_n|^p}{\sum\limits_{k\in\mathbb{N}} |x_k|^p}\,, \quad \beta = \frac{|y_n|^p}{\sum\limits_{k\in\mathbb{N}} |y_k|^p}\,, \quad \lambda = \frac{1}{p}\,, \quad 1-\lambda = \frac{1}{q}$$

into inequality (1), we have,

$$\begin{aligned}
&\frac{\sum\limits_{n\in\mathbb{N}} |x_n y_n|}{\left[\sum\limits_{k\in\mathbb{N}} |x_k|^p\right]^{1/p} \left[\sum\limits_{k\in\mathbb{N}} |y_k|^q\right]^{1/q}} \\
&= \sum_{n\in\mathbb{N}} \frac{|x_n|}{\left[\sum\limits_{k\in\mathbb{N}} |x_k|^p\right]^{1/p}} \cdot \frac{|y_n|}{\left[\sum\limits_{k\in\mathbb{N}} |y_k|^q\right]^{1/q}} \\
&\le \frac{1}{p} \sum_{n\in\mathbb{N}} \frac{|x_n|^p}{\sum\limits_{k\in\mathbb{N}} |x_k|^p} + \frac{1}{q} \sum_{n\in\mathbb{N}} \frac{|y_n|^q}{\sum\limits_{k\in\mathbb{N}} |y_k|^q} \\
&= \frac{1}{p} \frac{\sum\limits_{n\in\mathbb{N}} |x_n|^p}{\sum\limits_{k\in\mathbb{N}} |x_k|^p} + \frac{1}{q} \frac{\sum\limits_{n\in\mathbb{N}} |y_n|^q}{\sum\limits_{k\in\mathbb{N}} |y_k|^q} = \frac{1}{p} + \frac{1}{q} = 1\,.
\end{aligned}$$

Hence (2). In particular, as $x \in \ell_p$ and $y \in \ell_q$, it follows from inequality (1) that $z \in \ell_1$.

(c) If $p = 1$, Minkowski Inequality (3) follows immediately from the usual triangle inequality. So we assume $p > 1$. By triangle inequality, we have

$$\begin{aligned}
|x_k + y_k|^p &= |x_k + y_k|\,|x_k + y_k|^{p-1} \\
&\le |x_k|\,|x_k + y_k|^{p-1} + |y_k|\,|x_k + y_k|^{p-1}\,.
\end{aligned}$$

Let $q > 1$ be conjugate to p. Then by Hölder's Inequality, we have

$$\begin{aligned}
&\sum_{n\in\mathbb{N}} |x_k + y_k|^p \\
&\le \sum_{n\in\mathbb{N}} |x_n|\,|x_n + y_n|^{p-1} + \sum_{n\in\mathbb{N}} |y_n|\,|x_n + y_n|^{p-1} \\
&\le \left[\sum_{n\in\mathbb{N}} |x_n|^p\right]^{\frac{1}{p}} \left[\sum_{n\in\mathbb{N}} |x_n + y_n|^{(p-1)q}\right]^{\frac{1}{q}} \\
&\quad + \left[\sum_{n\in\mathbb{N}} |y_n|^p\right]^{\frac{1}{p}} \left[\sum_{n\in\mathbb{N}} |x_n + y_n|^{(p-1)q}\right]^{\frac{1}{q}}.
\end{aligned}$$

By observing that $(p-1)q = p$ and moving the common factor on the right hand side to the left hand side, we arrive at Minkowski Inequality (3).

Finally, to show that d is a metric, we need to show (M1), (M2), and (M3). But (M1) and (M2) are clear, and (M3) follows immediately from Minkowski Inequality (3).

1.2 Topology of Metric Spaces

Given just one single structure, namely, metric d, on a nonempty set X, a very rich theory called Topology could be developed. Roughly speaking, Topology is a branch of mathematics which studies the properties of geometric objects which are of the elastic type. That is, objects which are free to bend, twist, stretch, shrink, etc., that is, without closing up existing holes, opening up new holes, tearing, gluing, or passing through itself.

Elementary point-set Topology can be considered as a sub-branch of Analysis, as some methods used in Topology are very similar to those used in Analysis. Analysis is a science of approximation. In order to study the properties of an object, we study the properties of its surrounding objects first and then deduce the properties of the whole object under consideration. This makes sense even in real life. One way of understanding a person is to first understand his/her friends. So we will start by studying the so-called local properties of a metric space.

Definition 1.2.1. *Let $a \in X$ and $r > 0$. The open ball in X with center a and radius r is*

$$B(a,r) := \{x \in X : d(a,x) < r\}$$

and the closed ball in X with center a and radius r is

$$\overline{B}(a,r) := \{x \in X : d(a,x) \leq r\} .$$

Example 1.2.2.

(i) In $\mathbb{R}$, $B(0,1) = (-1,1)$.

(ii) In $\mathbb{R}^2$, $B(0,1) = \{(x_1,x_2) \in \mathbb{R}^2 : x_1^2 + x_2^2 < 1\}$ = the disc centered at the origin with radius 1 without boundary; and $\overline{B}(0,1) = \{(x_1,x_2) \in \mathbb{R}^2 : x_1^2 + x_2^2 \leq 1\}$ = the disc centered at the origin with radius 1 including the boundary.

(iii) In $(\mathbb{R}^2, \tilde{d})$, where $\tilde{d}$ is the metric introduced in Example 1.1.2 (v), $B(0,1) = \{(x_1,x_2) \in \mathbb{R}^2 : 100x_1^2 + x_2^2 < 1\}$, the region strictly inside the ellipse centered at the origin with semi-axes $\frac{1}{10}$ and 1, boundary *not* included; and $\overline{B}(0,1) = \{(x_1, x_2) \in \mathbb{R}^2 : x_1^2 + 4x_2^2 \leq 1\}$, the region enclosed by the same ellipse, boundary included.

(iv) Let $X := [0, \infty)$ equipped with the Euclidean metric. Then $B(0,1) = \{x \in [0,\infty) : d(x,0) < 1\} = [0,1)$, while $\overline{B}(0,1) = \{x \in [0,\infty) : d(x,0) \leq 1\} = [0,1]$.

(iv) Let $X := [0,1] \cup (2,4]$ equipped with the Euclidean metric. Then $B(0,3) = [0,1] \cup (2,3)$ and $\overline{B}(0,3) = [0,1] \cup (2,3]$.

(vi) At first sight, $\overline{B}(a,r)$ is only slightly larger than $B(a,r)$. However, in general that may not be the case. For example, if X is a discrete metric space and $a \in X$, then $B(a,1) = \{a\}$ is a singleton; while $\overline{B}(a,1) = X$ is the entire metric space.

Remark. Observe that if $\phi \neq S \subset X$ is considered as a metric subspace, the open balls in S are

$$B_S(a,r) = \{x \in S : d(a,x) < r\} = B_X(a,r) \cap S.$$

Example 1.2.3. $B_{[0,\infty)}(0,1) = [0,1) = B_{\mathbb{R}}(0,1) \cap [0,\infty)$.

Definition 1.2.4. *For any $S \subset X$, a point $a \in S$ is called an interior point of S in X if there exists $r > 0$ such that $B(a,r) \subset S$.*

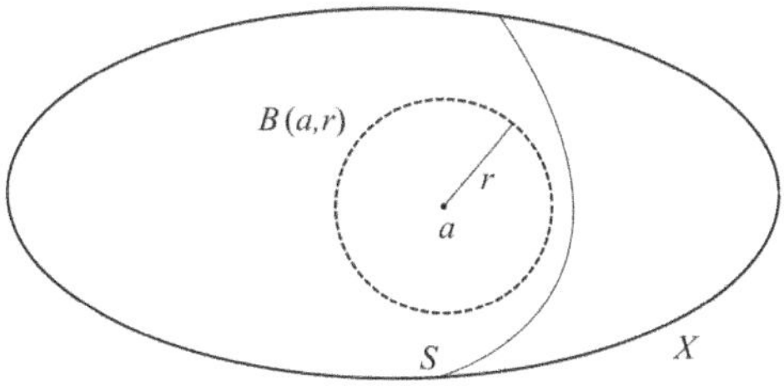

The collection of all interior points of S in X is called the interior of S in X and is denoted by S° or int S. S is said to be open in X if $S = S^\circ$, closed in X if $X \setminus S$ is open in X.

Observe that by definition, $S^\circ \subset S$. Hence $S = S^\circ$ if and only if $S \subset S^\circ$. Hence in order to show that S is open in X, all that is required is to show that every point in S is an interior point of S.

Example 1.2.5.

(i) If $X = \mathbb{R}$, $S = (0, 1)$, then $S^\circ = S$,

(ii) If $X := \mathbb{R}$, $T := [0, 1]$, then $T^\circ = (0, 1)$,

(iii) If $X := \mathbb{R}$, $P := [0, 1)$, then $P^\circ = (0, 1)$.

We have now the concepts of open balls, closed balls, open sets, and closed sets. Both "open balls" and "open sets" contain the word "open", so judging from the terminology it is natural to expect that they are somehow related, although up till this moment we have not investigated whether they have any relations at all. Similar for "closed balls" and "closed sets". The following example tells how they are related.

Example 1.2.6.

(i) ϕ, $X \subset X$ are both open and closed (in short, "*clopen*") in X.
Proof. Since nothing can go beyond X, every point in X is automatically an interior point of X and so X is open in X. Hence by definition, $\phi = X \setminus X$ is closed in X. On the other hand, as $S^\circ \subset S$ for any $S \subset X$, we have $\phi^\circ \subset \phi$. But then the only possible subset of ϕ is ϕ itself, so $\phi^\circ = \phi$ and so by definition, ϕ is open in X. Hence $X = X \setminus \phi$ is closed in X.

(ii) Every open ball $B(a, r)$ in X is open in X.
Proof. To show that $B(a, r)$ is open in X, we need to show that every point $x \in B(a, r)$ is an interior point of $B(a, r)$. So let $x \in B(a, r)$. By definition, we have $d(x, a) < r$. Hence $\delta := r - d(x, a) > 0$ and so $B(x, \delta)$ is a well-defined open ball in X. It is not hard to see that $B(x, \delta) \subset B(a, r)$. In fact, for any $y \in B(x, \delta)$, we have $d(y, x) < \delta$. Hence $d(y, a) \leq d(y, x) + d(x, a) < \delta + d(x, a) = r$ and so $y \in B(a, r)$.

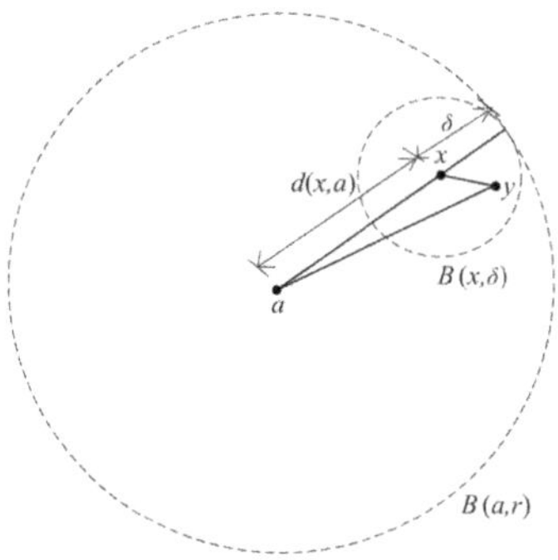

Since $y \in B(x, \delta)$ is arbitrary, we conclude that $B(x, \delta) \subset B(a, r)$ and hence by the definition of interior points, $x \in B(a, r)^{\circ}$.

(iii) Every closed ball $\overline{B}(a, r)$ in X is closed in X.

Proof. This can be proven by similar argument as that used in (ii).

(iv) If X is a discrete metric space, every subset of X is clopen.

Proof. Let $S \subset X$ be an arbitrary subset. If $S = \phi$, then it follows from (i) above that S is clopen. So assume that there exists $x \in S \neq \phi$. By Example 1.2.2 (vi), $B(x, 1) = \{x\} \subset S$ and so $x \in S^{\circ}$. Hence S is open in X. Therefore, all subsets of X are open in X. Since open subsets and closed subsets are complements of each other, all subsets of X are also closed in X.

Example 1.2.7. Let $X := \mathbb{R}$, $Y := [0, \infty)$, $S := [0, 1)$. Then for any $r > 0$, $B_X(0, r) = (-r, r) \not\subset S$ and so $0 \in S$ is *not* an interior point of S in X. Hence in particular, S is *not* open in X.

On the other hand, for any $r < 1$, $B_Y(0, r) = [0, r) \subset S$ and so $0 \in S$ *is* an interior point of S in Y. Furthermore, it is easy to see that all points in S other than the point 0 are interior points of S in Y. Hence in particular, S is open in Y. Alternatively, that S is open in Y can also be seen by the fact that $S = B_Y(0, 1)$.

Remarks.

(i) From Example 1.2.7, we see that a point $p \in S \subset Y \subset X$ is an interior point of S in Y does not necessarily imply that it is an interior point of S in X. Hence in particular, S is open in Y does not necessarily imply S is open in X, etc. So in general, when we talk about openness, closedness, interior, etc., we have to specify which metric space we are referring to. That is why we need to use the seemingly clumsy terminologies "open in X", "closed in Y", "interior point of S in X", "interior of S in X", etc.

(ii) However, in general, for any $S \subset Y \subset X$, it is clear that if S is open in X, then it is also open in Y.

Definition 1.2.8. *Let $\phi \neq S \subset X$ be a nonempty subset of X and $a \in S$. An open neighborhood of a in S is an open set U in S which contains a. More generally, a neighborhood of a in S is a set U in S which contains a in its interior. By Example 1.2.7 above, we see that in general, an open neighborhood of a in S may not be an open neighborhood of a in X.*

Remarks.

(i) By definition, S is open in X if and only if for any $x \in S$, there exists $r > 0$ such that the open ball $B(x, r)$ is lying entirely in S. Observe that this is equivalent to the condition that for any $x \in S$, there exists an open ball B in X such that $x \in B \subset S$, where we do not require that the center of B must be the point x. The point is, if such an open ball B exists, then for sure we can find a (smaller if necessary) open ball $B(x, r)$ centered at x such that $B(x, r) \subset B$, hence $B(x, r) \subset S$.

(ii) In the sequel, for the sake of simplicity, the statement "S is open (or closed) in T" will be abbreviated as "$S \subset T$ is open (or closed)".

Theorem 1.2.9. *Let $S \subset X$ be a nonempty subset and $U \subset S$. Then U is open in S if and only if $U = V \cap S$ for some $V \subset X$ which is open in X.*

Proof. ($\Rightarrow$) If U is open in S, for every $x \in U$, there exists $r_x > 0$ such that $B_S(x, r_x) \subset U$. Then $U = \bigcup_{x\in U}\{x\} \subset \bigcup_{x\in U} B_S(x, r_x) \subset U$ and so $U = \bigcup_{x\in U} B_S(x, r_x)$. Observe that $B_S(x, r) = B_X(x, r) \cap S$. So this suggests that we set $V := \bigcup_{x\in U} B_X(x, r)$ and so $U = V \cap S$. Observe that V must be open in X. In fact, for any $y \in V$, $y \in B_X(x, r_x)$ for some $x \in U$. By construction, $B_X(x, r_x) \subset V$. Hence by definition, V is open in X.

($\Leftarrow$) If $U = V \cap S$ for some open set V in X, then for any $x \in U \subset V$, there exists $r > 0$ such that $B_X(x, r) \subset V$. But then $B_S(x, r) = B_X(x, r) \cap S \subset V \cap S = U$. Hence U is open in S. □

Note that from the proof of Theorem 1.2.9, we implicitly showed that the union of any collection of open balls is open. In fact, this will be generalized to general open sets in Theorem 1.2.11 below. But before that, we observe that since closed sets and open sets are complements of each other, whenever we have a result on open sets, we have a corresponding results on closed sets for free.

Corollary 1.2.10. *Let $S \subset X$ be a nonempty subset and $D \subset S$. Then D is closed in S if and only if $D = E \cap S$ for some $E \subset X$ which is closed in X.*

Proof. This follows immediately from Theorem 1.2.9. In fact,

$$
\begin{aligned}
& D \subset S \text{ is closed} \\
\Leftrightarrow\ & S \setminus D \subset S \text{ is open} \\
\Leftrightarrow\ & S \setminus D = V \cap S \quad \text{for some open set } V \subset X \\
\Leftrightarrow\ & D = S \setminus (S \setminus D) = S \setminus (V \cap S) = S \setminus V = S \cap (X \setminus V) = S \cap E, \\
& \qquad \text{where } E := X \setminus V \subset X \text{ is closed.}
\end{aligned}
$$

□

Theorem 1.2.11.

(a) *The union of any collection of open sets is open.*

(b) *The intersection of any finite collection of open sets is open.*

Proof.

(a) Let $S = \bigcup_{\lambda \in I} S_\lambda$, where I is any index set and S_λ is open in X for every $\lambda \in I$. Then for any $a \in S$, $a \in S_\lambda$ for some λ and so there exists $r > 0$ such that $B(a, r) \subset S_\lambda \subset S$. Hence S is open in X.

(b) Let $S = \bigcap_{i=1}^{n} S_i$ where S_i is open in X for each $i = 1, \ldots, n$. Let $a \in S$. Then $a \in S_i$ for all i and so there exists $r_1, \ldots, r_n > 0$ such that $B(a, r_i) \subset S_i$, $i = 1, \ldots, n$. Let $r = \min\{r_1, \ldots, r_n\}$. Then $r > 0$ and $B(a, r) \subset B(a, r_i) \subset S_i$ for all i and so $B(a, r) \subset S$. Hence S is open in X. □

Corollary 1.2.12.

(a) *The union of finitely many closed sets is closed.*

(b) *The intersection of any collection of closed sets is closed.*

Proof. It is immediate from Theorem 1.2.11. □

Remark. The intersection of infinitely many open sets may not be open. Analogously, the union of infinitely many closed sets may not be closed. For example, in the real line $\mathbb{R}$, $\bigcap_{n=1}^{\infty} \left(-\frac{1}{n}, \frac{1}{n}\right) = \{0\}$ is not open in $\mathbb{R}$, while $\bigcup_{n=1}^{\infty} \left[0, 1 - \frac{1}{n}\right] = [0, 1)$ is not closed in $\mathbb{R}$.

Theorem 1.2.13. *If $A \subset X$ is open and $B \subset X$ is closed, then $A \setminus B$ is open in X and $B \setminus A$ is closed in X.*

Proof. This is obvious because open sets and closed sets are complements of each other. □

Theorem 1.2.14. *$\phi \neq U \subset X$ is open if and only if it is a union of open balls.*

Proof. Suppose $U \neq \phi$ is open in X, then each $x \in U$ is contained in some open ball $B(x) \subset U$, thus $U = \bigcup_{x \in U}\{x\} \subset \bigcup_{x \in U} B(x) \subset U$ and so $U = \bigcup_{x \in U} B(x)$. Conversely, suppose $U = \bigcup_{\alpha \in \Lambda} B_\alpha$, where for each α, B_α is an open ball, then by Theorem 1.2.11, U is open in X. □

Theorem 1.2.14 gives a characterization of open sets in a general metric space X. In the special case that $X = \mathbb{R}^n$, we can say more.

Theorem 1.2.15 (Lindelöf). *If $U \subset \mathbb{R}^n$ is an open set such that $U = \bigcup_{\lambda \in \Lambda} U_\lambda$, where Λ is an index set and for each $\lambda \in \Lambda$, U_λ is an open set in $\mathbb{R}^n$, then there is a countable subset $\{U_i\}_{i \in \mathbb{N}} \subset \{U_\lambda\}_{\lambda \in \Lambda}$ such that $U = \bigcup_{i=1}^{\infty} U_i$.*

Proof. First observe that the family

$$\mathcal{U} = \left\{B\left(r, \frac{1}{k}\right) : r \in \mathbb{Q}^n,\ k \in \mathbb{N}\right\}$$

is countable. Now if $U \subset \mathbb{R}^n$ is open and $U = \bigcup_{\lambda \in \Lambda} U_\lambda$, then for any $x \in U$, $x \in U_\lambda$ for some λ and there exists $B_x \in \mathcal{U}$ such that $x \in B_x \subset U_\lambda$. Since $\{B_x : x \in U\} \subset \mathcal{U}$ and $\mathcal{U}$ is countable, $\{B_x : x \in U\}$ is also countable and so we could rename it as $\{B_i\}_{i \in \mathbb{N}}$. Note that for each i, there exists $U_i \in \{U_\lambda\}_{\lambda \in \Lambda}$ such that $B_i \subset U_i$. Thus $U \subset \bigcup_{i \in \mathbb{N}} B_i \subset \bigcup_{i \in \mathbb{N}} U_i \subset \bigcup_{\lambda \in \Lambda} U_\lambda \subset U$ and so $U = \bigcup_{i \in \mathbb{N}} U_i$. □

Remark. Combining Theorem 1.2.14 and Theorem 1.2.15, we see that if $U \subset \mathbb{R}^n$ is open, then $U = \bigcup_{i=1}^{\infty} B_i$ for some countable family of open balls $\{B_i\}$. Furthermore, when $n = 1$, we can say a bit more.

Theorem 1.2.16. *Every nonempty open subset of $\mathbb{R}$ is the union of countably many pairwisely disjoint open intervals in $\mathbb{R}$. Furthermore, such a collection of intervals is unique.*

Proof. Let U be a nonempty open subset of $\mathbb{R}$. For any $x \in U$, there exists $r > 0$ such that the open interval $B(x,r) \subset U$. Let

$$I_x := \bigcup\{B : B \text{ is an open interval with } x \in B \subset U\}.$$

Then clearly $x \in I_x$ and I_x is an interval in $\mathbb{R}$. Furthermore, since union of open sets is open, we see that I_x is an open interval in $\mathbb{R}$. Repeat this construction for all $y \in U$. Observe that if $y \in I_x$, then $I_x \cup I_y$ is an open interval containing x, y and is contained in U, thus we must have $I_y \subset I_x$ and $I_x \subset I_y$. This forces $I_x = I_y$. Similarly, if $x \in I_y$, we also have $I_x = I_y$. Therefore, for any two distinct points $x, y \in U$, we must have either $I_x \cap I_y = \phi$ or $I_x = I_y$, for if they have a point z in common, then $I_x = I_z = I_y$. Hence

$$\mathcal{I} := \{\text{the distinct } I_x\text{'s, } x \in U\}$$

is a collection of pairwisely disjoint open intervals whose union is U. By Lindelöf's theorem, U equals the union of a countable sub-collection of $\mathcal{I}$. But since elements in $\mathcal{I}$ are pairwisely disjoint, no proper sub-collection of $\mathcal{I}$ could have union equals U. So this implies that $\mathcal{I}$ is itself already countable, and we can write $\mathcal{I} = \{I_i : i \in \mathbb{N}\}$. Finally, for uniqueness. If $U = \bigcup_{j=1}^{\infty} J_j$ with the J_j's being pairwisely disjoint open intervals, then each I_i must be lying entirely in some unique J_j, and vice versa. Hence the collections $\{I_i\}_{i\in\mathbb{N}}$ and $\{J_j\}_{j\in\mathbb{N}}$ are identical up to order. □

Remark. Theorem 1.2.16 is false in $\mathbb{R}^n$. Note however, that we still can show by similar arguments that every open set $U \subset \mathbb{R}^n$ can be decomposed into a disjoint union of countably many maximal open sets in U called the *connected components* of U. We will get into that in Chapter 3 below.

Definition 1.2.17. *Let $S \subset X$ be a subset. A point $x \in X$ is called an adherent point of S if $B(x, r) \cap S \neq \phi$ for all $r > 0$. The set of all adherent points of S in X is called the* closure *of S in X, and is denoted by $\overline{S}$. Clearly, $S \subset \overline{S}$.*

Example 1.2.18.

(i) If $X = \mathbb{R}$, $S = (0, 1) \cup \{2\}$, then $\overline{S} = [0, 1] \cup \{2\}$.

(ii) If $X =$ a discrete metric space, then $\overline{S} = S$ for all $S \subset X$.

Theorem 1.2.19. *$\overline{S}$ is the smallest closed subset of X containing S.*

Proof. For any $x \in X \setminus \overline{S}$, there exists $r > 0$ such that $B(x, r) \cap S = \phi$. That is, $B(x, r)$ stays away from S. But it is not entirely clear whether it also stays away from $\overline{S}$. So the picture *may* look like this:

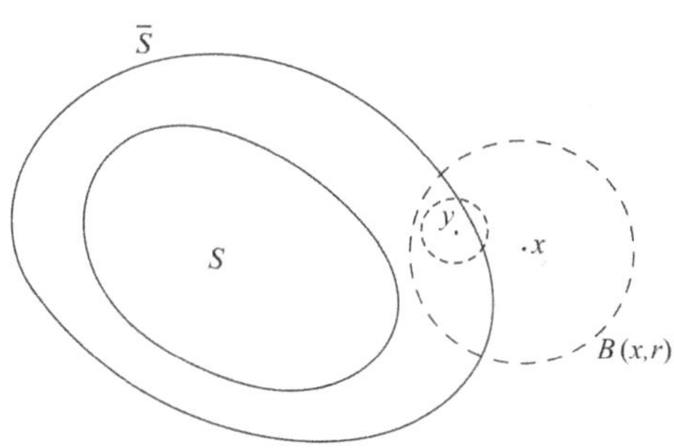

We will show that the diagram above is incorrect, that is, $B(x, r)$ actually stays away from $\overline{S}$, or $B(x, r) \subset X \setminus \overline{S}$. For this, take $y \in B(x, r)$. Then $B\big(y, r - d(x, y)\big) \subset B(x, r) \subset X \setminus S$. That is, $B\big(y, r - d(x, y)\big) \cap S = \phi$. Hence $y \in X \setminus \overline{S}$. Since $y \in B(x, r)$ is arbitrary, we have $B(x, r) \subset X \setminus \overline{S}$ and so $X \setminus \overline{S}$ is open in X and so the correct picture should look like this:

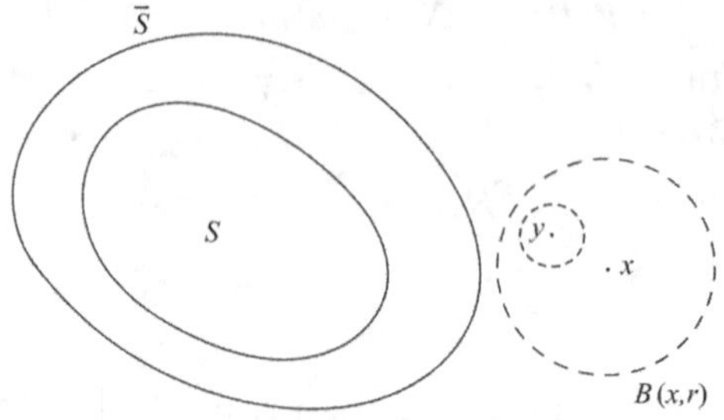

Thus $\overline{S}$ is closed in X.
Next, if $S \subset C$ and C is closed in X, then $X \setminus C \subset X \setminus S$. For any $x \in X \setminus C$, since $X \setminus C$ is open, there exists an open ball B in X such that $x \in B \subset X \setminus C \subset X \setminus S$. Hence $B \cap S = \phi$. Therefore $x \notin \overline{S}$ or $x \in X \setminus \overline{S}$. Since $x \in X \setminus C$ is arbitrary, we have $X \setminus C \subset X \setminus \overline{S}$.

□

Corollary 1.2.20. *For any $S \subset X$, $X \setminus S^\circ = \overline{X \setminus S}$.*

Proof. For any $S \subset X$, as $S^\circ \subset S$, we have $X \setminus S^\circ \supset X \setminus S$. But then since $X \setminus S^\circ$ is closed, it contains the smallest closed subset of X containing $X \setminus S$, i.e., $X \setminus S^\circ \supset \overline{X \setminus S}$. Conversely, for any $x \in X \setminus S^\circ$, $B(x,r) \not\subset S$ for any $r > 0$, i.e., $B(x,r) \cap (X \setminus S) \neq \phi$ for any $r > 0$. This implies $x \in \overline{X \setminus S}$. □

Definition 1.2.21. *Let $S \subset T \subset X$ be subsets of X. S is said to be dense in T if $\overline{S} \supset T$. In particular, $S \subset X$ is dense in X if and only if $\overline{S} = X$. Geometrically, S is dense in T if arbitrarily close to every point of T, there are points of S.*

Example 1.2.22. In the metric space $\mathbb{R}$,

(i) $(0,1) \cup (1,2)$ is dense in $(0,2)$.

(ii) $\mathbb{Q}$ is dense in $\mathbb{R}$.

(iii) $\{\frac{1}{n} : n \in \mathbb{N}\}$ is dense in $\{\frac{1}{n} : n \in \mathbb{N}\} \cup \{0\}$.

Definition 1.2.23. *Let $S \subset X$ be a subset. A point $x \in X$ is called an* accumulation point (*or,* a limit point) *of S if it is an adherent point of $S \setminus \{x\}$, that is, if $B(x,r) \cap S \setminus \{x\} \neq \phi$ for all $r > 0$. Hence $x \in X$ is an accumulation point of $S \subset X$ if and only if there are points in S which are "arbitrarily close" to x. The set of all accumulation points of S in X is called the* derived set of S in X, *and is denoted by S'. Clearly we have $S' \subset \overline{S}$ but in general, equality is not expected.*

Definition 1.2.24. *Let $S \subset X$ be a subset. A point $x \in S \setminus S'$ is called an* isolated point of S. *Hence a point $x \in S$ is an isolated point of S if and only if there exists $r > 0$ such that $B(x,r) \cap S \setminus \{x\} = \phi$, or, equivalently, $B(x,r) \cap S = \{x\}$.*

Example 1.2.25.

(i) If $X = \mathbb{R}$, $S = [a,b]$, then $S' = \overline{S} = S$.

(ii) If $X = \mathbb{R}$, $S = (0,1) \cup \{2\}$, then $S' = [0,1]$.

(iii) If $X = \mathbb{R}$, $S = \mathbb{Q}$, then $S' = \overline{S} = \mathbb{R} \supset S$.

(iv) If $X =$ discrete, then $S' = \phi$ for all $S \subset X$.

Theorem 1.2.26. *Let $S \subset X$ be any subset. Then $\overline{S} = S' \cup S$.*

Proof. It is straight forward by definition. In fact, it is obvious by definitions that $S' \subset \overline{S}$ and $S \subset \overline{S}$. Hence $S' \cup S \subset \overline{S}$. Conversely, for any $x \in \overline{S}$. If $x \in S$, then $x \in S' \cup S$. If $x \notin S$, then $B(x,r) \cap S \setminus \{x\} = B(x,r) \cap S \neq \phi$ for all $r > 0$. Hence $x \in S' \subset S' \cup S$. □

Theorem 1.2.27. *Let $S \subset X$ be any subset and $x \in X$. Then $x \in S'$ if and only if $\#\{B(x,r) \cap S\} = \infty$ for all $r > 0$, where for any nonempty set A, $\#(A)$ denotes the cardinal number of A.*

Proof. Obvious by definition. □

Remark. By Theorem 1.2.27, a necessary condition for S to have an accumulation point is that it contains infinitely many distinct points. Note that obviously the converse of this is false. For instance, $S = \mathbb{N}$ has infinitely many points but it has no accumulation point.

Theorem 1.2.28. *Let $S \subset X$ be any subset. Then the following are equivalent:*

(a) *S is closed in X.*

(b) *$S' \subset S$.*

(c) *$\overline{S} = S$.*

Proof. (a) $\Leftrightarrow$ (c): By Theorem 1.2.19.
(b) $\Leftrightarrow$ (c): By Theorem 1.2.26. $\square$

So far we have, associated to any subset $S \subset X$, studied several types of points, namely, interior points, adherent points, and accumulation points. The readers will see in the sequel that they provide important information for various properties of the subset S. We have yet another type of points: boundary points. At first sight it seems to be so obvious that there is no need to formally define it. But in reality it is not the case. For simple geometric objects it is easy to "see" what the boundary points of a set should be. However, for more complicated sets, and even more so, for subsets of an abstract metric space, it is not entirely clear what boundary points are. For instance, for the sets $\left\{\frac{1}{n} : n \in \mathbb{N}\right\} \subset \mathbb{R}$ and $\mathbb{Q} \subset \mathbb{R}$, etc., it is not intuitively clear what their boundary points should be. So a formal definition is due. It turns out that it is nothing more than our intuition on simple subsets $S \subset X$.

Definition 1.2.29. *Let $S \subset X$ be a subset. A point $x \in X$ is said to be a boundary point of S if*

$$\begin{cases} B(x,r) \cap S \neq \phi \\ B(x,r) \cap (X \setminus S) \neq \phi \end{cases}$$

for all $r > 0$. The set of all boundary points of S is called the boundary of S, and is denoted by ∂S. Note that $\partial S \subset \overline{S}$ and $S^\circ \cap \partial S = \phi$.

Example 1.2.30.

(i) In $\mathbb{R}^n$, $\partial B(a,r) = S(a,r) = \{x \in \mathbb{R}^n : \|x-a\| = r\}$.

(ii) In $\mathbb{R}$, $\partial (0,\infty) = \{0\}$.

(iii) In $\mathbb{R}$, $\partial \mathbb{Q} = \mathbb{R}$.

(iv) In $\mathbb{R}$, if $S := \{\frac{1}{n} : n \in \mathbb{N}\}$, then $\partial S = S \cup \{0\}$.

Remarks.

(i) By definition, we have $\partial S = \overline{S} \cap \overline{X \setminus S}$.

(ii) Also by definition, we have $\partial S = \partial (X \setminus S)$.

(iii) In particular, we see that ∂S is always closed for any $S \subset X$.

Theorem 1.2.31. *For any subset $S \subset X$, $\overline{S} = \partial S \uplus S^\circ$, where $A \uplus B := A \cup B$ with an extra piece of information that A and B are mutually disjoint.*

Proof. It is rather obvious from definition:
For any $x \in S^\circ$, there exists $r > 0$ such that $B(x,r) \subset S$. Hence $B(x,r) \cap X \setminus S = \phi$ and so $x \notin \partial S$. Therefore, $S^\circ \cap \partial S = \phi$. On the other hand, for any $x \in \overline{S}$, if $x \notin S^\circ$, then $B(x,r) \cap X \setminus S \neq \phi$ for all $r > 0$ and so $x \in \partial S$. □

Exercise 1.2

Unless otherwise specified, X will stand for a general metric space and S, T, etc., are arbitrary subsets of X.

Part A: True or False Questions

For each of the following statements, determine if it is true or false. If it is true, prove it. If it is false, give a counterexample or provide proper justification.

1. Let (X, d) be a metric space and S, T be subsets of X.
   (a) $S \subset T \implies S^\circ \subset T^\circ$;
   (b) $(S^\circ)^\circ = S^\circ$;
   (c) $S^\circ \subset T \subset S \Longrightarrow T^\circ = S^\circ$;
   (d) $S^\circ \cap T^\circ \subset (S \cap T)^\circ$;
   (e) $S^\circ \cap T^\circ \supset (S \cap T)^\circ$;
   (f) $S^\circ \cup T^\circ \subset (S \cup T)^\circ$;
   (g) $S^\circ \cup T^\circ \supset (S \cup T)^\circ$.

   Answer:

   (a) True.

   Proof. Let $s \in S^\circ$. Then there exists $r > 0$ such that $B(s, r) \subset S \subset T$. So $s \in T^\circ$ and thus $S^\circ \subset T^\circ$.

   (b) True.

   Proof. Suffices to show that S° is open. For any $x \in S^\circ$, there exists $r > 0$ such that $B(x, r_x) \subset S$. Note that actually $B(x, r_x) \in S^\circ$. In fact, for any $y \in B(x, r_x)$, since $B(x, r_x)$ is open, there is $\delta > 0$ such that $B(y, \delta) \subset B(x, r_x)$. Hence $B(y, \delta) \subset S$. By definition, y is also an interior point of S, that is, $y \in S^\circ$.

   (c) True.

   Proof. By Parts (a) and (b), $S^\circ \subset T \subset S \Rightarrow S^\circ = (S^\circ)^\circ \subset T^\circ \subset S^\circ$ and so we must have $T^\circ = S^\circ$.

   (d) True.

Proof. For any $x \in S^\circ \cap T^\circ$, there exist r_1, $r_2 > 0$ such that $B(x, r_1) \subset S$ and $B(x, r_2) \subset T$. Let $r = \min\{r_1, r_2\} > 0$. Then $B(x, r) \subset S \cap T$. Hence $x \in (S \cap T)^\circ$ and so $S^\circ \cap T^\circ \subset (S \cap T)^\circ$.

(e) True.

Proof. By Part (a), $S \cap T \subset S \Rightarrow (S \cap T)^\circ \subset S^\circ$. Similarly, $S \cap T \subset T \Rightarrow (S \cap T)^\circ \subset T^\circ$. Hence $(S \cap T)^\circ \subset S^\circ \cap T^\circ$.

(f) True.

Proof. In the proof of Part (b), we see that S° and T° are open, so the same is true for their union. Hence $S^\circ \cup T^\circ = (S^\circ \cup T^\circ)^\circ$. Since $S^\circ \cup T^\circ \subset S \cup T$, by (a) we have $S^\circ \cup T^\circ = (S^\circ \cup T^\circ)^\circ \subset (S \cup T)^\circ$.

(g) False.

Example: Let $X := \mathbb{R}$, $S := [0, 1]$, and $T := [1, 2]$. Then $(S \cup T)^\circ = (0, 2)$ but $S^\circ \cup T^\circ = (0, 1) \cup (1, 2)$.

2. For any $\phi \neq S$, $T \subset \mathbb{R}$, define

$$\begin{aligned} S \cdot T &:= \{st : s \in S, t \in T\} , \\ S + T &:= \{s + t : s \in S, t \in T\} . \end{aligned}$$

Then

(a) S is open $\Longrightarrow S \cdot T$ is open.

(b) S is open $\Longrightarrow S + T$ is open.

Answer:

(a) False.

Example: Take $S := \mathbb{R}$ which is open, and $T := \{0\}$. Then $S \cdot T = \{0\}$ which is not open.

(b) True.

Proof. For any $x \in S + T$, there exist $s \in S$ and $t \in T$ such that $x = s + t$. Since S is open, there exists $r_s > 0$ such that $B(s, r_s) \subset S$. Hence for any $y \in B(x, r_s)$,

$$|(y - t) - s| = |y - (s + t)| = |y - x| < r_s .$$

So $y - t \in B(s, r_s) \subset S$. As a result,

$$y = (y - t) + t \in S + T \ .$$

Since $y \in B(x, r_s)$ is arbitrary, we conclude that $B(x, r_s) \subset S + T$ and we are done.

3. Let (X, d) be a metric space and S, T be subsets of X.
   (a) $S \subset T \Longrightarrow \overline{S} \subset \overline{T}$.
   (b) $\overline{(\overline{S})} = \overline{S}$.
   (c) $S \subset T \subset \overline{S} \Longrightarrow \overline{T} = \overline{S}$.
   (d) $\overline{S} \cup \overline{T} \subset \overline{S \cup T}$.
   (e) $\overline{S} \cup \overline{T} \supset \overline{S \cup T}$.
   (f) $\overline{S} \cap \overline{T} \subset \overline{S \cap T}$.
   (g) $\overline{S} \cap \overline{T} \supset \overline{S \cap T}$.

   Answer:

   (a) True.
   Proof. For any $x \in \overline{S}$ and any $r > 0$, we have $B(x, r) \cap T \supset B(x, r) \cap S \neq \phi$. Hence $x \in \overline{T}$.

   (b) True.
   Proof. It follows trivially from the fact that $\overline{S}$ is closed and so the smallest closed set in X containing $\overline{S}$ is $\overline{S}$ itself. Hence $\overline{(\overline{S})} = \overline{S}$.

   (c) True.
   Proof. By (a) and (b), $S \subset T \subset \overline{S} \Rightarrow \overline{S} \subset \overline{T} \subset \overline{(\overline{S})} = \overline{S} \Rightarrow \overline{T} = \overline{S}$.

   (d) True.
   Proof. As $S \subset S \cup T$ and $T \subset S \cup T$, we have by (a), $\overline{S} \subset \overline{S \cup T}, \overline{T} \subset \overline{S \cup T}$ and so $\overline{S} \cup \overline{T} \subset \overline{S \cup T}$.

   (e) True.
   Proof. For any $x \in \overline{S \cup T}$ and any $r > 0$, we have

$$(B(x, r) \cap S) \cup (B(x, r) \cap T) = B(x, r) \cap (S \cup T) \neq \phi \ . \ (*)$$

If $B(x,r) \cap S \neq \phi$ for all $r > 0$, then $x \in \overline{S} \subset \overline{S} \cup \overline{T}$. Otherwise, there exists $r_0 > 0$ such that $B(x, r_0) \cap S = \phi$ and so $B(x,r) \cap S = \phi$ for all $0 < r \leq r_0$. But then by (*), we must have $B(x,r) \cap T \neq \phi$ for all $0 < r < r_0$. Thus $B(x,r) \cap T \neq \phi$ for all $r > 0$ and so $x \in \overline{T} \subset \overline{S} \cup \overline{T}$.

(f) False.

Example: Take $X := \mathbb{R}$, $S := [0,1)$ and $T := (1,2]$. Then $\overline{S} \cap \overline{T} = \{1\}$ but $\overline{S \cap T} = \phi$.

(g) True.

Proof. As $S \cap T \subset S$ and $S \cap T \subset T$, by (a) we have $\overline{S \cap T} \subset \overline{S}$ and $\overline{S \cap T} \subset \overline{T}$. Thus $\overline{S \cap T} \subset \overline{S} \cap \overline{T}$.

4. Let (X,d) be a metric space and S, T be subsets of X.
   (a) $S \subset T \implies S' \subset T'$.
   (b) $(S')' \subset S'$.
   (c) $(S')' \supset S'$.
   (d) $S' \cup T' \subset (S \cup T)'$.
   (e) $S' \cup T' \supset (S \cup T)'$.
   (f) $S' \cap T' \subset (S \cap T)'$.
   (g) $S' \cap T' \supset (S \cap T)'$.

Answer:

(a) True.

Proof. For any $x \in S'$ and any $r > 0$, by Theorem 1.2.27, we have

$$\#\{B(x,r) \cap T\} \supset \#\{B(x,r) \cap S\} = \infty .$$

Hence by Theorem 1.2.27 again, $x \in T'$.

(b) True.

Proof. Let $x \in (S')'$ and $r > 0$ be arbitrary. Then there exists $y \in B(x,r) \cap S' \setminus \{x\}$. Put $\delta := r - d(x,y) > 0$. Then $B(y,\delta) \subset B(x,r)$. As $y \in S'$, $\#(B(y,\delta) \cap S \setminus \{y\}) = \infty$. Hence there exists $z \in B(y,\delta) \cap S \setminus \{x,y\} \subset B(x,r) \cap S \setminus \{x\}$. Therefore $x \in S'$ and hence $(S')' \subset S'$.

(c) False.

Example: Let $X := \mathbb{R}$, $S := \left\{\frac{1}{n} : n \in \mathbb{N}\right\}$. Then $S' = \{0\}$ and $(S')' = \phi$.

(d) True.

Proof. As $S \subset S \cup T$ and $T \subset S \cup T$, by (a) we have $S' \subset (S \cup T)'$, $T' \subset (S \cup T)'$ and so $S' \cup T' \subset (S \cup T)'$.

(e) True.

Proof. For any $x \in (S \cup T)'$ and any $r > 0$, we have

$$\begin{aligned}&(B(x,r) \cap S \setminus \{x\}) \cup (B(x,r) \cap T \setminus \{x\}) \\ &= B(x,r) \cap (S \cup T) \setminus \{x\} \neq \phi\,. \qquad (**)\end{aligned}$$

If $B(x,r) \cap S \setminus \{x\} \neq \phi$ for all $r > 0$, then $x \in S' \subset S' \cup T'$. Otherwise, there exists $r_0 > 0$ such that $B(x, r_0) \cap S \setminus \{x\} = \phi$ and so $B(x,r) \cap S \setminus \{x\} = \phi$ for all $0 < r \leq r_0$. But then by (**), we must have $B(x,r) \cap T \setminus \{x\} \neq \phi$ for all $0 < r < r_0$. Thus $B(x,r) \cap T \setminus \{x\} \neq \phi$ for all $r > 0$ and so $x \in T' \subset S' \cup T'$.

(f) False.

Example: Take $X := \mathbb{R}$, $S := [0,1)$, and $T := (1,2]$. Then $S' \cap T' = [0,1] \cap [1,2] = \{1\}$ but $(S \cap T)' = \phi$.

(g) True.

Proof. As $S \cap T \subset S$ and $S \cap T \subset T$, by (a) we have $(S \cap T)' \subset S'$ and $(S \cap T)' \subset T'$ and so $(S \cap T)' \subset S' \cap T'$.

5. Let (X, d) be a metric space and $S, T \subset X$.

(a) $S \subset T \implies \partial S \subset \partial T$.

(b) $\partial(\partial S) \subset \partial S$.

(c) $\partial(\partial S) \supset \partial S$.

(d) $\partial S \subset \partial(S^\circ)$.

(e) $\partial S \supset \partial(S^\circ)$.

(f) $\partial S \subset \partial(S')$.

(g) $\partial S \supset \partial(S')$.

(h) $\partial S \subset \partial(\overline{S})$.

(i) $\partial S \supset \partial(\overline{S})$.
(j) $\partial(S') \subset \partial(S^\circ)$.
(k) $\partial(S') \supset \partial(S^\circ)$.
(l) $\partial(\overline{S}) \subset \partial(S')$.
(m) $\partial(\overline{S}) \supset \partial(S')$.
(n) $\partial(S^\circ) \subset \partial(\overline{S})$.
(o) $\partial(S^\circ) \supset \partial(\overline{S})$.
(p) $(\partial S)^\circ = \phi$.
(q) $(\partial S)' = \phi$.

Answer:

(a) False.
Example: Take $X := \mathbb{R}$, $S := [0, 3]$, $T := [1, 2]$. Then $\partial S = \{0, 3\}$ while $\partial T = \{1, 2\}$.

(b) True.
Proof. Since ∂S is closed, we have $\partial(\partial S) \subset \overline{\partial S} = \partial S$.

(c) False.
Example: Take $X := \mathbb{R}$, $S := \mathbb{Q}$, then $\partial S = \mathbb{R}$ while $\partial(\partial S) = \partial\,\mathbb{R} = \phi$.

(d) False.
Example: Take $X := \mathbb{R}$, $S := (0, 1) \cup \{2\}$. Then $\partial S = \{0, 1, 2\} \not\subset \{0, 1\} = \partial\big((0, 1)\big) = \partial(S^\circ)$.

(e) True.
Prove. For any $x \in \partial(S^\circ)$ and any $r > 0$, we have by Corollary 1.2.20,

$$\begin{cases} B(x,r) \cap S & \supset B(x,r) \cap S^\circ \qquad \neq \phi \qquad \text{(i)} \\ B(x,r) \cap \overline{X \setminus S} & = B(x,r) \cap (X \setminus S^\circ) \neq \phi\,. \qquad \text{(ii)} \end{cases}$$

Observe that from (ii) we actually have $B(x, r) \cap (X \setminus S) \neq \phi$. In fact, by (ii), there exists $y \in B(x, r) \cap \overline{X \setminus S}$. Let $\delta > 0$ be s.t. $B(y, \delta) \subset B(x, r)$. As $y \in \overline{X \setminus S}$, there is a point $z \in B(y, \delta) \cap (X \setminus S) \subset B(x, r) \cap (X \setminus S)$. Thus $B(x, r) \cap (X \setminus S) \neq \phi$. This together with (i) implies $x \in \partial S$.

(f) False.

Example: Take $X := \mathbb{R}$, $S := (0,1) \cup \{2\}$. Then $\partial S = \{0,1,2\} \not\subset \{0,1\} = \partial([0,1]) = \partial(S')$.

(g) False.

Example: Let $X := S := \{\frac{1}{n} : n \in \mathbb{N}\} \cup \{0\}$. Then $\partial S = \phi \not\supset \{0\} = \partial(\{0\}) = \partial(S')$.

(h) False.

Example: Let $X := \mathbb{R}$, $S := (0,1) \cup (1,2)$. Then $\partial S = \{0,1,2\} \not\subset \{0,2\} = \partial([0,2]) = \partial(\overline{S})$.

(i) True.

Proof: $\partial\overline{S} = \overline{(\overline{S})} \cap \overline{X \setminus \overline{S}} = \overline{S} \cap \overline{X \setminus \overline{S}} \subset \overline{S} \cap \overline{X \setminus S} = \partial S$.

(j) False.

Example: Let $X := \mathbb{R}$, $S := \{\frac{1}{n} : n \in \mathbb{N}\}$. Then $\partial(S') = \partial(\{0\}) = \{0\} \not\subset \phi = \partial(\phi) = \partial(S^\circ)$.

(k) False.

Example: Let $X := \mathbb{R}$, $S := (0,1) \cup (1,2)$. Then $\partial(S^\circ) = \partial S = \{0,1,2\} \not\subset \{0,2\} = \partial([0,2]) = \partial(S')$.

(l) False.

Example: Let $X := \mathbb{R}$, $S := \{0\}$. Then $\partial(\overline{S}) = \partial S = \{0\} \not\subset \phi = \partial\phi = \partial(S')$.

(m) False.

Example: Let $X := S := \{\frac{1}{n} : n \in \mathbb{N}\} \cup \{0\}$. Then $\partial(\overline{S}) = \partial(X) = \phi \not\supset \{0\} = \partial\{0\} = \partial(S')$.

(n) False.

Example: Let $X := \mathbb{R}$, $S := (0,1) \cup (1,2)$. Then $\partial(S^\circ) = \partial S = \{0,1,2\} \not\subset \{0,2\} = \partial([0,2]) = \partial(\overline{S})$.

(o) False.

Example: Let $X := \mathbb{R}$, $S := \{\frac{1}{n} : n \in \mathbb{N}\}$. Then $\partial(S^\circ) = \partial(\phi) = \phi \not\supset \overline{S} = \partial(\overline{S})$.

(p) False.

Example: Take $X := \mathbb{R}$, $S := \mathbb{Q}$. Then $(\partial S)^\circ = \mathbb{R}^\circ = \mathbb{R}$.

(q) False.

Example: Take $X := \mathbb{R}^2$, $S := \{(x,y) : x^2 + y^2 \leq 1\}$. Then $(\partial S)' = (S^1)' = S^1$.

6. Let $S, T \subset \mathbb{R}^n$.
   (a) $\partial(S \cup T) \subset \partial S \cup \partial T$.
   (b) $\partial(S \cup T) \supset \partial S \cup \partial T$.
   (c) $\partial(S \cap T) \subset \partial S \cap \partial T$.
   (d) $\partial(S \cap T) \supset \partial S \cap \partial T$.
   (e) $\partial(S \cap T) \subset \partial S \cup \partial T$.
   (f) $\partial(S \cap T) \supset \partial S \cup \partial T$.
   (g) $\partial(S \setminus T) \subset \partial S \cup \partial T$.
   (h) $\partial(S \setminus T) \supset \partial S \cup \partial T$.

   Answer:

   (a) True.

   Proof. For any $x \in \partial(S \cup T)$, we have

$$\big(B(x,r) \cap S\big) \cup \big(B(x,r) \cap T\big) = B(x,r) \cap (S \cup T) \neq \phi$$

for all $r > 0$ and

$$B(x,r) \cap (X \setminus S) \cap (X \setminus T) = B(x,r) \cap \big(X \setminus (S \cup T)\big) \neq \phi$$

for all $r > 0$. Note that the latter implies that we always have

$$B(x,r) \cap (X \setminus S) \neq \phi \quad \text{and} \quad B(x,r) \cap (X \setminus T) \neq \phi$$

for all $r > 0$.

If $B(x,r) \cap S \neq \phi$ for all $r > 0$, then we have

$$\begin{cases} B(x,r) \cap S \neq \phi \\ B(x,r) \cap \big(X \setminus S\big) \neq \phi \end{cases} \qquad \forall\, r > 0\,.$$

Thus $x \in \partial S$.

If there is an $r_0 > 0$ such that $B(x,r_0) \cap S = \phi$, then $B(x,r) \cap S = \phi$ for all $0 < r < r_0$. This forces $B(x,r) \cap T \neq \phi$ for all $0 < r < r_0$ and so $B(x,r) \cap T \neq \phi$ for all $0 < r$. Hence

$$\begin{cases} B(x,r) \cap T \neq \phi \\ B(x,r) \cap \big(X \setminus T\big) \neq \phi \end{cases} \qquad \forall\, r > 0\,.$$

Thus $x \in \partial T$.

(b) False.

Example: Let $X := \mathbb{R}$, $S := [0,2]$, $T := \{1\}$.
Then $\partial(S \cup T) = \partial([0,2]) = \{0,2\} \not\supset \{0,1,2\} = \partial([0,2]) \cup \partial(\{1\}) = \partial S \cup \partial T$.

(c) False.

Example: Let $X := \mathbb{R}$, $S := [0,3]$, $T := [1,2]$.
Then $\partial(S \cap T) = \partial([1,2]) = \{1,2\} \not\subset \phi = \{0,3\} \cap \{1,2\} = \partial S \cap \partial T$.

(d) False.

Example: Let $X := \mathbb{R}$, $S := [0,1)$, $T := (1,2]$.
Then $\partial(S \cap T) = \partial \phi = \phi \not\supset \{1\} = \{0,1\} \cap \{1,2\} = \partial S \cap \partial T$.

(e) True.

Proof. For any $x \in \partial(S \cap T)$, we have

$$B(x,r) \cap S \cap T = B(x,r) \cap (S \cap T) \neq \phi$$

for all $r > 0$ and

$$\begin{aligned} &\big(B(x,r) \cap (X \setminus S)\big) \cup \big(B(x,r) \cap (X \setminus T)\big) \\ &= B(x,r) \cap \big(X \setminus (S \cap T)\big) \neq \phi \end{aligned}$$

for all $r > 0$. Note that the former implies that we always have

$$B(x,r) \cap S \neq \phi \quad \text{and} \quad B(x,r) \cap T \neq \phi$$

for all $r > 0$.
If $B(x,r) \cap (X \setminus S) \neq \phi$ for all $r > 0$, then we have

$$\begin{cases} B(x,r) \cap S \neq \phi \\ B(x,r) \cap (X \setminus S) \neq \phi \end{cases} \qquad \forall\, r > 0\,.$$

Thus $x \in \partial S$.
If there is an $r_0 > 0$ such that $B(x,r_0) \cap (X \setminus S) = \phi$, then $B(x,r) \cap (X \setminus S) = \phi$ for all $0 < r < r_0$. This forces $B(x,r) \cap$

$(X \setminus T) \neq \phi$ for all $0 < r < r_0$ and so $B(x,r) \cap (X \setminus T) \neq \phi$ for all $0 < r$. Hence

$$\begin{cases} B(x,r) \cap T \neq \phi \\ B(x,r) \cap (X \setminus T) \neq \phi \end{cases} \qquad \forall\, r > 0\,.$$

Thus $x \in \partial T$.

(f) False.

Example: Let $X := \mathbb{R}$, $S := [0,1]$, $T := \{2\}$.
Then $\partial(S \cap T) = \partial\phi = \phi \not\supset \{0,1,2\} = \partial S \cup \partial T$.

(g) True.

Proof. Recall that $\partial S = \partial(X \setminus S)$ for any $S \subset X$. We have by (e),

$\partial(S \setminus T) = \partial\big(S \cap (X \setminus T)\big) \subset \partial S \cup \partial(X \setminus T) = \partial S \cup \partial T$.

(h) False.

Example: Let $X := \mathbb{R}$, $S := [0,1]$, $T := \{2\}$.
Then $\partial(S \setminus T) = \partial\big([0,1]\big) = \{0,1\} \not\supset \{0,1,2\} = \partial S \cup \partial T$.

7. Let (X,d) be a metric space and S, T be subsets of X.
   (a) $(S \cup T)^\circ = S^\circ \cup T^\circ$.
   (b) $(S \cap T)^\circ = S^\circ \cap T^\circ$.
   (c) $\overline{S \cup T} = \overline{S} \cup \overline{T}$.
   (d) $\overline{S \cap T} = \overline{S} \cap \overline{T}$.
   (e) $(S \cup T)' = S' \cup T'$.
   (f) $(S \cap T)' = S' \cap T'$.
   (g) S' is closed.
   (h) $S' \neq \phi \implies S \cap S' \neq \phi$.
   (i) $S^\circ \subset S'$
   (j) $S' \subset (S')'$.
   (k) $(S')' \subset S'$.
   [This is identical with Exercise 1.2, Part A, Problem #4(b). It is included here to facilitate the readers to make comparison between S' and $(S')'$.]
   (l) $\overline{S'} \subset \overline{S}$.

(m) $\overline{S'} \supset \overline{S}$.
(n) $\overline{S}\,' \subset \overline{S}$.
(o) $\overline{S}\,' \supset \overline{S}$.
(p) $\overline{S^\circ} \subset \overline{S}$.
(q) $\overline{S^\circ} \supset \overline{S}$.
(r) $(\overline{S})^\circ \subset S^\circ$.
(s) $(\overline{S})^\circ \supset S^\circ$.
(t) $(S^\circ)' \subset S'$.
(u) $(S^\circ)' \supset S'$.
(v) $(S')^\circ \subset S^\circ$.
(w) $(S')^\circ \supset S^\circ$.
(x) $(\overline{S})^\circ \subset \overline{S^\circ}$.
(y) $(\overline{S})^\circ \supset \overline{S^\circ}$.
(z) $(\overline{S})' \subset S'$.
(aa) $(\overline{S})' \supset S'$.
(ab) $\partial(S') \subset \partial S \Longleftrightarrow S' \cap S^\circ \subset (S')^\circ$.

Answer:

(a) False.

Justification. By Exercise 1.2, Part A, Problem #1(g).

(b) True.

Proof. By Exercise 1.2, Part A, Problem #1(d), (e).

(c) True.

Proof. By Exercise 1.2, Part A, Problem #3(d), (e).

(d) False.

Justification. By Exercise 1.2, Part A, Problem #3(f).

(e) True.

Proof. By Exercise 1.2, Part A, Problem #4(d), (e).

(f) False.

Justification. By Exercise 1.2, Part A, Problem #4(f).

(g) True.

Proof. By Theorem 1.2.26 and Exercise 1.2, Part A, Problem #4(b), $\overline{S'} = (S')' \cup S' = S'$ and so S' is closed.

(h) False.

Example: Let $X := \mathbb{R}$, $S := \left\{\frac{1}{n} : n \in \mathbb{N}\right\}$. Then $S' = \{0\} \neq \phi$ but $S \cap S' = \phi$.

(i) False.

Example: Let $X =$ any discrete metric space and $S \subset X$ be nonempty. Then $S^\circ = S \not\subset \phi = S'$.

(j) False.

Example: Let $X := \mathbb{R}$, $S := \left\{\frac{1}{n} : n \in \mathbb{N}\right\}$. Then $S' = \{0\} \not\subset \phi = (S')'$.

(k) True.

Proof. The readers are referred to the Proof for Exercise 1.2, Part A, Problem #4(b).

(l) True.

Proof. By definition, $S' \subset \overline{S}$. Hence by Exercise 1.2, Part A, Problem #3(a), (b), we have $\overline{S'} \subset \overline{(\overline{S})} = \overline{S}$.

(m) False.

Example: Take $X := \mathbb{R}$, $S := (0, 1) \cup \{2\}$.
Then $\overline{S'} = [0, 1] \not\supset [0, 1] \cup \{2\} = \overline{S}$.

(n) True.

Proof. By definition and Exercise 1.2, Part A, Problem #3(b), $\overline{S}\,' \subset \overline{(\overline{S})} = \overline{S}$.

(o) False.

Example: Take $X := \mathbb{R}$, $S := (0, 1) \cup \{2\}$.
Then $\overline{S}\,' = [0, 1] \not\supset [0, 1] \cup \{2\} = \overline{S}$.

(p) True.

Proof. $S^\circ \subset S \Longrightarrow \overline{S^\circ} \subset \overline{S}$.

(q) False.

Example: Take $X := \mathbb{R}$, $S := \{0\}$.
Then $\overline{S^\circ} = \phi \not\supset \{0\} = \overline{S}$.

(r) False.

Example: Take $X := \mathbb{R}$, $S := \mathbb{Q}$.
Then $(\overline{S})^\circ = \mathbb{R} \not\subset \phi = S^\circ$.

(s) True.

Proof. $S \subset \overline{S} \Longrightarrow S^\circ \subset (\overline{S})^\circ$.

(t) True.

Proof. For any $x \in (S^\circ)'$ and any $r > 0$, there exists $y \in B(x,r) \cap S^\circ \setminus \{x\}$. As $y \in S^\circ$, there exists $0 < \delta < \min\{d(y,x), r-d(y,x)\}$ small enough such that $B(y,\delta) \subset S$. Then $B(x,r) \cap S \setminus \{x\} \supset B(y,\delta) \cap S \setminus \{x\} = B(y,\delta) \neq \phi$ and so $x \in S'$.

(u) False.

Example: Take $X := \mathbb{R}$, $S := \mathbb{Q}$.
Then $(S^\circ)' = \phi \not\supset \mathbb{R} = S'$.

(v) False.

Example: Take $X := \mathbb{R}$, $S := (0,1) \cup (1,2)$.
Then $(S')^\circ = (0,2) \not\subset (0,1) \cup (1,2) = S^\circ$.

(w) False.

Example: Let X be discrete and $S \subset X$ be any nonempty subset. Then $(S')^\circ = \phi \not\supset S = S^\circ$.

(x) False.

Example: Take $X := \mathbb{R}$, $S := \mathbb{Q}$.
Then $(\overline{S})^\circ = \mathbb{R} \not\subset \phi = \overline{S^\circ}$.

(y) False.

Example: Take $X := \mathbb{R}$, $S := (0,1)$.
Then $(\overline{S})^\circ = (0,1) \not\supset [0,1] = \overline{S^\circ}$.

(z) True.

Proof. For any $x \in (\overline{S})'$ and any $r > 0$, there exists $y \in B(x,r) \cap \overline{S} \setminus \{x\}$. Let $\delta := \min\{d(y,x), r - d(y,x)\} > 0$. As $y \in \overline{S}$, we have

$$B(x,r) \cap S \setminus \{x\} \supset B(y,\delta) \cap S \setminus \{x\} = B(y,\delta) \cap S \neq \phi$$

and hence $x \in S'$.

(aa) True.

Proof. By Exercise 1.2, Part A, Problem #4(a), $S \subset \overline{S} \implies S' \subset (\overline{S})'$.

(ab) True.

Proof. We have

$$\begin{aligned} S' &= S' \cap \overline{S} \\ &= S' \cap (\partial S \cup S^\circ) \\ &= (S' \cap \partial S) \cup (S' \cap S^\circ) . \end{aligned} \qquad (*)$$

On the other hand, as S' is closed, we have $\partial(S') \subset S'$ and

$$S' = \overline{S'} = \partial(S') \cup (S')^\circ . \qquad (**)$$

Thus

$$\begin{aligned} \partial(S') \subset \partial S &\iff \partial(S') \subset S' \cap \partial S \\ &\iff S' \setminus \partial(S') \supset S' \setminus (S' \cap \partial S) \\ &\iff (S')^\circ \supset S' \cap S^\circ \qquad \text{by } (**) \text{ and } (*) . \end{aligned}$$

8. Recall that a nonempty set $S \subset \mathbb{R}^n$ is said to be *convex* if for any $x, y \in S$ and any $t \in [0, 1]$, we have $(1 - t)x + ty \in S$. Geometrically, S is convex if the straight line segment joining two points of S lies entirely in S. Suppose $S \subset \mathbb{R}^n$.
   (a) S is convex $\Rightarrow S^\circ$ is convex.
   (b) S° is convex $\Rightarrow S$ is convex.
   (c) S is convex $\Rightarrow \overline{S}$ is convex.
   (d) $\overline{S}$ is convex $\Rightarrow S$ is convex.

   Answer:

   (a) True.

   Proof. Let $S \subset \mathbb{R}^n$ be a convex set, $a, b \in S^\circ$, and $t \in [0, 1]$. It suffices to show that $c := (1 - t)a + tb \in S^\circ$. Since a, $b \in S^\circ$, there exist $r_a > 0$ and $r_b > 0$ such that $B(a, r_a) \subset S$ and $B(b, r_b) \subset S$. Define $r := \min\{r_a, r_b\} > 0$. Then $B(a, r) \subset S$, $B(b, r) \subset S$. For any $p \in B(c, r)$, let $x := a+(p-c) \in B(a, r) \subset S$ and $y := b+(p-c) \in B(b, r) \subset S$. It is easy to verify that $p = (1-t)x+ty$. Since S is convex and

$x, y \in S$, this implies $p \in S$. Since $p \in B(c, r)$ is arbitrary, we have $B(c, r) \subset S$ and so $c \in S^\circ$.

(b) False.

Example: Let $X := \mathbb{R}$, $S := (0, 1) \cup \{2\}$. Then $S^\circ = (0, 1)$ is convex but S is not.

(c) True.

Proof. For any $a, b \in \overline{S}$ and any $t \in [0, 1]$, it suffices to show that $c := (1 - t)a + tb \in \overline{S}$. To see this, let $r > 0$. Then there exist $x \in B(a, r) \cap S$ and $y \in B(b, r) \cap S$. Write $z := (1 - t)x + ty$. As S is convex and $x, y \in S$, we have $z \in S$. Furthermore, since

$$\begin{aligned} |z - c| &= |(1 - t)(x - a) + t(y - b)| \\ &\leq (1 - t)|x - a| + t|y - b| \\ &\leq (1 - t)r + tr \\ &= r\ , \end{aligned}$$

we have $z \in B(c, r) \cap S$ and so $c \in \overline{S}$. Hence $\overline{S}$ is convex.

(d) False.

Example: Let $X := \mathbb{R}$, $S := \mathbb{Q}$. Then $\overline{S} = \mathbb{R}$ is convex but S is not.

Part B: Problems

In what follows, unless otherwise specified, X is a general metric space.

1. Describe the open unit balls $B(0, 1)$ in $\mathbb{R}^2$ with metrics $\tilde{d}$, d_S and d_∞ which are defined in Example 1.1.2 (v), (viii) and (ix), respectively.

   Solution:

   For $\tilde{d}$:

   $$B(0, 1) = \left\{(x_1, x_2) \in \mathbb{R}^2 : \frac{(x_1 - y_1)^2}{(\frac{1}{10})^2} + (x_2 - y_2)^2 < 1\right\} < 1,$$

   which is the region inside the ellipse centered at the origin with semi x-axis $\frac{1}{10}$ and semi y-axis 1, excluding the boundary.

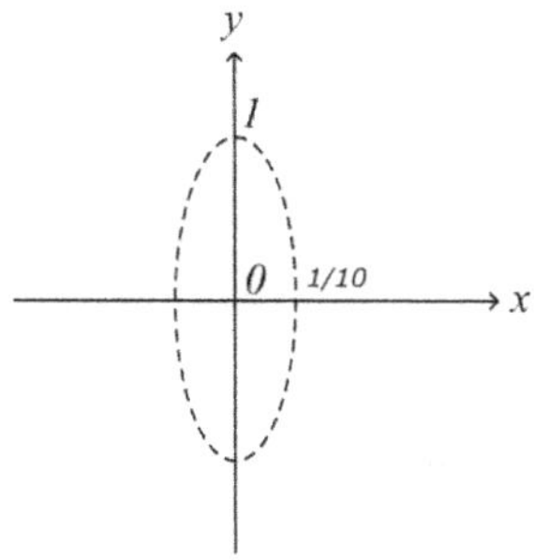

For d_S:
$B(0,1) = \{(x_1, x_2) \in \mathbb{R}^2 : |x_1| + |x_2| < 1\}$, which is the rectangular region centered at 0 with vertices $(\pm 1, 0)$, $(0, \pm 1)$, excluding the boundary.

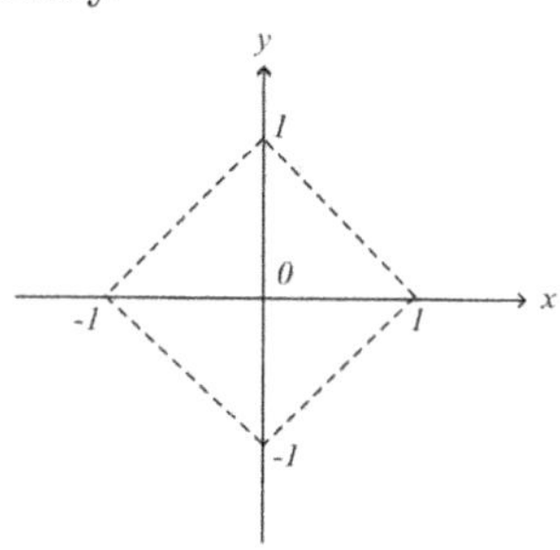

For d_∞:
$B(0,1) = \{(x_1, x_2) \in \mathbb{R}^2 : |x_1| < 1 \text{ and } |x_2| < 1\}$, which is the rectangular region centered at 0 with vertices $(\pm 1, \pm 1)$, excluding the boundary.

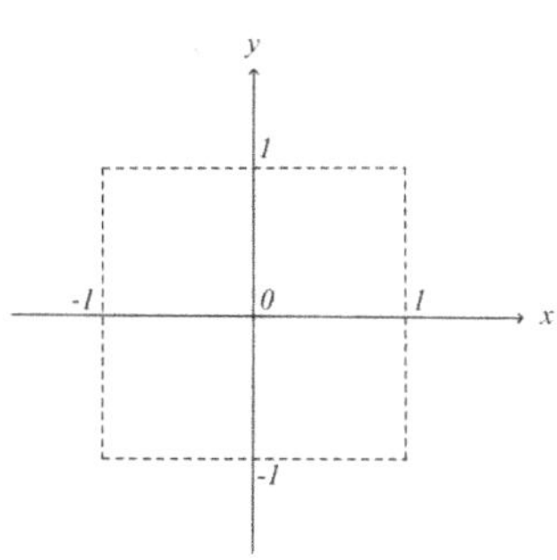

2. Compute
   (a) interior of $[0, \infty)$ in $\mathbb{R}$;
   (b) interior of $[0, 1)$ in $\mathbb{R}$;
   (c) interior of $[0, 1)$ in $[0, \infty)$;
   (d) interior of $\mathbb{Q}$ in $\mathbb{R}$.

Solution: Recall first that interior points of a set are automatically points in the given set. So to find all interior points of a set, we only need to consider the points in the given set and study whether they are interior points or not.

(a) *Answer*: $(0, \infty)$.
In fact, for any $x \in (0, \infty)$, $B_{\mathbb{R}}(x, \frac{x}{2}) = (\frac{x}{2}, \frac{3x}{2}) \subset [0, \infty)$ and so every point in $(0, \infty)$ is an interior point of $[0, \infty)$ in $\mathbb{R}$. It remains to consider the point $0 \in [0, \infty)$. As $B_{\mathbb{R}}(0, r) = (-r, r) \not\subset [0, \infty)$ for any $r > 0$, we see that 0 is not an interior point of $[0, \infty)$ in $\mathbb{R}$. Thus the interior of $[0, \infty)$ in $\mathbb{R}$ is $(0, \infty)$.

(b) *Answer*: $(0, 1)$.
In fact, for any $x \in (0, 1)$, if we write $r := \min\{x, 1 - x\} > 0$, then $B_{\mathbb{R}}(x, r) \subset [0, 1)$. Hence every point in $(0, 1)$ is an interior point of $[0, 1)$ in $\mathbb{R}$. It remains to consider the point $0 \in [0, 1)$. As $B_{\mathbb{R}}(0, r) = (-r, r) \not\subset [0, 1)$ for any $r > 0$, we see that 0 is not an interior point of $[0, 1)$ in $\mathbb{R}$. Thus the interior of $[0, 1)$ in $\mathbb{R}$ is $(0, 1)$.

(c) *Answer*: $[0, 1)$.
Similar to part (b) above, for any $x \in (0, 1)$, write $r := \min\{x, 1 - x\} > 0$. We have $B_{\mathbb{R}}(x, r) \subset [0, 1)$ and so $B_{[0,\infty)}(x, r) = B_{\mathbb{R}}(x, r) \cap [0, \infty) \subset [0, 1) \cap [0, \infty) = [0, 1)$. So every point in $(0, 1)$ is an interior point of $[0, 1)$ in $[0, \infty)$. It remains to consider the point $0 \in [0, 1)$. As $B_{[0,\infty)}(0, r) = B_{\mathbb{R}}(0, r) \cap [0, \infty) = (-r, r) \cap [0, \infty) = [0, r) \subset [0, 1)$ for any $0 < r < 1$, we see that 0 is an also interior point of $[0, 1)$ in $[0, \infty)$. Therefore, the whole set $[0, 1)$ is the interior of $[0, 1)$ in $[0, \infty)$.

(d) *Answer*: ϕ.

In fact, for any $x \in \mathbb{Q}$ and any $r > 0$, $B_{\mathbb{R}}(x, r) = (x-r, x+r)$ is bound to contain irrational numbers. Hence $B_{\mathbb{R}}(x, r) \not\subset \mathbb{Q}$. Therefore, no point in $\mathbb{Q}$ can be an interior point of $\mathbb{Q}$ in $\mathbb{R}$.

3. Let $S = \{(-1)^n + \frac{1}{n} : n \in \mathbb{N}\} \subset \mathbb{R}$ and $T = \{1 + \frac{1}{2n} : n \in \mathbb{N}\} \subset S$.
   (a) Is S open in $\mathbb{R}$?
   (b) Is S closed in $\mathbb{R}$?
   (c) Is T open in $\mathbb{R}$?
   (d) Is T closed in $\mathbb{R}$?
   (e) Is T open in S?

   Solution:
   (a) False.

   Justification. It is obvious that $B_{\mathbb{R}}\left((-1)^n + \frac{1}{n}, r\right) \not\subset S$ for any $n \in \mathbb{N}$ and any $r > 0$. Hence no point in S is an interior point of S in $\mathbb{R}$ and so S is not open in $\mathbb{R}$.

   (b) False.

   Justification. Consider $1 \in \mathbb{R} \setminus S$. For any $r > 0$, $1 + \frac{1}{2r} \in S \cap B_{\mathbb{R}}(1, r)$. So $1 \in \mathbb{R} \setminus S$ is not an interior point of $\mathbb{R} \setminus S$ and so $\mathbb{R} \setminus S$ is not open in $\mathbb{R}$. Hence S is not closed in $\mathbb{R}$.

   (c) False.

   Justification. Similar to (a). Details are left with the readers.

   (d) False.

   Justification. Similar to (b). Details are left with the readers.

   (e) True.

   Proof. Observe first that T contains all the positive elements of S. For any $t \in T$, write $t = 1 + \frac{1}{2n}$, where $n \in \mathbb{N}$. Note that all elements in $B_S\left(t, \frac{1}{2n}\right)$ are positive and so $B_S\left(t, \frac{1}{2n}\right) \subset T$. Hence T is open in S.

4. Show that S° is the largest subset of S which is open in X.

Proof.
S° is open: For any $x \in S^\circ$, there exists $r > 0$ such that $B(x, r_x) \subset S$. Note that actually $B(x, r_x) \in S^\circ$. In fact, for any $y \in B(x, r_x)$, since $B(x, r_x)$ is open, there is $\delta > 0$ such that $B(y, \delta) \subset B(x, r_x)$. Hence $B(y, \delta) \subset S$ and so $y \in S^\circ$. Since $y \in B(x, r_x)$ is arbitrary, we have $B(x, r_x) \subset S^\circ$ and so S° is open.
S° is the largest open set contained in S: Let $U \subset S$ be an open set in X. For any $u \in U$, there exists $r > 0$ such that $B(u, r) \subset U$. Hence $B(u, r) \subset S$ and so $u \in S^\circ$. Since $u \in U$ is arbitrary, we have $U \subset S^\circ$.

5. Show that $S^{\circ\circ} = S^\circ$ for any subset S of X.

   Proof. By definition, a subset $A \subset X$ is open if and only if $A^\circ = A$. Now by Exercise 1.2, Part B, Problem #4, $S^\circ \subset X$ is open, so we have $S^{\circ\circ} = S^\circ$.

6. Let $S = \{\frac{1}{n} : n \in \mathbb{N}\} \subset \mathbb{R}$. What are $\overline{S}$, S', $(S')'$, $\overline{S \cup \{0\}}$, $(S \cup \{0\})'$, and ∂S?

   Solution:
   $\overline{S} = \{\frac{1}{n} : n \in \mathbb{N}\} \cup \{0\}$,
   $S' = \{0\}$,
   $(S')' = \phi$,
   $\overline{S \cup \{0\}} = \{\frac{1}{n} : n \in \mathbb{N}\} \cup \{0\}$,
   $(S \cup \{0\})' = \{0\}$,
   $\partial S = \{\frac{1}{n} : n \in \mathbb{N}\} \cup \{0\}$.

7. Let S and T be two subsets of a metric space (X, d).
   (a) Show that $d(x, S) = 0$ if and only if $x \in \overline{S}$.
   (b) Prove or disprove that if S and T are two disjoint closed subsets of X, then $d(S, T) > 0$.
   (c) Prove or disprove that if S and T are two disjoint closed subsets of X, then there exist two open sets U and V with $S \subset U$, $T \subset V$, and $U \cap V = \phi$.

Solution:

(a)
$$\begin{aligned} d(x,S)=0 &\iff \inf_{s\in S} d(x,s)=0 \\ &\iff \forall\ r>0,\ \exists\ s_r\in S \text{ s.t. } d(x,s_r)<r \\ &\iff \forall\ r>0, B(x,r)\cap S\neq\phi \\ &\iff x\in\overline{S}\,. \end{aligned}$$

(b) False.

Example: In $X := \mathbb{R}$, consider the subsets $S := \mathbb{N}$ and $T := \{n+\frac{1}{2^n} : n\in\mathbb{N}\}$. Clearly, S and T are disjoint closed subsets of $\mathbb{R}$. But as $d(n, n+\frac{1}{2^n}) = \frac{1}{2^n}$ is arbitrarily small as n is arbitrarily large, we have $d(S,T)=0$.

(c) True.

Proof. Since S and T are closed and disjoint, we have by (a), $d(s,T)>0$ and $d(t,S)>0$ for all $s\in S$ and all $t\in T$. Let

$$U := \bigcup_{s\in S} B\Big(s, \frac{d(s,T)}{3}\Big), \qquad V := \bigcup_{t\in T} B\Big(t, \frac{d(t,S)}{3}\Big).$$

Clearly both U and V are open in X, $S\subset U$, and $T\subset V$. It remains to show that $U\cap V=\phi$. Suppose not, then there exists $z\in U\cap V$. There are $s_0\in S$ and $t_0\in T$ such that $z\in B(s_0, \frac{d(s_0,T)}{3})\cap B(t_0, \frac{d(t_0,S)}{3})$. Without loss of generality, assume $d(s_0,T)\geq d(t_0,S)$. Then we have

$$\begin{aligned} d(s_0,T) &\leq d(s_0,t_0)\leq d(s_0,z)+d(z,t_0) \\ &< \frac{d(s_0,T)}{3}+\frac{d(t_0,S)}{3}\leq\frac{2d(s_0,T)}{3}, \end{aligned}$$

which is impossible.

8. For any metric space X and any subset $\phi\neq E\subset X$, determine whether each of the following statements is true:
   (a) For any $a\in X$, $d(a,E)=d(a,\overline{E})$.
   (b) $d(E)=d(\overline{E})$.

Answer:

(a) True.

Proof. As $E \subset \overline{E}$, we have

$$\{d(a,y) : e \in E\} \subset \{d(a,y) : y \in \overline{E}\}$$

and so

$$d(a,E) = \inf_{y \in E}\{d(a,y)\} \geq \inf_{y \in \overline{E}}\{d(a,y)\} = d(a,\overline{E})\ .$$

Conversely, let $\varepsilon > 0$ be given. For any $x \in \overline{E}$, by Exercise 1.2, Part B, Problem #7(a), $d(x,E) = 0$. Hence there exists $e \in E$ such that $d(x,e) < \varepsilon$ and so

$$d(a,E) \leq d(a,e) \leq d(a,x) + d(x,e) < d(a,x) + \varepsilon\ .$$

Since this is true for all $x \in \overline{E}$, we have

$$d(a,E) \leq d(a,\overline{E}) + \varepsilon\ .$$

Since $\varepsilon > 0$ is arbitrary, we conclude that $d(a,E) \leq d(a,\overline{E})$ and so the result follows.

(b) True.

Proof. Since $E \subset \overline{E}$, we have

$$\{d(x,y) : x,y \in E\} \subset \{d(x,y) : x,y \in \overline{E}\}$$

and so

$$d(E) = \sup_{x,y \in E}\{d(x,y)\} \leq \sup_{x,y \in \overline{E}}\{d(x,y)\} = d(\overline{E})\ .$$

Conversely, let $\varepsilon > 0$ be given. For any x, $y \in \overline{E}$, there exist x', $y' \in E$ with $d(x,x') < \varepsilon$ and $d(y,y') < \varepsilon$. Therefore,

$$\begin{aligned} d(x,y) \leq&\ d(x,x') + d(x',y') + d(y',y) \\ <&\ d(x',y') + 2\varepsilon \leq d(E) + 2\varepsilon\ . \end{aligned}$$

Since x, $y \in \overline{E}$ are arbitrary, this implies

$$d(\overline{E}) = \sup\{d(x,y) : x,y \in \overline{E}\} \leq d(E) + 2\varepsilon\ .$$

Finally, since $\varepsilon > 0$ is arbitrary, we conclude that $d(\overline{E}) \leq d(E)$ and so the result follows.

9. (a) Let S be a nonempty subset of X. Show that $\overline{S}$ is the intersection of all closed sets of X containing S.
   (b) Is it always true that $(\bigcup_{n=1}^{\infty} A_n)^{\circ} = \bigcup_{n=1}^{\infty} A_n^{\circ}$?
   [Compare with Exercise 1.2, Part A, Problem #1(d), (e) and Problem #7(a).]
   (c) Is it always true that $\overline{\bigcup_{n=1}^{\infty} A_n} = \bigcup_{n=1}^{\infty} \overline{A_n}$?
   [Compare with in Exercise 1.2, Part A, Problem #3(d), (e) and Problem #7(c).]
   (d) Is it always true that $(\bigcup_{n=1}^{\infty} A_n)' = \bigcup_{n=1}^{\infty} A_n'$?
   [Compare with Exercise 1.2, Part A, Problem #4(d), (e) and Problem #7(e).]
   (e) Is it always true that $(\bigcap_{n=1}^{\infty} A_n)^{\circ} = \bigcap_{n=1}^{\infty} A_n^{\circ}$?
   [Compare with Exercise 1.2, Part A, Problem #1(d), (e) and Problem #7(b).]
   (f) Give examples of a metric space in which $\overline{B}(a,r) \neq \overline{B(a,r)}$, i.e., the closed ball $\overline{B}(a,r)$ is not the closure of the open ball $B(a,r)$.

Solution:

(a) First of all, as $\overline{S} \in \{A \subset X : S \subset A \text{ and } A \text{ is closed in } X\}$, we have

$$\bigcap\{A \subset X : S \subset A \text{ and } A \text{ is closed in } X\} \subset \overline{S}.$$

On the other hand, since

$$\bigcap\{A \subset X : S \subset A \text{ and } A \text{ is closed in } X\}$$

is closed in X and contains S, and $\overline{S}$ is the smallest such set, we have

$$\overline{S} \subset \bigcap\{A \subset X : S \subset A \text{ and } A \text{ is closed in } X\}.$$

(b) No.
Example: Let $X := \mathbb{R}$ and $A_n := [n, n+1]$, $n \in \mathbb{N}$. Then $(\bigcup_{n=1}^{\infty} A_n)^{\circ} = [1,\infty)^{\circ} = (1,\infty) \neq \bigcup_{n=1}^{\infty}(n, n+1) = \bigcup_{n=1}^{\infty} A_n^{\circ}$

(c) No.

Example: Let $X := \mathbb{R}$. Since $\mathbb{Q}$ is countable, we can write $\mathbb{Q} = \{r_n : n \in \mathbb{N}\}$. Take $A_n := \{r_n\}$, $n \in \mathbb{N}$. Then $\overline{\bigcup_{n=1}^{\infty} A_n} = \overline{\mathbb{Q}} = \mathbb{R} \neq \mathbb{Q} = \bigcup_{n=1}^{\infty} \overline{A_n}$.

(d) No.

Using the same example as that in (c), we have $(\bigcup_{n=1}^{\infty} A_n)' = \mathbb{Q}' = \mathbb{R} \neq \phi = \bigcup_{n=1}^{\infty} A_n'$.

(e) No.

Example: Let $X := \mathbb{R}$ and $A_n := \left(-1, \frac{1}{n}\right)$, $n \in \mathbb{N}$. Then

$$\left(\bigcap_{n=1}^{\infty} A_n\right)^{\circ} = \left(\bigcap_{n=1}^{\infty}\left(-1, \frac{1}{n}\right)\right)^{\circ} = (-1, 0]^{\circ} = (-1, 0)\,,$$

while

$$\bigcap_{n=1}^{\infty} A_n^{\circ} = \bigcap_{n=1}^{\infty}\left(-1, \frac{1}{n}\right) = (-1, 0]\,.$$

(f) *Example 1*:

Let X be a discrete metric space with $\#(X) > 1$. Let $a \in X$. Then $\overline{B}(a, 1) = X \neq \{a\} = \overline{B(a, 1)}$.

Example 2: Let $X := \mathbb{Z}$ with the Euclidean metric. Then $\overline{B}(0, 1) = \{-1, 0, 1\} \neq \{0\} = \overline{B(0, 1)}$.

Example 3: Let $X := (-\infty, 0] \cup [1, +\infty)$ with the Euclidean metric. Then $\overline{B}(0, 1) = [-1, 0] \cup \{1\} \neq [-1, 0] = \overline{B(0, 1)}$.

10. Show that if $\phi \neq S \subset \mathbb{R}$ is both open and closed, then $S = \mathbb{R}$.

Proof. Suppose to the contrary that $S \neq \mathbb{R}$. Let $T := \mathbb{R} \setminus S$. Then T is also nonempty, proper, and both open and closed in $\mathbb{R}$. Let $\alpha > 0$ be such that both $(-\alpha, \alpha) \cap S$ and $(-\alpha, \alpha) \cap T$ are nonempty (why exists?). As S is open in $\mathbb{R}$, so is the intersection $(-\alpha, \alpha) \cap S$. By Theorem 1.2.16, there is a countable family of pairwisely disjoint open intervals $\{I_n\}$ such that

$$(-\alpha, \alpha) \cap S = \bigcup_n I_n.$$

Write $I_1 = (a, b)$. If $b \neq \alpha$, by the definition of T, $b \in T$. However, as T is open, there is $0 < \delta < b - a$ such that $(b - \delta, b + \delta) \subset T$. But it means that $S \cap T \supset (b - \delta, b) \neq \phi$ and contradiction arises. Therefore $b = \alpha$. Similarly, one has $a = -\alpha$. That means the whole family $\{I_n\}$ reduces to one single open interval, namely, $(-\alpha, \alpha)$, i.e., $(-\alpha, \alpha) \cap S = (-\alpha, \alpha)$, or $(-\alpha, \alpha) \subset S$. But that contradicts to $T \cap (-\alpha, \alpha) \neq \phi$. Hence $S = \mathbb{R}$.

11. Let X be a nonempty set. A metric d on X is said to be an *ultrametric* if $d(x, y) \leq \max(d(x, z), d(y, z))$ for all $x, y, z \in X$.
    (a) Give an example of ultrametric.
    (b) Suppose d is an ultrametric. Show that
        (i) for any $a, b, x \in X$, if $d(a, x) < d(a, b)$, then $d(x, b) = d(a, b)$;
        (ii) closed balls in (X, d) are open;
        (iii) open balls in (X, d) are closed;
        (iv) the sphere $\{x \in X : d(x, a) = r\}$ is clopen for any $r > 0$.

Solution:

(a) *Example 1*: The discrete metric is an ultrametric.

Example 2: Let p be a prime number. Every non-zero rational number x can be expressed as $p^k \cdot \frac{r}{s}$ for some k, $r \in \mathbb{Z}$ and $s \in \mathbb{N}$, with neither r nor s is divisible by p. Define

$$|x|_p := \begin{cases} p^{-k} & \text{if } x \neq 0 \\ 0 & \text{if } x = 0, \end{cases}$$

and $d(x, y) := |x - y|_p$ for any x, $y \in \mathbb{Q}$. It turns out that this is a well-defined metric on $\mathbb{Q}$ and is known as the "p-adic metric" on $\mathbb{Q}$. In fact, it is not difficult to check that it is non-negative, symmetric, and $d(x, y) = 0$ if and only if $x = y$. Note that the triangle inequality would follow once we have $d(x, y) \leq \max\{d(x, z), d(y, z)\}$. Hence it only remains to show this inequality. To do this, we write $|x - z|_p = p^{-n_1}$

and $|z - y|_p = p^{-n_2}$. Without loss of generality, we assume $n_1 \le n_2$. By the definition of $|\cdot|_p$, we have $x - z = p^{n_1} \cdot \frac{u_1}{v_1}$ and $z - y = p^{n_2} \cdot \frac{u_2}{v_2}$ with $p \nmid u_1, u_2, v_1, v_2$, which implies

$$\begin{aligned} x - y &= p^{n_1}\left(\frac{u_1}{v_1} + p^{n_2 - n_1}\frac{u_2}{v_2}\right) \\ &= p^{n_1}\left(\frac{u_1 v_2 + p^{n_2-n_1} u_2 v_1}{v_1 v_2}\right) . \end{aligned}$$

Since $p \nmid v_1 v_2$ and $p \nmid u_1 v_2$, we have

$$\begin{aligned} d(x,y) = |x-y|_p \le p^{-n_1} &= \max(p^{-n_1}, p^{-n_2}) \\ &= \max(|x-z|_p, |y-z|_p) \\ &= \max\{d(x,z), d(y,z)\} . \end{aligned}$$

(b) (i) Since $d(a,x) < d(a,b)$, we have

$$\begin{aligned} d(x,b) \le \max\{d(x,a), d(a,b)\} &= d(a,b) \\ &\le \max\{d(a,x), d(x,b)\} . \end{aligned}$$

But then again because $d(a,x) < d(a,b)$, the last inequality forces $d(x,b) \ge d(a,b)$. Combining, we conclude that $d(x,b) = d(a,b)$.

(ii) For any $a \in X$ and any $r > 0$, let $b \in \overline{B}(a,r)$. For any $c \in B(b,r)$, we have

$$d(c,a) \le \max\{d(c,b), d(b,a)\} \le r ,$$

thus $c \in \overline{B}(a,r)$. Hence $B(b,r) \subset \overline{B}(a,r)$ and so $\overline{B}(a,r)$ is open.

(iii) For any $a \in X$ and any $r > 0$, let $b \in X \setminus B(a,r)$. If there is $c \in B(b,r) \cap B(a,r)$, we have

$$d(a,b) \le \max\{d(a,c), d(c,b)\} < r ,$$

that is, $b \in B(a,r)$, which is absurd. Therefore, $B(b,r) \cap B(a,r) = \phi$, or equivalently, $B(b,r) \subset X \setminus B(a,r)$. Hence $X \setminus B(a,r)$ is open.

(iv) Note that

$$\{x \in X : d(x,a) = r\} = \overline{B}(a,r) \cap (X \setminus B(a,r)).$$

By (ii) and (iii) above, both sets on the right hand side are clopen, hence the sphere is also clopen.

12. Let $S \subset \mathbb{R}^n$. Show that there are at most countably many isolated points in S. Hence, show that if S is uncountable then S' is uncountable.

Proof. Let $\Lambda \subset S$ be the set of isolated points of S. For any $\lambda \in \Lambda$, there is an open set $U_\lambda \subset \mathbb{R}^n$ such that $U_\lambda \cap S = \{\lambda\}$. So we have $\Lambda = \bigcup_{\lambda\in\Lambda} U_\lambda$. By Lindelöf's theorem, there exists a countable subcollection $\{U_i\}_{i\in\mathbb{N}} \subset \{U_\lambda\}_{\lambda\in\Lambda}$ such that $\Lambda = \bigcup_{i\in\mathbb{N}} U_i$.

On the other hand, as each U_λ contains exactly one point in Λ, no proper subcollection of $\{U_\lambda\}_{\lambda\in\Lambda}$ could cover Λ. So $\{U_\lambda\}_{\lambda\in\Lambda} = \{U_i\}_{i\in\mathbb{N}}$ and hence Λ is countable.

Finally, if S is uncountable, $S \setminus \Lambda$ is uncountable and so $S' \supset S \setminus \Lambda$ is uncountable.

13. A collection $\mathcal{A}$ of subsets of X is said to have the "*countable intersection property*" if every countable intersection of elements of $\mathcal{A}$ is nonempty. Show that every family of closed nonempty subsets of $\mathbb{R}^n$ which has the countable intersection property has a nonempty intersection.

Proof. Let $\mathcal{A}$ be a family of closed subsets of $\mathbb{R}^n$ with the countable intersection property. Suppose $\bigcap_{A\in\mathcal{A}} A = \phi$. Consider $\mathcal{U} := \{\mathbb{R}^n \setminus A : A \in \mathcal{A}\}$. We have

$$\bigcup_{U\in\mathcal{U}} U = \bigcup_{A\in\mathcal{A}} \left(\mathbb{R}^n \setminus A\right) = \mathbb{R}^n \setminus \left(\bigcap_{A\in\mathcal{A}} A\right) = \mathbb{R}^n \setminus \phi = \mathbb{R}^n .$$

Hence by Lindelöf Theorem, there is a countable subcollection $\mathcal{U}' \subset \mathcal{U}$ such that

$$\bigcup_{U \in \mathcal{U}'} U = \mathbb{R}^n .$$

Let $\mathcal{A}' := \{\mathbb{R}^n \setminus U : U \in \mathcal{U}'\}$. Then $\mathcal{A}'$ is countable and

$$\phi = \mathbb{R}^n \setminus \Big(\bigcup_{U \in \mathcal{U}'} U \Big) = \bigcap_{U \in \mathcal{U}'} \Big(\mathbb{R}^n \setminus U \Big) = \bigcap_{A \in \mathcal{A}'} A ,$$

contradicting the countable intersection property of $\mathcal{A}$. Hence $\bigcap_{A \in \mathcal{A}} A \neq \phi$.

14. Would Lindelöf Theorem remain valid when $\mathbb{R}^n$ is equipped with other metrics instead of the Euclidean one?

    Answer: Not in general.
    Justification. Let $X := \mathbb{R}$ with the discrete metric. Then $\mathcal{F} := \big\{\{x\} : x \in X\big\}$ is an infinite collection of open subsets of X which covers X. But then it is obvious that no countable subcollection of $\mathcal{F}$ can cover X.

15. Show that a collection of nonempty disjoint open sets in $\mathbb{R}^n$ must be countable. What will happen if we replace open sets by closed sets?

    Solution: Let $\mathcal{F}$ be a collection of nonempty disjoint open sets in $\mathbb{R}^n$. Since $\mathbb{Q}^n$ is countable, we can write $\mathbb{Q}^n = \{r_k : k \in \mathbb{N}\}$. For any $S \in \mathcal{F}$, since S is an open in $\mathbb{R}^n$ and $\mathbb{Q}^n$ is dense in $\mathbb{R}^n$, $S \cap \mathbb{Q}^n \neq \phi$. Hence there exists some $r_S \in S \cap \mathbb{Q}^n$. Do this for every $S \in \mathcal{F}$ and we form a set $\{r_S : S \in \mathcal{F}\} \subset \mathbb{Q}^n$. As $\mathbb{Q}^n$ is countable, the same is true for the set $\{r_S : S \in \mathcal{F}\}$. Observe that if $r_S = r_{S'}$ for some $S, S' \in \mathcal{F}$, then we have $r_S \in S \cap \mathbb{Q}^n$ and $r_S \in S' \cap \mathbb{Q}^n$. By the pairwise disjointness of elements in $\mathcal{F}$, this forces $S = S'$. Hence it follows that $\mathcal{F}$ and $\{r_S : S \in \mathcal{F}\}$ have the same cardinality, and so $\mathcal{F}$ is countable.

Finally, the same statement is not true if we replace open sets by closed sets, that is, there are uncountable collections of disjoint closed sets in $\mathbb{R}^n$. For example, $\{\{x\} : x \in \mathbb{R}^n\}$.

16. (a) If X is a discrete metric space and $S \subset X$, what is ∂S?
    (b) For any subset $S \subset X$, is ∂S always closed?

Solution:

(a) $\partial S = \phi$.

(b) True.

Proof. Note that $\partial S = \overline{S} \setminus S^\circ = \overline{S} \cap (X \setminus S^\circ)$. Being the intersection of two closed sets, ∂S is always closed.

17. Let $X := \mathbb{R}$, $Y := [a, \infty)$, $S =: [a, b)$, $a < b$. Compute
    (a) the boundary of Y in X;
    (b) the boundary of S in X;
    (c) the boundary of S in Y.

Solution:

(a) $\partial Y = \{a\}$.

(b) $\partial S = \{a, b\}$.

(c) The boundary of S in Y is $\{b\}$.

18. Let S, $T \subset X$.
    (a) Show that if S open, then $(\partial S)^\circ = \phi$.
    (b) We have seen in Exercise 1.2, Part A, Problem #6(a), (b) that in general, $\partial(S \cup T) \subset \partial S \cup \partial T$ but $\partial(S \cup T) \not\supset \partial S \cup \partial T$. Show that if $\overline{S} \cap \overline{T} = \phi$, then $\partial S \cup \partial T = \partial(S \cup T)$.

Proof.

(a) For any $x \in (\partial S)^\circ$, there exists $r_0 > 0$ such that $B(x, r_0) \subset \partial S$. But as $x \in (\partial S)^\circ \subset \partial S$, we have $\phi \neq B(x, r_0) \cap S \subset \partial S \cap S = \partial S \cap S^\circ = \phi$, which is absurd.

(b) It suffices to show that $\partial S \cup \partial T \subset \partial(S \cup T)$. Let $x \in \partial S \cup \partial T$.

Without loss of generality, assume $x \in \partial S$. Then

$$\left\{\begin{array}{ll} B(x,r)\cap(S\cup T)\supset B(x,r)\cap S & \neq \phi \\ \qquad\qquad\qquad\qquad B(x,r)\cap(X\setminus S) & \neq\phi \end{array}\right. \quad \forall\, r>0.$$

On the other hand, as $\overline{S}\cap\overline{T}=\phi$, we have $x\in\partial S\subset\overline{S}\subset X\setminus\overline{T}$. Hence there exists $r_0>0$ such that

$$B(x,r_0)\subset X\setminus\overline{T}$$

and so

$$B(x,r)\subset X\setminus\overline{T}\quad \forall\, 0<r\leq r_0\,.$$

Hence for all $0<r\leq r_0$, we have

$$\begin{aligned} B(x,r)\cap(X\setminus(S\cup T)) &= B(x,r)\cap(X\setminus T)\cap(X\setminus S)\\ &\supset B(x,r)\cap(X\setminus\overline{T})\cap(X\setminus S)\\ &= B(x,r)\cap(X\setminus S)\\ &\neq\phi\,, \end{aligned}$$

and so

$$B(x,r)\cap(X\setminus(S\cup T))\neq\phi\qquad\forall\, r>0\,.$$

Therefore, $x\in\partial(S\cup T)$.

19. Let $X:=C[0,1]$. For any $f,\,g\in X$, define

$$\begin{aligned} d_\infty(f,g) &:= \sup_{x\in[0,1]}|f(x)-g(x)|\,,\\ d_1(f,g) &:= \int_0^1 |f(x)-g(x)|\,dx\,,\\ d_2(f,g) &:= \left[\int_0^1 (f(x)-g(x))^2\,dx\right]^{1/2}. \end{aligned}$$

Let $S:=\{f\in X\,:\,f(0)=0\}$ and $T=\{f\in S\,:\,f(1)\neq 0\}$.

(a) Show that d_∞, d_1, and d_2 are metrics.
(b) Determine whether S is closed in (X, d_1).
(c) Determine whether S is closed in (X, d_2).
(d) Determine whether S is closed in (X, d_∞).
(e) Determine whether T is open in (X, d_∞).
(f) Determine whether T is open in (S, d_∞).
(g) Are there two positive real numbers α and β such that

$$\alpha\, d_\infty(f, g) \le d_2(f, g) \le \beta\, d_\infty(f, g)$$

for all $f,\ g \in X$?
[In general, if d and d' are two metrics on the same underlying set X and there are constants $\alpha,\ \beta > 0$ such that

$$\alpha\, d(x, y) \le d'(x, y) \le \beta\, d(x, y) \qquad \forall\ x, y \in X\ ,$$

then the two metrics d and d' are said to be *equivalent*. For example, the metrics d_1, d_2 and d_∞ on $\mathbb{R}^n$ discussed in Exercise 1.1, Part B, Problem #6 are all equivalent. Note that when two metrics d and d' on a set X are equivalent, the collections of open sets in (X, d) and (X, d') are identical. So here the problem is asking whether d_∞ and d_2 are equivalent.]
(h) Show that for all $f, g \in X$ with $f \ne g$ and $d_2(f, 0) = d_2(g, 0) = 1$, we have $d_2(f + g, 0) < 2$. What if d_2 is replaced by d_∞?

Solution:

(a) d_∞ is a metric:
Clearly, $d_\infty(f, g) \ge 0$ and $d_\infty(f, f) = 0$ for all $f,\ g \in X$. Moreover, if $d_\infty(f, g) = 0$, then for all $t \in [0, 1]$,

$$0 \le |f(t) - g(t)| \le \sup_{x\in[0,1]} |f(x) - g(x)| = d_\infty(f, g) = 0\ .$$

It follows that $f = g$ on $[0, 1]$ and hence (M1) holds. Obviously, (M2) also holds. Finally, let f, g, $h \in X$. For all $x \in [0, 1]$, we have

$$\begin{aligned}|f(x) - g(x)| &\leq |f(x) - h(x)| + |h(x) - g(x)| \\ &\leq d_\infty(f, h) + d_\infty(h, g) \ .\end{aligned}$$

Thus

$$d_\infty(f, g) = \sup_{x \in [0,1]} |f(x) - g(x)| \leq d_\infty(f, h) + d_\infty(h, g)$$

and so (M3) also holds.

d_1 is a metric:

(M1) and (M2) are clear. It remains to verify (M3). For this we let f, g, $h \in X$. For any $x \in [0, 1]$, by the usual triangle inequality for the Euclidean metric in $\mathbb{R}$, we have

$$|f(x) - h(x)| \leq |f(x) - g(x)| + |g(x) - h(x)| \ .$$

Since x is arbitrary, we have

$$\begin{aligned}d_1(f, h) &= \sup_{x \in [0,1]} |f(x) - h(x)| \\ &\leq \sup_{x \in [0,1]} \left(|f(x) - g(x)| + |g(x) - h(x)|\right) \\ &\leq \sup_{x \in [0,1]} |f(x) - g(x)| + \sup_{x \in [0,1]} |g(x) - h(x)| \\ &\quad \text{(why?)} \\ &= d_1(f, g) + d_1(g, h) \ .\end{aligned}$$

Hence (M3).

d_2 is a metric:

(M1) and (M2) are clear. It remains to verify (M3). For this we let f, g, $h \in X$. By Cauchy–Schwarz inequality,

$$\begin{aligned}
&d_2^2(f,g)\\
&= \int_0^1 \big(f(x)-g(x)\big)^2 dx\\
&= \int_0^1 \big(f(x)-h(x)+h(x)-g(x)\big)^2 dx\\
&= \int_0^1 \big(f(x)-h(x)\big)^2 dx + \int_0^1 \big(h(x)-g(x)\big)^2 dx\\
&\quad + 2\int_0^1 \big(f(x)-h(x)\big)\big(h(x)-g(x)\big)dx\\
&\le \int_0^1 \big(f(x)-h(x)\big)^2 dx + \int_0^1 \big(h(x)-g(x)\big)^2 dx\\
&\quad + 2\Big[\int_0^1 \big(f(x)-h(x)\big)^2 dx\Big]^{\frac{1}{2}} \Big[\int_0^1 \big(h(x)-g(x)\big)^2 dx\Big]^{\frac{1}{2}}\\
&= \big(d_2(f,h)+d_2(h,g)\big)^2 .
\end{aligned}$$

Hence (M3) holds.

(b) *Answer*: No, S is not closed in (X, d_1).

Justification. It suffices to show that $X \setminus S$ is not open in (X, d_1). This boils down to finding an $f \in X \setminus S$ such that for any $r > 0$ there is some $g \in B(f, r) \cap S$. For this we consider $f :\equiv 1$. Then clearly, $f \in X \setminus S$. For any $r > 0$, assuming without loss of generality that $0 < r < 1$, we define

$$g(x) := \begin{cases} \dfrac{x}{r} & \text{if } 0 \le x \le r \\ 1 & \text{if } x > r\ . \end{cases}$$

Then g is continuous and $g(0) = 0$. Hence, $g \in S$. Note that

$$\begin{aligned}
d_1(f,g) &= \int_0^1 |f(x)-g(x)|\, dx\\
&= \int_0^r \Big(1-\frac{x}{r}\Big)\, dx = \frac{r}{2} < r
\end{aligned}$$

and so $g \in B(f, r)$. It follows that $X \setminus S$ is not open and hence S is not closed.

(c) *Answer*: No, S is not closed in (X, d_2).
Justification. It suffices to show that $X \setminus S$ is not open in (X, d_2). For this, take $f :\equiv 1$ on $[0, 1]$. Clearly, $f \in X \setminus S$. For any $\delta > 0$, set

$$f_\delta(x) := \begin{cases} \frac{1}{\delta}x & \text{if } 0 \le x \le \delta \\ 1 & \text{if } \delta < x \le 1 \,. \end{cases}$$

Note that $f_\delta(0) = 0$ and hence $f_\delta \in S$. Furthermore,

$$d_2(f_\delta, f) = \left(\int_0^\delta \left(1 - \frac{x}{\delta}\right)^2 dx \right)^{\frac{1}{2}} = \sqrt{\frac{\delta}{3}} < \sqrt{\delta} \,.$$

So for any $r > 0$, let $\delta := r^2$. Then we have

$$d_2(f_\delta, f) < \sqrt{\delta} = r \,,$$

that is, $f_\delta \in B_{d_2}(f, r) \cap S$. In other words, for any $r > 0$, $B_{d_2}(f, r) \not\subset X \setminus S$. Hence $f \in X \setminus S$ is not an interior point of $X \setminus S$ and so $X \setminus S$ is not open in (X, d_2).

(d) *Answer*: Yes, S is closed in (X, d_∞).
Proof. It suffices to show that $X \setminus S$ is open in (X, d_∞). Let $f \in X \setminus S$, so $f(0) \ne 0$. Then for any $g \in B_{d_\infty}(f, |f(0)|/2)$, we have

$$\begin{aligned} \Big| |f(0)| - |g(0)| \Big| &\le |f(0) - g(0)| \le \sup_{x \in [0,1]} |f(x) - g(x)| \\ &= d_\infty(f, g) < |f(0)|/2 \end{aligned}$$

and so $|g(0)| > |f(0)|/2 > 0$. Thus $g \in X \setminus S$. Since $g \in B_{d_\infty}(f, |f(0)|/2)$ is arbitrary, we have $B_{d_\infty}(f, |f(0)|/2) \subset X \setminus S$ and so $X \setminus S$ is open in (X, d_∞).

(e) *Answer*: T is not open in (X, d_∞).
Justification. Consider the function $f(x) := x$, $x \in [0,1]$. Since $f(0) = 0$ and $f(1) = 1 \neq 0$, $f \in T$. For each $0 < \delta < 1$, set $f_\delta := f + \frac{\delta}{2}$. Then

$$d_\infty(f_\delta, f) = \sup_{x\in[0,1]} |f_\delta(x) - f(x)| = \delta/2 < \delta \ ,$$

and so $f_\delta \in B_{d_\infty}(f, \delta)$. However, as $f_\delta(0) = \delta/2 > 0$, $f_\delta \notin T$. Thus $B_{d_\infty}(f, \delta) \not\subset T$ and so T is not open in (X, d_∞).

(f) *Answer*: T is open in (S, d_∞).
Proof. In fact, note that $T = S \cap V$ where

$$V := \{f \in X : f(1) \neq 0\}.$$

An argument exactly the same as that in (d) shows that $X \setminus V = \{f \in X : f(1) = 0\}$ is closed in (X, d_∞). Hence V is open in (X, d_∞) and so $T = S \cap V$ is open in (S, d_∞).

(g) *Answer*: No, d_∞ and d_2 are not equivalent.
Justification. Suppose there is a positive real number α such that

$$\alpha\, d_\infty(f, g) \le d_2(f, g) \qquad \forall\ f, g \in X\ . \tag{*}$$

Take $n > \max\left\{\frac{1}{\alpha}, 1\right\}$, $f_n(x) := x^{\frac{n^2-1}{2}}$ and $g := 0$. Then

$$\begin{aligned} d_\infty(f_n, g) &= \sup_{x\in[0,1]} |f_n(x) - 0| = 1\ , \\ d_2(f_n, g) &= \left[\int_0^1 \left(x^{\frac{n^2-1}{2}}\right)^2 dx\right]^{1/2} = \frac{1}{n}\ . \end{aligned}$$

By (*), we have $\alpha \le \frac{1}{n}$, which contradicts to the choice of n. Therefore there is no such α. Similar arguments show that there is no β satisfying $d_2(f, g) \le \beta\, d_\infty(f, g)$ for all f, $g \in X$. Hence we conclude that the metrics d_∞ and d_2 are not equivalent.

(h) Simple algebra yields

$$d_2^2(f+g,0)+d_2^2(f-g,0)=2[d_2^2(f,0)+d_2^2(g,0)]=4$$

from which the inequality $d_2(f+g,0)<2$ follows immediately. However, the corresponding result for d_∞ fails to hold. For example, consider $f:\equiv 1$ and $g(x):=x$. Then

$$d_\infty(f,0)=d_\infty(g,0)=1$$

but $d_\infty(f+g,0)=2$.

1.3 Compactness

Definition 1.3.1. *Let S be a subset of X. A family $\mathcal{F}$ of subsets of X is said to be a cover of S (or, equivalently, $\mathcal{F}$ is said to cover S) if $S \subset \bigcup_{F\in\mathcal{F}} F$. If all elements of $\mathcal{F}$ are open sets in X, $\mathcal{F}$ is called an open cover of S.*

Example 1.3.2.

(i) $X = \mathbb{R}$, $\{(\frac{1}{n}, \frac{2}{n})\}_{n=1}^{\infty}$ is an open cover of $(0, 1)$.

(ii) $X = \mathbb{R}$, $\big\{(-n, n)\big\}_{n=1}^{\infty}$ is an open cover of $\mathbb{R}$.

(iii) $X = \mathbb{R}$, $\big\{(n, n+2)\big\}_{n\in\mathbb{Z}}$ is an open cover of $\mathbb{Q}$.

(iv) $X = \mathbb{R}$, $\big\{(r, \infty)\big\}_{r>0}$ is an open cover of $(0, \infty)$.

(v) $X = [1, \infty)$, $\left\{\left[1, 3 - \frac{1}{n}\right)\right\}_{n\in\mathbb{N}}$ is an open cover of $[1, 3)$.

Here, note that the open covers in (i), (ii), (iii) and (v) are *countable*, while the one in (iv) is *uncountable*.

Definition 1.3.3. *A subset $S \subset X$ is said to be compact if every open cover of S has a finite sub-cover, that is, for any open cover $\mathcal{F} = \{U_\alpha\}_{\alpha\in\Lambda}$ of S, where Λ is an index set, there is a finite subcollection $\{U_1, \dots, U_n\} \subset \mathcal{F}$ such that $S \subset \bigcup_{i=1}^{n} U_\alpha$.*

Example 1.3.4.

(i) In $\mathbb{R}$, $\mathcal{F} := \big\{B(x, \frac{1}{n}) : x \in [0, 1], n \in \mathbb{N}\big\}$ is an open cover of $[0, 1]$. It is clear that $\big\{B(0, 1), B(1, 1)\big\}$ is a finite subcover of $\mathcal{F}$. But then of course we cannot conclude just with this observation that $[0, 1]$ is compact. The point is, to show that $[0, 1]$ is compact, we need to show that *every* open cover of X has a finite subcover. Showing that one particular open cover has a finite subcover is not sufficient.

(ii) $\{(n-1, n+1) : n \in \mathbb{Z}\}$ is an open cover of $\mathbb{R}$ with no finite subcover. Note that since we have exhibited that one specific open cover of $\mathbb{R}$ has no finite subcover, we can conclude that $\mathbb{R}$ is not compact.

(iii) $\left\{\left(\frac{1}{n}, 1-\frac{1}{n}\right) : n \in \mathbb{N}\right\}$ is an open cover of (0, 1) with no finite subcover. Hence $(0, 1)$ is not compact.

From these examples, we can imagine that if a metric space is not bounded (like $\mathbb{R}$), or it is not closed (like $(0, 1)$), then there is no hope that the space is compact. In fact, we have the following necessary condition for compactness.

Theorem 1.3.5. *Every compact subset of X is closed and bounded.*

Proof. Let $S \subset X$ be compact. If $S = \phi$, the theorem is clear. So suppose $S \neq \phi$ and so we have a point $p \in S$. Then $\{B(p, n) : n \in \mathbb{N}\}$ is an open cover of S and so it has a finite sub-cover, say, $\{B(p, n_i)\}_{i=1}^{k}$. Note that the $B(p, n_i)$'s are concentric open balls. Let $M := \max\{n_1, \dots, n_k\}$. Then we have $S \subset \bigcup_{i=1}^{k} B(p, n_i) = B(p, M)$ and so in particular, S is bounded. Next, for any $p \in X \setminus S$, since S is compact, the open cover $\left\{B\left(x, \frac{1}{2}d(x, p)\right) : x \in S\right\}$ of S has a finite sub-cover, say $\{B(x_i, r_i) : i = 1, \dots, n\}$. Let $r = \min\{r_1, \dots, r_n\}$. Then $r > 0$ and we have

$$B(p, r) \cap B(x_i, r_i) = \phi \qquad \text{for all } i = 1, \dots, n\,.$$

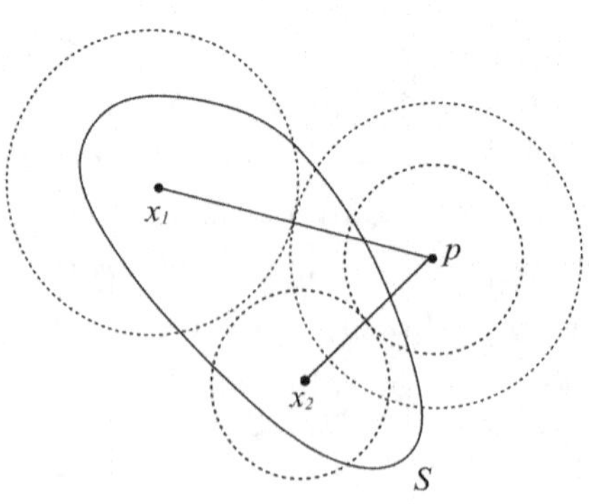

Since $S \subset \bigcup_{i=1}^{n} B(x_i, r_i)$, we have $B(p,r) \cap S = \phi$ and therefore, S is closed. $\square$

Another useful necessary condition for compactness is the *Bolzano-Weierstrass Property.*

Theorem 1.3.6. *Every compact subset S of X has the Bolzano-Weierstrass property, that is, every infinite subset of S has an accumulation point in S.*

Proof. Suppose $T \subset S$ is an infinite set with no accumulation point in S. Then every $x \in S$ has an open neighborhood $B(x)$ such that $\#\{B(x) \cap T\} \leq 1$. Since S is compact, the open cover $\{B(x) : x \in S\}$ has a finite subcover. In particular, this finite sub-cover also covers T, but this is impossible since T is infinite but each $B(x)$ contains at most 1 point of T. $\square$

Theorem 1.3.7. *For any $C \subset S \subset X$, C is compact in S if and only if C is compact in X.*

Proof. ($\Rightarrow$) Let $\{U_\alpha\}_{\alpha \in \Lambda}$ be an open cover of C in X. For each $\alpha \in \Lambda$, U_α is open in X and so $U_\alpha \cap S$ is open in S. Since

$$\bigcup_{\alpha \in \Lambda} (U_\alpha \cap S) = \left(\bigcup_{\alpha \in \Lambda} U_\alpha \right) \cap S \supset C \cap S = C \ ,$$

$\{U_\alpha \cap S\}_{\alpha \in \Lambda}$ is an open cover of C in S. As C is compact in S, $\{U_\alpha \cap S\}_{\alpha \in \Lambda}$ has a finite subcover, say, $\{U_i \cap S\}_{i=1}^{n}$. Hence

$$C \subset \bigcup_{i=1}^{n} (U_i \cap S) = \left(\bigcup_{i=1}^{n} U_i \right) \cap S \subset \bigcup_{i=1}^{n} U_i \ ,$$

that is, $\{U_i\}_{i=1}^{n}$ is a finite subcover of $\{U_\alpha\}_{\alpha \in \Lambda}$.

($\Leftarrow$) Let $\{U_\alpha\}_{\alpha\in\Lambda}$ be an open cover of C in S. Then for each $\alpha \in \Lambda$, $U_\alpha = V_\alpha \cap S$ for some open set $V_\alpha \subset X$. As

$$C \subset \bigcup_{\alpha\in\Lambda} U_\alpha = \bigcup_{\alpha\in\Lambda} (V_\alpha \cap S) = \left(\bigcup_{\alpha\in\Lambda} V_\alpha\right) \cap S \subset \bigcup_{\alpha\in\Lambda} V_\alpha ,$$

$\{V_\alpha\}_{\alpha\in\Lambda}$ is an open cover of C in X. Since C is compact in X, it has a finite subcover, say, $\{V_i\}_{i=1}^n$. Hence

$$C \subset \left(\bigcup_{i=1}^n V_i\right) \cap S = \bigcup_{i=1}^n (V_i \cap S) = \bigcup_{i=1}^n U_i ,$$

that is, $\{U_i\}_{i=1}^n$ is a finite subcover of $\{U_\alpha\}_{\alpha\in\Lambda}$. □

Remark. By Theorem 1.3.7, we see that unlike openness and closedness, compactness is independent of the ambient space we are working on. So when we talk about compactness, we do not need to specify which metric space it is referred to.

A first sufficient condition for compactness is the following.

Theorem 1.3.8. *Closed subsets of compact sets are compact. More precisely, if $S \subset X$ is compact and $T \subset S$ is closed in S, then T is compact.*

Proof. In view of Theorem 1.3.7 and the preceding Remark, we only need to show that T is compact as a subset of S. Let $\{U_\alpha\}_{\alpha\in\Lambda}$ be an open cover of T in S. As $S \setminus T$ is open in S, $\{U_\alpha, S\setminus T : \alpha \in \Lambda\}$ is an open cover of S in the metric space S and so it has a finite sub-cover. Without loss of generality, we may assume that the open set $S \setminus T$ is in this finite subcover and so it takes the form $\{U_1, \dots, U_n, S \setminus T\}$. Hence

$$T \subset \left(\bigcup_{i=1}^n U_i\right) \cup (S \setminus T) .$$

But then as it is clear that $T \cap (S \setminus T) = \phi$, we have $T \subset \bigcup_{i=1}^{n} U_i$ and so $\{U_i\}_{i=1,\dots,n}$ is a finite subcover of $\{U_\alpha\}_{\alpha \in \Lambda}$. Hence T is compact. □

Remark. It is clear that if $T \subset S \subset X$ and T is closed in X, then T is also closed in S. Hence the same assertion of Theorem 1.3.8 holds in case $S \subset X$ is compact and $T \subset S$ is closed in X.

Exercise 1.3

Unless otherwise specified, X will stand for a general metric space and S, T, etc., are arbitrary subsets of X.

Part A: True or False Questions

For each of the following statements, determine if it is true or false. If it is true, prove it. If it is false, give a counterexample or provide proper justification.

1. Every finite subset $S \subset X$ is compact.

   Answer: True.

   Proof. Let $S \subset X$ be a finite set. Write $S = \{x_1, \dots, x_n\}$. If $\{U_\alpha\}_{\alpha\in\Lambda}$ is an open cover of S in X, then for any $i = 1, \dots, n$, there is an $\alpha_i \in \Lambda$ such that $x_i \in U_{\alpha_i}$ and so $\{U_{\alpha_i} : i = 1, \dots, n\}$ is a finite subcover of $\{U_\alpha\}_{\alpha\in\Lambda}$.

2. $\mathbb{N} \subset \mathbb{R}$ is compact.

   Answer: False.

   Justification. Since $\mathbb{N}$ is unbounded. By Theorem 1.3.5, $\mathbb{N}$ is not compact.

   Alternatively, $\left\{\left(n - \frac{1}{2}, n + \frac{1}{2}\right) : n \in \mathbb{N}\right\}$ is an open cover of $\mathbb{N}$ without finite subcover. Hence $\mathbb{N}$ is not compact.

3. $S := \{\frac{1}{n} : n \in \mathbb{N}\} \subset \mathbb{R}$ is compact.

   Answer: False.

   Justification. Since $\overline{S} = S \cup \{0\} \supsetneqq S$, S is not closed and so by Theorem 1.3.5, S is not compact.

   Alternatively, observe that $\left\{B\left(\frac{1}{n}, \frac{1}{n(n+1)}\right) : n \in \mathbb{N}\right\}$ is an open cover of S. It is easy to see that

$$\frac{1}{k} \in B\left(\frac{1}{n}, \frac{1}{n(n+1)}\right) \quad \text{if and only if } k = n\,.$$

Hence for each $n \in \mathbb{N}$, $B\left(\frac{1}{n}, \frac{1}{n(n+1)}\right)$ contains one and only one point of S and so any finitely many of them could not cover S. Thus S is not compact.

4. $S := \{\frac{1}{n} : n \in \mathbb{N}\} \cup \{0\} \subset \mathbb{R}$ is compact.

   Answer: True.

   Proof. Let $\mathcal{U} = \{U_\alpha\}_{\alpha \in \Lambda}$ be any open cover of S. Then there is an element, say $U_0 \in \mathcal{U}$, such that $0 \in U_0$. Since U_0 is open in $\mathbb{R}$, there exists $r > 0$ such that $B(0, r) \subset U_0$. In particular, $\frac{1}{n} \in B(0, r) \subset U_0$ for all $n > \frac{1}{r}$. On the other hand, for every $n \leq \frac{1}{r}$, there exists $U_n \in \mathcal{U}$ such that $\frac{1}{n} \in U_n$. Hence $\{U_0, U_n : n \leq \frac{1}{r}\}$ is a finite subcover of $\mathcal{U}$.

5. Every discrete metric space is non-compact.

   Answer: False. The statement is true only when the discrete metric space is infinite. In fact, in this case, $\{\{x\} : x \in X\}$ is an open cover without finite subcover.

   Justification. In case the discrete metric space is finite, by Exercise 1.3, Part A, Problem #1, it must be compact.

6. If $C_i \subset X$ is compact for every $i = 1, \ldots, N$, then
   (a) $\bigcup_{i=1}^N C_i$ is compact.
   (b) $\bigcap_{i=1}^N C_i$ is compact.

   Answer:

   (a) True.

   Proof. Let $\mathcal{U} := \{U_\alpha\}_{\alpha \in I}$ be an open cover of $\bigcup_{i=1}^N C_i$ in X. Then for every fixed $i = 1, \ldots, N$, C_i is covered by $\mathcal{U}$ and so there is a finite subcover $\mathcal{U}_i \subset \mathcal{U}$. Collectively,

   $$\mathcal{U}_1 \cup \mathcal{U}_2 \cup \cdots \cup \mathcal{U}_N \subset \mathcal{U}$$

   is a finite subcover of $\mathcal{U}$.

(b) True.

Proof. For every i, since C_i is compact, it is closed in X. Hence $\bigcap_{i=1}^{N} C_i$ is closed in X. In particular, $\bigcap_{i=1}^{N} C_i$ is a closed subset of the compact set C_1 and so by Theorem 1.3.8, it must be compact.

7. The intersection of any collection of closed sets of X is compact if at least one of them is compact.

Answer: True.

Proof. Since the intersection of an arbitrary collection of closed sets of X is closed in X, if one of these closed sets is compact, the intersection of them is a closed subset of this compact set. Hence by Theorem 1.3.8, it is compact.

8. The intersection of any collection of compact sets of X is compact.

Answer: True.

Proof. This follows immediately from Exercise 1.3, Part A, Problem #7.

9. $A := \left\{\left(x, \sin\frac{1}{x}\right) : x \in (0,1]\right\} \subset \mathbb{R}^2$ is compact.

Answer: False.

Justification. It suffices to show that A is not closed, or equivalently, $\mathbb{R}^2 \setminus A$ is not open. To achieve this, all we need is to show that $(0,0) \in \mathbb{R}^2 \setminus A$ is not an interior point of $\mathbb{R}^2 \setminus A$. So let $\varepsilon > 0$ be given. Choose $n \in \mathbb{N}$ such that $n > \frac{1}{2\pi\varepsilon}$. Then

$$\left\|\left(\frac{1}{2n\pi}, \sin 2n\pi\right) - (0,0)\right\| = \left\|\left(\frac{1}{2n\pi}, 0\right)\right\| = \frac{1}{2n\pi} < \varepsilon\,.$$

That is,

$$\left(\frac{1}{2n\pi}, \sin 2n\pi\right) \in B((0,0),\varepsilon) \cap A$$

and so $B((0,0),\varepsilon) \not\subset \mathbb{R}^2 \setminus A$. Hence $(0,0) \in \mathbb{R}^2 \setminus A$ is not an interior point of $\mathbb{R}^2 \setminus A$.

10. Consider the metric space $X = C[0,1]$ with metric

$$d(f,g) := \sup\{|f(x) - g(x)| : x \in [0,1]\} .$$

(a) $A := \{f \in X : f = \text{constant}\}$ is compact.
(b) $B := \{f \in X : f(0) = 0\}$ is compact.
(c) $C := \left\{f \in X : \int_0^1 f(x)\, dx = 1\right\}$ is compact.
(d) $\overline{B}(0,1) \subset X$ is compact.

Answer:

(a) False.

Justification. For every $n \in \mathbb{N}$, let f_n be the constant function $f_n :\equiv n$ on $[0,1]$. Then $f_n \in A$ for all $n \in \mathbb{N}$, while we have trivially the constant zero function $0 \in A$. Hence

$$d(A) \geq d(f_n, 0) = n \qquad \text{for all } n \in \mathbb{N}.$$

That means $d(A) = \infty$, that is, A is not bounded and hence not compact.

(b) False.

Justification. For every $n \in \mathbb{N}$, let $g_n : [0,1] \to \mathbb{R}$ be given by $g_n(x) := nx$, $x \in [0,1]$. Then $g_n \in B$ for all $n \in \mathbb{N}$, while we have trivially the constant zero function $0 \in B$. Hence

$$d(B) \geq d(g_n, 0) = n \qquad \text{for all } n \in \mathbb{N} .$$

That means $d(B) = \infty$ and so B is not bounded, hence not compact.

(c) False.

Justification. For every $n \in \mathbb{N}$, let $h_n : [0,1] \to \mathbb{R}$ be defined by

$$h_n(x) := \begin{cases} -2n^2x + 2n & \text{if } 0 \leq x \leq 1/n , \\ 0 & \text{otherwise.} \end{cases}$$

Then

$$\int_0^1 h_n(x)\,dx = \int_0^{\frac{1}{n}} (-2n^2x + 2n)dx = 1$$

and so $f_n \in C$ for all $n \in \mathbb{N}$. Hence we have

$$d(C) \geq d(h_{2n}, h_n) \geq |h_{2n}(0) - h_n(0)| = 2n$$

for all $n \in \mathbb{N}$. Hence $d(C) = \infty$ and so C is not bounded, therefore not compact.

(d) False.

Justification. For any $n \in \mathbb{N}$, define $f_n \in C[0,1]$ by

$$f_n(x) := \begin{cases} 1 & \text{if } 0 \leq x < \frac{1}{n+1}; \\ -n(n+1)x + n + 1 & \text{if } \frac{1}{n+1} \leq x < \frac{1}{n}; \\ 0 & \text{if } \frac{1}{n} \leq x \leq 1\,. \end{cases}$$

Then

$$d(f_n, 0) = 1 \quad \text{for every } n \in \mathbb{N}$$

and so $f_n \in \overline{B}(0,1)$ for every n. However, it is easy to verify that

$$d(f_n, f_m) = 1 \quad \text{for all } n, m \in \mathbb{N} \text{ with } \; m \neq n$$

and so the infinite set $\{f_n : n \in \mathbb{N}\} \subset \overline{B}(0,1)$ cannot have any accumulation point in $\overline{B}(0,1)$. Thus $\overline{B}(0,1)$ does not possess the Bolzano-Weierstrass property and so it is not compact.

Part B: Problems

1. The collection $\mathcal{F}$ of open intervals of the form $(\frac{1}{n}, \frac{2}{n})$, where $n = 2, 3, \ldots$, is an open cover of the open interval $(0,1)$. Show that $\mathcal{F}$ has no finite subcover.

Proof. Any finite sub-collection of $\mathcal{F}$ is of the form

$$\mathcal{F}' = \left\{ \left(\frac{1}{n_1}, \frac{2}{n_1} \right), \dots, \left(\frac{1}{n_k}, \frac{2}{n_k} \right) \right\}.$$

Take any $p \in (0,1)$ such that $p < \min\{\frac{1}{n_1}, \dots, \frac{1}{n_k}\}$. Then $p \notin \left(\frac{1}{n_i}, \frac{2}{n_i}\right)$ for any $1 \le i \le k$. Thus $\mathcal{F}'$ cannot be a subcover.

2. Give an example of a metric space in which the closed unit ball is not compact.

Solution:

Example 1: Let X be a discrete metric space which consists of infinitely many elements. Note that $B(x,1) = \{x\}$ for every $x \in X$. Pick any $a \in X$. Then $\mathcal{F} := \{\{x\} : x \in X\}$ is an open cover of $X = \overline{B}(a,1)$ but clearly no finite subcollection of $\mathcal{F}$ can cover X.

Example 2: Part (d) of Exercise 1.3, Part A, Problem #10 is another interesting non-trivial example.

3. (a) Show that for any $x \in X$ and any open set $U \subset X$ containing x, there is an open set $V \subset X$ such that $\{x\} \subset V \subset \overline{V} \subset U$.

   (b) Let S be a compact subset of X and U be an open subset of X containing S, show that there is an open set $V \subset X$ such that $S \subset V \subset \overline{V} \subset U$. What if the compactness of S is missing?

   (c) Let S be a compact subset of X and T be a closed subset of X with $S \cap T = \phi$. Show that there are open subsets U_S and U_T of X such that $S \subset U_S$, $T \subset U_T$ and $U_S \cap U_T = \phi$. [Compare with Exercise 1.2 Part B, Problem #7(c).]

Solution:

(a) Since $x \in U$ which is open, there exists $r > 0$ such that $B(x,r) \subset U$. Then the set $V := B(x, \frac{r}{2})$ will serve the purpose. In fact, we already have $\{x\} \subset V \subset B(x,r) \subset U$ and $V \subset \overline{V}$. Hence it only remains to show $\overline{V} \subset B(x,r)$. Let

$y \in \overline{V}$. Then there is $z \in B(y, \frac{r}{2}) \cap V = B(y, \frac{r}{2}) \cap B(x, \frac{r}{2})$. It follows that $d(y, x) \leq d(y, z) + d(z, x) < \frac{r}{2} + \frac{r}{2} = r$ which means $y \in B(x, r)$. Hence $\overline{V} \subset B(x, r)$, as desired.

(b) For each $a \in S$, by (a), there is an open set V_a such that $\{a\} \subset V_a \subset \overline{V_a} \subset U$. Then the collection $\{V_a : a \in S\}$ is an open cover of S. By the compactness of S, $\{V_a : a \in S\}$ has a finite subcover, say $\{V_{a_1}, \cdots, V_{a_n}\}$ for some $n \in \mathbb{N}$. Take $V := \bigcup_{i=1}^n V_{a_i}$. Then V is open and satisfies $S \subset V \subset \overline{V} = \overline{\bigcup_{i=1}^n V_{a_i}} = \bigcup_{i=1}^n \overline{V_{a_i}} \subset U$.

In case the compactness of S is missing, the assertion will no longer hold.

Example: Let $X := \mathbb{R}$, $S := (0, 1)$, $U := (0, 1)$. Then S is not compact. Clearly, $(0, 1)$ is the only possible subset of X that contains S and is contained in U, but $\overline{(0, 1)} = [0, 1] \not\subset U$.

(c) Since T is closed, $X \setminus T$ is an open subset of X containing S and so by (b), there exists an open set U_S in X such that $S \subset U_S \subset \overline{U_S} \subset X \setminus T$. Since $\overline{U_S}$ is closed, $U_T := X \setminus \overline{U_S}$ is open and contains T. Obviously $U_S \cap U_T = \phi$.

4. Let S, T be nonempty subsets of X.

(a) If $S \subset X$ is compact, show that there exists a point $x \in S$ such that $d(x, T) = d(S, T)$. In this case we say that S has the *nearest point property*.

(b) If $S \subset X$ is compact and $T \subset X$ is closed such that $S \cap T = \phi$, show that $d(S, T) > 0$.

[Compare with Exercise 1.2 Part B, Problem #7(b).]

Solution:

(a) If S is finite, then statement is obviously true.

So assume S is an infinite set. Write $d(S, T) = a$. For any $n \in \mathbb{N}$, there exists $x_n \in S$ such that $a \leq d(x_n, T) < a + \frac{1}{n}$. If $\#\{x_n : n \in \mathbb{N}\} < \infty$, then there exists $x \in \{x_n : n \in \mathbb{N}\}$

such that

$$a \le d(x,T) < a + \frac{1}{n}$$

for infinitely many n. Hence $d(x,T) = a$.

So suppose $\#\{x_n : n \in \mathbb{N}\} = \infty$. Since S is compact, it possesses the Bolzano-Weierstrass Property. Hence the infinite set $\{x_n : n \in \mathbb{N}\} \subset S$ has an accumulation point $x \in S$. Therefore, for any $\varepsilon > 0$, there is $N > \frac{1}{\varepsilon}$ such that $x_N \in B(x,\varepsilon)$ and so $d(x,T) \le d(x,x_N)+d(x_N,T) < \varepsilon+a+\frac{1}{N} < a+2\varepsilon$. This forces $d(x,T) = a$.

(b) Suppose $d(S,T) = 0$. By (a), there exists $x \in S$ such that $d(x,T) = 0$. By Exercise 1.2 Part B, Problem #7(a), $x \in \overline{T} = T$. So $x \in S \cap T$, contradicting the assumption that $S \cap T = \phi$.

5. Let S, T be disjoint subsets of a metric space X, S is compact, and T is closed. Show that there exists $\delta > 0$ such that $d(s,t) > \delta$ for any $s \in S$ and any $t \in T$.

Proof. By Exercise 1.3, Part B, Problem #4(a), there exists $s_0 \in S$ such that

$$d(s_0,T) = d(S,T) \; .$$

Since $S \cap T = \phi$, we have $s_0 \notin T = \overline{T}$ and so by Exercise 1.2, Part B, Problem #7, we have

$$d(S,T) = d(s_0,T) > 0 \; .$$

Hence for any $0 < \delta < d(S,T)$, we have

$$d(s,t) \ge d(S,T) > \delta > 0 \qquad \text{for any } s \in S,\ t \in T \; .$$

6. For any $\phi \ne A$, $B \subset \mathbb{R}^n$, define

$$A + B := \{a + b : a \in A, b \in B\} \; .$$

If A is compact and B is closed, show that $A + B$ is closed. What if the compactness of A is missing?

Solution: For any $x \notin A+B$, put $C = x - B := \{x - b : b \in B\}$. As B is closed, it is clear that C is also closed and $C \cap A = \phi$. By Exercise 1.3, Part B, Problem #5, there exists $\delta > 0$ such that $d(a,c) > \delta$ for any $a \in A$ and any $c \in C$. We claim that $B(x,\delta) \cap (A+B) = \phi$. In fact, if there exists $y \in B(x,\delta) \cap (A+B)$, then $y = a_0 + b_0$ for some $a_0 \in A$ and $b_0 \in B$. But then as $x - b_0 \in C$, we have

$$\begin{aligned} \delta > d(x,y) &= |x-y| = |x - a_0 - b_0| \\ &= |(x-b_0) - a_0| = d(x-b_0, a_0) > \delta , \end{aligned}$$

which is absurd. Thus $B(x,\delta) \cap (A+B) = \phi$ and hence $A+B$ is closed.

In case A is not compact, the assertion no longer holds.

Example: In $\mathbb{R}^2$, consider

$$\begin{aligned} A &:= \{(x, \tan^{-1} x) : x \in \mathbb{R}\} \\ B &:= \{(x,0) : x \in \mathbb{R}\} . \end{aligned}$$

It is clear that both A, B are closed in $\mathbb{R}^2$ but A is not compact. Observe that $A+B = \{(x,y) : x \in \mathbb{R}, -1 < y < 1\}$ is not closed in $\mathbb{R}^2$.

7. (a) Show that every compact space has a countable dense subset. That is, if X is compact, then there exists a countable subset $S \subset X$ such that $\overline{S} = X$.
   (b) Without compactness, is (a) still true?

Solution:

(a) Since X is compact, for each fixed $n \in \mathbb{N}$, the open cover $\left\{B\left(x, \frac{1}{n}\right) : x \in X\right\}$ has a finite subcover

$$\left\{B\left(x_1^n, \frac{1}{n}\right), B\left(x_2^n, \frac{1}{n}\right), \ldots, B\left(x_{k_n}^n, \frac{1}{n}\right)\right\} .$$

Write

$$S_n := \left\{x_1^n, x_2^n, \ldots, x_{k_n}^n\right\} .$$

Do this for every $n \in \mathbb{N}$ and define

$$S := \bigcup_{n \in \mathbb{N}} S_n \ .$$

Then S is countable. On the other hand, for any $x \in X$ and any $r > 0$, let $n \in \mathbb{N}$ be such that $n > \frac{1}{r}$. Since $\left\{B\left(x_1^n, \frac{1}{n}\right), B\left(x_2^n, \frac{1}{n}\right), \dots, B\left(x_{k_n}^n, \frac{1}{n}\right)\right\}$ covers X, there is $j \in \{1, \dots, k_n\}$ such that $x \in B\left(x_j^n, \frac{1}{n}\right)$. But that means

$$x_j^n \in B\left(x, \frac{1}{n}\right)$$

and so $B\left(x, \frac{1}{n}\right) \cap S \neq \phi$. Hence $x \in \overline{S}$.

(b) No, (a) will not hold any more in case X is not compact.

Example: Let X be a discrete metric space consisting of uncountably many elements. Since $\overline{S} = S$ for any $S \subset X$, no countable subset of X could be dense in X. Note that in this case X is not compact.

8. Recall (Exercise 1.1, Part B, Problem #13) that ℓ^2 is the space of sequences of real numbers $x = \{x_n\}_{n \in \mathbb{N}}$ with $\sum_{n \in \mathbb{N}} |x_n|^2 < \infty$, and the ℓ^2-metric is defined as

$$d(x, y) := \left(\sum_{n \in \mathbb{N}} |x_n - y_n|^2 \right)^{1/2} \qquad \text{for all } x, y \in \ell^2 \ .$$

Determine whether $\overline{B}(0, 1)$ is compact. Here as usual, 0 stands for the zero sequence in ℓ^2.

Solution: It is non-compact.

Denote by $e^n \in \ell^2$ the sequence defined by $e_k^n := \delta_{nk}$. Here we recall that the *Kronecker Delta Function* δ_{ij} is defined by

$$\delta_{ij} := \begin{cases} 1 & \text{if } i = j \\ 0 & \text{if } i \neq j \ . \end{cases}$$

So e^n is simply the sequence whose nth term is 1 and all other terms are 0. Clearly, $d(e^n, 0) = 1$ and so $e^n \in \overline{B}(0,1)$ for all $n \in \mathbb{N}$. Furthermore, for any $m \neq n$, $d(e^m, e^n) = \sqrt{2}$. Now it is clear that $\mathcal{U} := \left\{B(x, \frac{1}{2}) : x \in \ell^2\right\}$ is an open cover of ℓ^2. Fix any $x \in \ell^2$. Note that if there are distinct intergers $n \neq m$ in $\mathbb{N}$ such that both e^m, $e^n \in B(x, \frac{1}{2})$, then we have

$$\sqrt{2} = d(e^m, e^n) \leq d(e^m, x) + d(e^n, x) < \frac{1}{2} + \frac{1}{2} = 1 ,$$

which is absurd. Hence each $B(x, \frac{1}{2})$ can contain at most one e^n. Therefore, no finite subcollection of $\mathcal{U}$ could cover all the e^n's, and thus no finite subcollection of $\mathcal{U}$ could cover $\overline{B}(0,1)$. Hence $\overline{B}(0,1)$ is not compact.

9. A metric space X is said to be *noetherian* if every nonempty family of open subsets has a maximal element. It is known that X is noetherian if and only if it satisfies the *ascending chain condition*, that is, every increasing sequence of open subsets

$$U_1 \subset U_2 \subset \cdots$$

of X stabilizes, i.e., there is some $N \in \mathbb{N}$ such that $U_n = U_N$ whenever $n \geq N$.

(a) Show that a space X is noetherian if and only if every open subset is compact.

(b) Hence show that a noetherian metric space must be a finite set.

Proof.

(a) ($\Rightarrow$): Let $U \subset X$ be open. Let $\{U_\alpha\}_{\alpha\in\Lambda}$ be any open cover of U of nonempty sets, and $\mathcal{W}$ be the collection of all finite unions of the U_α's. By definition, the family $\mathcal{W}$ has a maximal element, say W. If $W \not\supset U$, then there is an element $x \in U \backslash W$. Since $\{U_\alpha\}_{\alpha\in\Lambda}$ covers U, there exists some $\alpha_0 \in \Lambda$ such that $x \in U_{\alpha_0}$. But then we have

$$W \subsetneq W \cup U_{\alpha_0} ,$$

which contradicts the maximality of W. So $W \supset U$. Note that W represents a finite subcover of $\{U_\alpha\}_{\alpha\in\Lambda}$ for U and so U is compact.

($\Leftarrow$): In view of the given equivalent condition for noetherian metric spaces, it suffices to show that X has the ascending chain condition. So let

$$U_1 \subset U_2 \subset \cdots$$

be an ascending chain of open sets in X. By assumption, the open set $W := \bigcup_{n\in\mathbb{N}} U_n$ is compact and so the open cover $\{U_n\}_{n\in\mathbb{N}}$ of W has a finite sub-cover, say, $\{U_{n_i} : 1 \le i \le m\}$. Let $N := \max\{n_i : 1 \le i \le m\}$. Then we have $W = U_N$ and hence

$$U_N \subset U_n \subset W = U_N \quad \text{for all } n \ge N\ ,$$

which forces $U_n = U_N$ for all $n \ge N$.

(b) Suppose X is noetherian. By (a), every open set in X is compact and hence closed. In particular, complements of singletons are closed and so singletons are open. Therefore, $\mathcal{U} := \big\{\{x\} : x \in X\big\}$ is an open cover of X. Note that X is open in itself and hence is compact, and so $\mathcal{U}$ must have a finite subcover. This can happen only when X is finite.

10. A metric space X is said to be *totally bounded* if for every $\varepsilon > 0$, there exist $k > 0$ and $x_1, \ldots, x_k \in X$ such that $\{B(x_i, \varepsilon)\}_{i=1}^k$ is an open cover of X. Equivalently, X is totally bounded if for any $\varepsilon > 0$, X can be decomposed into finitely many subsets each of which has diameter less than ε. A subset $S \subset X$ is *totally bounded* if it is totally bounded when considered as a metric space, with metric induced by that of X.
    (a) Show that if $S \subset X$ is totally bounded, then S is bounded.
    (b) Show that if $S \subset X$ is compact, then S is totally bounded.

(c) (i) Let $X = C[0,1]$ with

$$d(f,g) := \sup\{|f(x) - g(x)| : x \in [0,1]\}.$$

Show that $\overline{B}(0,1)$ is *not* totally bounded.

(ii) Show that every bounded subset of $\mathbb{R}$ is totally bounded. [Hence, together with (a), a subset in $\mathbb{R}$ is bounded if and only if it is totally bounded.]

(iii) Let $X = \mathbb{R}$ with the metric

$$d(x,y) := \min\{1, |x-y|\} .$$

Find a subset of X that is bounded but not totally bounded.

(d) Does closedness + total boundedness imply compactness? Either prove it or give a counterexample.

Solution:

(a) If $S \subset X$ is totally bounded, then S is covered by finitely many open balls of radius 1. Hence S is bounded.

(b) Let $S \subset X$ be compact. For any $\varepsilon > 0$, the open cover $\{B(s,\varepsilon)\}_{s\in S}$ of S has a finite subcover. That is, S is covered by finitely many ε-balls and so it is totally bounded.

(c) (i) For any $n \in \mathbb{N}$, define $f_n \in X = C[0,1]$ by

$$f_n(x) := \begin{cases} nx & \text{if } x \in [0, \frac{1}{n}] \\ 1 & \text{if } x \in [\frac{1}{n}, 1] . \end{cases}$$

It is easy to see that $f_n \in \overline{B}(0,1)$ for all $n \in \mathbb{N}$ and $d(f_n, f_m) = \frac{m-n}{m}$ for all $m > n$. Consider the open cover $\mathcal{U} := \{B(f, \frac{1}{4}) : f \in \overline{B}(0,1)\}$ of $\overline{B}(0,1)$. Fix an element $B(f, \frac{1}{4}) \in \mathcal{U}$. Suppose $B(f, \frac{1}{4})$ constains f_k for some $k \in \mathbb{N}$. Then for any $m \neq k$,

$$f_m \in B\left(f, \frac{1}{4}\right) \Longrightarrow d(f_m, f_k) \leq d(f_m, f) + d(f_k, f)$$
$$< \frac{1}{4} + \frac{1}{4} = \frac{1}{2} .$$

In case $m > k$, this implies

$$\frac{m-k}{m} < \frac{1}{2} \qquad \text{or} \qquad k < m < 2k \ ;$$

and in case $m < k$, this implies

$$\frac{k-m}{k} < \frac{1}{2} \qquad \text{or} \qquad \frac{k}{2} < m < k \ .$$

Combining the two cases, we see that each $B(f, \frac{1}{4})$ can contain at most finitely many f_n's. In other words, finitely many $\frac{1}{4}$-balls cannot cover all the f_n's, hence cannot cover $\overline{B}(0,1)$. Therefore, $\overline{B}(0,1)$ is not totally bounded.

(ii) Let $S \subset \mathbb{R}$ be bounded. Then $S \subset [-M, M]$ for some $M > 0$. For any $\varepsilon > 0$, let $N := \left[\frac{M}{\varepsilon}\right] + 1$. Then the ε-balls

$$\{(k\varepsilon, (k+2)\varepsilon) : k = -N-1, -N, \ldots, N-1\}$$

is a finite cover of $[-M, M]$, hence a finite cover of S.

(iii) Note first that under the given metric d, every subset of $\mathbb{R}$ is bounded. So in particular, $\mathbb{N} \subset \mathbb{R}$ is bounded. However, as $B(n, \frac{1}{4}) \cap \mathbb{N} = \{n\}$ for all $n \in \mathbb{N}$, any finite collection of $\frac{1}{4}$-balls $\{B(n, \frac{1}{4}) : n \in \mathbb{N}\}$ cannot cover $\mathbb{N}$. Hence N is not totally bounded.

(d) False.

Example: Let $X := (0,1)$ with the usual Euclidean metric. Similar to (c)(ii), we see X it is totally bounded. It is also trivial that X is closed in X. However, X is not compact, as the open cover $\{(\frac{1}{n}, 1 - \frac{1}{n}) : n \in \mathbb{N}\}$ has no finite subcover.

11. Let (X, d) be a metric space and $\mathcal{U}$ a cover of X. $\mathcal{U}$ is said to have a *Lebesgue number* $\lambda > 0$ if every subset of X of diameter less than λ is contained in some element in $\mathcal{U}$, i,e,, for any $E \subset X$, if $d(E) < \lambda$, there is some $U \in \mathcal{U}$ such that $E \subset U$.

(a) Give an example of a metric space in which an open cover may not have a Lebesgue number.

(b) Show that if X is a compact metric space, then every open cover of X has a Lebesgue number. [This is known as *Lebesgue Covering Lemma*.]

Solution:

(a) Consider $X = (0,1)$ with the usual Euclidean metric. Then $\{(\frac{1}{n},1) : n \in \mathbb{N}\}$ is an open cover of X without a Lebesgue number. In fact, for any $\lambda > 0$, the open set $(0,\frac{\lambda}{2}) \cap (0,1) \subset X$ is of diameter $< \lambda$ but $(0,\frac{\lambda}{2}) \cap (0,1) \not\subset (\frac{1}{n},1)$ for any $n \in \mathbb{N}$.

(b) Let X be compact and $\mathcal{U}$ be an open cover of X. For any $x \in X$, there exists an open set $U_x \in \mathcal{U}$ such that $x \in U_x$. As U_x is open, there exists $r_x > 0$ such that $B(x, r_x) \subset U_x$. Clearly,

$$\mathcal{B} := \left\{B\left(x, \frac{r_x}{2}\right) : x \in X\right\}$$

is an open cover of X. As X is compact, there is a finite subcover, say

$$\mathcal{B}' = \left\{B\left(x_i, \frac{r_{x_i}}{2}\right) : i = 1, 2, ..., n\right\} .$$

Consider

$$\lambda := \min\left\{\frac{r_{x_i}}{2} : i = 1, 2, ..., n\right\} > 0 .$$

Then λ is a Lebesgue number for $\mathcal{U}$. To see this, let E be any subset of X with $d(E) < \lambda$. Fix $p \in E$. As $\mathcal{B}'$ is a cover of X, there exists $k \in \{1, 2, ..., n\}$ such that $p \in B\left(x_k, \frac{r_{x_k}}{2}\right)$. Hence for any $q \in E$, we have

$$d(q, x_k) \leq d(q, p) + d(p, x_k) < \lambda + \frac{r_{x_k}}{2} \leq r_{x_k}$$

and so $q \in B(x_k, r_{x_k}) \subset U_{x_k}$. Thus $E \subset U_{x_k}$.

12. A collection of subsets $\{U_\lambda\}_{\lambda \in \Lambda}$ of a metric space X is said to have the *finite intersection property* (*f.i.p.*) if every finite subcollection of $\{U_\lambda\}_{\lambda \in \Lambda}$ has nonempty intersection.

Show that X is compact if and only if every collection of closed subset of X having f.i.p. has nonempty intersection.

Proof.

($\Rightarrow$): Let X be compact and $\{U_\lambda\}_{\lambda\in\Lambda}$ be any collection of closed subsets of X having f.i.p. Assume to the contrary that $\{U_\lambda\}_{\lambda\in\Lambda}$ has empty intersection, that is, $\bigcap_{\lambda\in\Lambda} U_\lambda = \phi$. Then

$$\bigcup_{\lambda\in\Lambda}(X \setminus U_\lambda) = X \setminus \bigcap_{\lambda\in I} U_\lambda = X$$

and so $\{X \setminus U_\lambda\}_{\lambda\in\Lambda}$ is an open cover of X. Since X is compact, $\{X \setminus U_\lambda\}_{\lambda\in\Lambda}$ has a finite sub-cover, say $\{X \setminus U_{\lambda_i}\}_{i=1}^N$. Hence

$$X = \bigcup_{i=1}^{N}(X \setminus U_{\lambda_i}) = X \setminus \bigcap_{i=1}^{N} U_{\lambda_i}$$

and so $\bigcap_{i=1}^N U_{\lambda_i} = \phi$, contradicting with the f.i.p. of $\{U_\lambda : \lambda \in \Lambda\}$.
($\Leftarrow$): We prove by contrapositive argument. Suppose X is not compact. Then there exists an open cover $\{V_\lambda\}_{\lambda\in\Lambda}$ of X which has no finite subcover. For each $\lambda \in \Lambda$, $U_\lambda := X \setminus V_\lambda$ is closed in X. Furthermore, for any finite subcollection $\{U_{\lambda_i}\}_{i=1}^N$ of $\{U_\lambda\}_{\lambda\in\Lambda}$, we have

$$\bigcap_{i=1}^{N} U_{\lambda_i} = \bigcap_{i=1}^{N}(X \setminus V_{\lambda_i}) = X \setminus \bigcup_{i=1}^{N} V_{\lambda_i} \neq \phi$$

and so $\{U_\lambda\}_{\lambda\in\Lambda}$ is a family of closed subsets having f.i.p. However,

$$\bigcap_{\lambda\in\Lambda} U_\lambda = \bigcap_{\lambda\in\Lambda}(X \setminus V_\lambda) = X \setminus \bigcup_{\lambda\in\Lambda} V_\lambda = \phi\,,$$

hence $\{U_\lambda\}_{\lambda\in\Lambda}$ is a family of closed subsets having f.i.p. but with empty intersection. Hence the assertion.

13. Let $X = C[0,1]$ with metric $d(f,g) := \sup_{x\in[0,1]} |f(x) - g(x)|$.
    (a) If $S := \left\{f \in X : \int_0^1 f(x)\,dx = 1\right\}$, prove that S is closed and determine whether it is compact.

(b) If $T := \{f \in X : d(f,0) \leq 1, f(0) = 0\}$, prove that T is closed and bounded. By considering the set $\{f_1, f_2, \dots\}$, where

$$f_n(x) := \begin{cases} 0 & \text{if } 0 \leq x \leq 1 - \frac{1}{n} \\ 1 - n + nx & \text{if } 1 - \frac{1}{n} < x \leq 1\,, \end{cases}$$

show that T is not compact.

Solution:

(a) To show the closedness of S, it suffices to prove $S' \subset S$. So let $f \in S'$. For any $\varepsilon > 0$, there is $f_\varepsilon \in S$ such that $0 < d(f, f_\varepsilon) < \varepsilon$. Hence

$$f_\varepsilon(x) - \varepsilon < f(x) \ < \ f_\varepsilon(x) + \varepsilon \quad \text{for all } x \in [0,1]$$

and so

$$\begin{aligned} 1 - \varepsilon = \int_0^1 (f_\varepsilon(x) - \varepsilon)\,dx &\leq \int_0^1 f(x)\,dx \\ &\leq \int_0^1 (f_\varepsilon(x) + \varepsilon)\,dx = 1 + \varepsilon\,. \end{aligned}$$

Since $\varepsilon > 0$ is arbitrary, we conclude that $\int_0^1 f(x)\,dx = 1$, i.e., $f \in S$.

However, S is not compact. In fact, for any $n \in \mathbb{N}$, define

$$f_n(x) := \begin{cases} -2n^2x + 2n & \text{if } 0 \leq x \leq \frac{1}{n}\,, \\ 0 & \text{otherwise.} \end{cases}$$

It is clear that each f_n is continuous and $\int_0^1 f_n(x)\,dx = 1$. Hence $f_n \in S$ for all $n \in \mathbb{N}$. But then

$$d(f_{2n}, f_n) \geq |f_{2n}(0) - f_n(0)| = 2n \qquad \text{for all } n \in \mathbb{N}$$

and so S is unbounded, hence not compact.

(b) Since $T \subset \overline{B}(0,1) \subset X$, it is bounded. On the other hand, as

$$T = \overline{B}(0,1) \cap \{f \in X : f(0) = 0\}$$

and both $\overline{B}(0,1)$ and $\{f \in X : f(0) = 0\}$ are closed in X, T is closed in X.

However, T is not compact.

In fact, consider the subset $A := \{f_n : n \in \mathbb{N}\}$, where

$$f_n(x) := \begin{cases} 0 & \text{if } 0 \le x \le 1 - \frac{1}{n}\,, \\ 1 - n + nx & \text{if } 1 - \frac{1}{n} < x \le 1\,. \end{cases}$$

It is easy to verify that $A \subset T$. If T were compact, then being a bounded infinite subset of T, A has an accumulation point $f \in T$. Hence for any $r > 0$, there is $N \in \mathbb{N}$ such that $f_N \in B(f, r)$, that is,

$$\begin{aligned} r > d(f_N, f) &= \sup_{x \in [0,1]} |f_N(x) - f(x)| \\ &\ge |f_N(1) - f(1)| \\ &= |1 - f(1)|\,. \end{aligned}$$

Since $r > 0$ is arbitrary, we must have $f(1) = 1$. On the other hand, since $f \in T$ is an accumulation point of A, near f there are infinitely many elements of A, that is, there are infinitely many f_n's. So for any $r > 0$ and any $x \in (0,1)$, there exists $M \in \mathbb{N}$ such that $1 - \frac{1}{M} > x$ and $f_M \in B(f, r)$. Hence $F_M(x) = 0$ and so

$$r > d(f_n, f) \ge |f_M(x) - f(x)| = |f(x)| \ge 0\,.$$

Since $r > 0$ is arbitrary, we have $f(x) = 0$. Since $x \in (0,1)$ is arbitrary, we have $f = 0$ on $(0,1)$. But then from above we have $f(1) = 1$ and so f is not continuous at 1, contradicting to the fact that $f \in T \subset C[0,1]$. Hence T is not compact.

1.4 Compactness in the Euclidean Space $\mathbb{R}^n$

As the readers may have already observed by now, in general, compactness in a general metric space is a rather intricate concept to study. In this section, we will restrict our attention to the case where the metric space is the Euclidean space $\mathbb{R}^n$. In fact, we will see that in this setting, neater and more fruitful results on compactness can be obtained. In particular, we shall establish the equivalence among the following conditions on subsets in $S \subset \mathbb{R}^n$:

(a) S is compact.

(b) S is closed and bounded.

(c) S has the Bolzano-Weierstrass property.

Observe that by Theorems 1.3.5 and 1.3.6, we already have (a) $\Rightarrow$ (b) and (a) $\Rightarrow$ (c). So in order to establish the equivalence relations, it suffices to show (b) $\Rightarrow$ (a) and (c) $\Rightarrow$ (b).

Theorem 1.4.1 (Bolzano-Weierstrass). *Every bounded infinite set $S \subset \mathbb{R}^n$ has an accumulation point in $\mathbb{R}^n$.*

Proof. Since S is bounded, $S \subset J^0 := \prod_{i=1}^n [-a, a]$ for some $a > 0$. The "mid-planes" $x_i = 0$, $i = 1, \ldots, n$, partition the cube J^0 into 2^n identical cubes each side of which has length a. Since S is an infinite set, at least one of these 2^n cubes, say J^1, contains infinitely many points of S. Repeating the same process with J^1 in place of J^0, we obtain a cube J^2 each side of which has length $\frac{a}{2}$ and contains infinitely many points of S. Inductively, we obtain a decreasing nest of cubes $\{J^k\}_{k=1}^{\infty}$, $J^{k+1} \subset J^k$, with the length of each side of J^k being $\frac{a}{2^{k-1}}$, and for each k, J^k contains infinitely many points of S. Observe that $\bigcap_{k=1}^{\infty} J^k \neq \phi$.

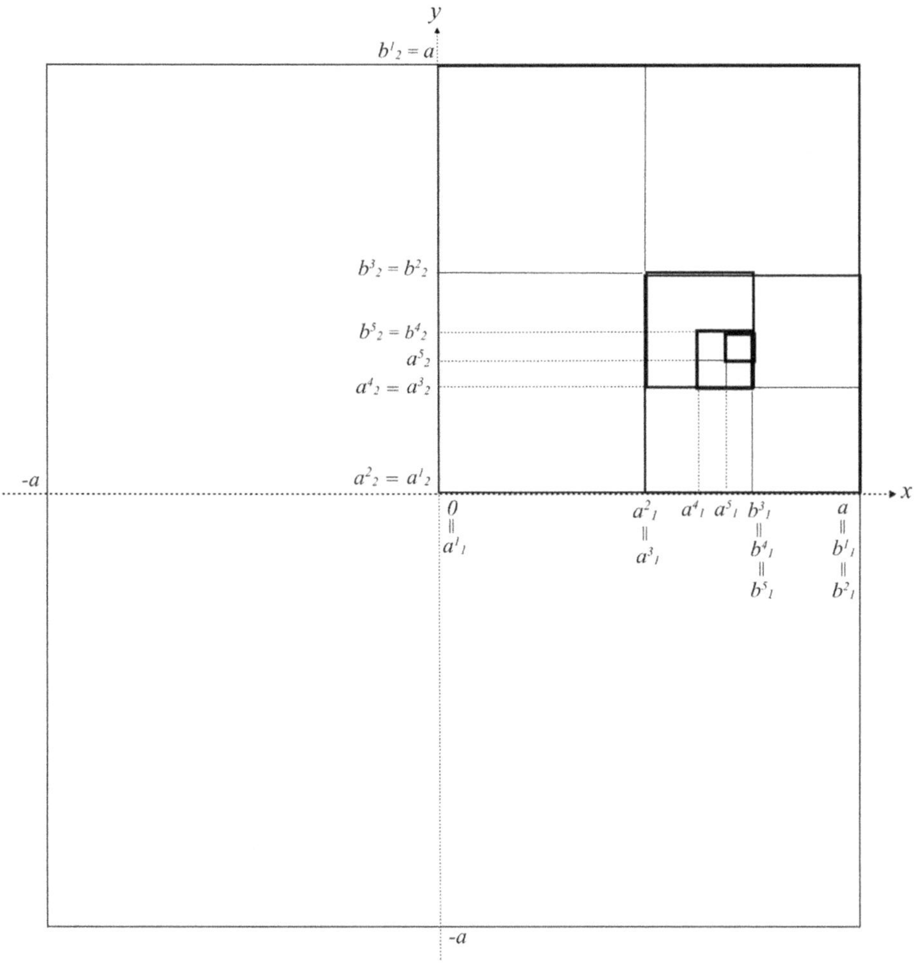

In fact, for each $k \in \mathbb{N}$, if we write

$$J^k = \prod_{i=1}^{n} [a_i^k, b_i^k] \ ,$$

then for each $i = 1, \ldots, n$,

$$b_i^k - a_i^k = \frac{a}{2^{k-1}}$$

and so as $k \to \infty$, $b_i^k - a_i^k \to 0$, thus

$$\lim_{k\to\infty} a_i^k = \lim_{k\to\infty} b_i^k = t_i \ . \qquad (*)$$

[Question: why do the limits exist? – see Remark (i) below.]
Let $t = (t_1, \dots, t_n)$. Then it is not hard to see that $t \in J^k$ for every k and so $t \in \bigcap_{k=1}^{\infty} J^k$. (**)

[Question: why? – see Remark (ii) below.]

It is then evident that t is an accumulation point of S. In fact, for any $r > 0$, $B(t, r)$ will contain the entire cube J^k whenever k is large enough so that $\frac{\sqrt{n}\, a}{2^{k-1}} < r$. Thus $B(t, r) \cap \big(S \setminus \{t\}\big) \neq \phi$ and so $t \in S'$. $\square$

Remarks.

(i) A common elementary mistake mathematics freshmen would make is

$$\text{“}\{x_k - y_k\} \to 0 \text{ as } k \to \infty$$
$$\implies \{x_k\} \text{ and } \{y_k\} \text{ converge to the same limit as } k \to \infty\text{”}.$$

A simple example for this statement to be false is $x_k := k$ and $y_k := k + \frac{1}{k}$, as in this case we have $\{x_k - y_k\} = \{-\frac{1}{k}\} \to 0$ as $k \to \infty$, while neither $\{x_k\}$ nor $\{y_k\}$ converges as $k \to \infty$.

The more matured readers could tell that the statement above is true only when both $\{x_k\}$ and $\{y_k\}$ are convergent as $k \to \infty$.

Hence to show that (*) is true, we need to show that for each fixed $i = 1, \dots, n$, both $\{a_i^k\}_{k\in\mathbb{N}}$ and $\{b_i^k\}_{k\in\mathbb{N}}$ are convergent as $k \to \infty$. But these should be immediate once we observe that for every fixed $i = 1, \dots, n$ (that is, at each coordinate), the sequence $\{a_i^k\}_{k\in\mathbb{N}}$ is nondecreasing and bounded above, while $\{b_i^k\}_{k\in\mathbb{N}}$ is nonincreasing and bounded below. Hence both $\{a_i^k\}_{k\in\mathbb{N}}$ and $\{b_i^k\}_{k\in\mathbb{N}}$ are convergent as $k \to \infty$.

(ii) By the observation in (i), we have, for every fixed $i = 1, \ldots, n$,

$$a_i^k \le a_i^{k+1} \le \cdots \le t_i \le \cdots \le b_i^{k+1} \le b_i^k \quad \text{for all } k \in \mathbb{N} .$$

Hence for every $i = 1, \ldots, n$, $t_i \in [a_i^k, b_i^k]$ for all $k \in \mathbb{N}$ and so $t = (t_1, \ldots, t_n) \in J^k = \prod_{i=1}^n [a_i^k, b_i^k]$ for every $k \in \mathbb{N}$. Hence (**).

Theorem 1.4.2 (Cantor Intersection Theorem). *Every decreasing nest of closed and bounded nonempty subsets of* $\mathbb{R}^n$ *has nonempty intersection. That is, if* $\{Q_k\}_{k=1}^\infty$ *is a collection of nonempty closed subsets of* $\mathbb{R}^n$ *satisfying*

(a) $Q_{k+1} \subset Q_k$ *for all* k, *and*

(b) Q_1 *is bounded,*

then $\bigcap_{k=1}^\infty Q_k \neq \phi$.

Proof. Let $S = \bigcap_{k=1}^\infty Q_k$. If Q_k is finite for some $k \in \mathbb{N}$, then there exists $r \ge k$ such that $Q_r = Q_{r+1} = Q_{r+2} = \cdots$ and hence in this case $\bigcap_{k=1}^\infty Q_k = Q_r \neq \phi$. So let us assume that Q_k is infinite for all k. Let

$$A = \{x_1, x_2, \ldots\},$$

where for each k, $x_k \in Q_k$ is an arbitrary point distinct from $x_1, \ldots, x_{k-1}$. This can be done since all Q_k's are infinite. By Bolzano-Weierstrass theorem, A has an accumulation point $x \in \mathbb{R}^n$. By Theorem 1.2.27 every open neighborhood U of x in $\mathbb{R}^n$ contains infinitely many points of A and thus contains infinitely many points of each Q_k. Hence x is an accumulation point of each Q_k. Now for each $k \in \mathbb{N}$, Q_k is closed. Hence we have $Q_k' \subset Q_k$ and so $x \in Q_k' \subset Q_k$ for all $k \in \mathbb{N}$. Therefore, $x \in \bigcap_{k=1}^\infty Q_k = S$. □

Theorem 1.4.3 (Heine-Borel Theorem). *Every closed and bounded subset of* $\mathbb{R}^n$ *is compact.*

Proof. Let $S \subset \mathbb{R}^n$ be closed and bounded. Let $\mathcal{U} = \{U_\alpha\}_{\alpha\in\Lambda}$ be an open cover of S in $\mathbb{R}^n$. By Lindelöf's covering theorem (Theorem 1.2.15), $\mathcal{U}$ has a countable sub-cover $\{U_k\}_{k=1}^\infty$. For any $m \geq 1$, define

$$S_m := \bigcup_{k=1}^{m} U_k \,,$$
$$Q_m := S \setminus S_m.$$

Then the S_m's are open, increasing and hence the Q_m's are closed and decreasing. Furthermore, since S is bounded, the Q_m's are also bounded. Now if $Q_m \neq \phi$ for all m, by Cantor Intersection Theorem (Theorem 1.4.2),

$$\phi \neq \bigcap_{k=1}^{\infty} Q_k = \bigcap_{k=1}^{\infty} \left(S \setminus S_k\right) = S \setminus \bigcup_{k=1}^{\infty} S_k \,,$$

which is absurd since $S \subset \bigcup_{k=1}^\infty S_k$. Thus there exists $m \geq 1$ such that $Q_m = \phi$ and hence $S \subset S_m = \bigcup_{k=1}^m U_k$. □

Corollary 1.4.4. *Let* $S \subset \mathbb{R}^n$ *be a subset. Then the following are equivalent:*

(a) S *is compact.*

(b) S *is closed and bounded.*

(c) S *has the Bolzano-Weierstrass property.*

Proof. By Theorems 1.3.5 and 1.4.3, (a) $\Leftrightarrow$ (b). By Theorem 1.3.6, (a) $\Rightarrow$ (c). It thus remains to show (c) $\Rightarrow$ (b). So assume that S has the Bolzano-Weierstrass property. Clearly S must be bounded for otherwise there exists a sequence $\{x_k\}_{k\in\mathbb{N}}$ in S with $\|x_k\| \geq k$ for all k. But then obviously such a sequence cannot have an accumulation point in S. Next, to show that S is closed, it sufficies to show that

$S' \subset S$. So let $x \in S'$. Since every open neighborhood of x contains infinitely many points of S, we can choose an infinite sequence of distinct points $\{x_k : k \in \mathbb{N}\} \subset S$ with $x_k \in B(x, \frac{1}{k})$ for each $k \in \mathbb{N}$. Since S has the Bolzano-Weierstrass property, $\{x_k : k \in \mathbb{N}\}$ has an accumulation point $y \in S$. We claim that $y = x$. To see this, observe that for any $k \in \mathbb{N}$,

$$\begin{aligned} \|y - x\| &\leq \|y - x_k\| + \|x_k - x\| \\ &< \|y - x_k\| + \frac{1}{k} \,. \end{aligned}$$

Now certainly the second term $\frac{1}{k}$ on the right is controllable, and is going to be small when k gets large. However, in general the first term $\|y - x_k\|$ need not be small for *all* $k \in \mathbb{N}$. In fact, we only know that there are infinitely many k's for which x_k is close to y, but it is not necessarily true that all x_k's are close to y. But that is already good enough for our purpose: For any $r > 0$, there are infinitely many $k \in \mathbb{N}$ such that $x_k \in B(y, r)$. Hence

$$\|y - x\| < r + \frac{1}{k} \qquad \text{for infinitely many } k \in \mathbb{N}\,.$$

This forces

$$\|y - x\| \leq r\,.$$

But then as $r > 0$ is arbitrary, this implies $x = y \in S$ and so $S' \subset S$. Hence S is closed. □

Remark. For a general metric space, Corollary 1.4.4 is only partially true, namely, we have

$$\begin{aligned} &\text{(a)} \Rightarrow \text{(b)} \qquad \text{by Theorem 1.3.5}\,, \\ &\text{(a)} \Rightarrow \text{(c)} \qquad \text{by Theorem 1.3.6}\,, \end{aligned}$$

furthermore, it can be proved without much difficulty that

$$\text{(c)} \Rightarrow \text{(a)}$$

but in general,

$$\text{(b)} \not\Rightarrow \text{(a)} .$$

For instance, let $X := \mathbb{Q}$ with the Euclidean metric, and $S := \big\{ r \in \mathbb{Q} : r \in (a, b) \big\}$, where a, $b \in \mathbb{R} \setminus \mathbb{Q}$, $a < b$. Then it is not difficult to see that S is closed and bounded in X but not compact (see Exercise 1.4, Part A, Problem #12 below). Hence for a general metric space, including the case where it is a proper subspace of $\mathbb{R}^n$, we only have

$$\text{(a)} \Leftrightarrow \text{(c)} \Rightarrow \text{(b)} .$$

Exercise 1.4

Unless otherwise specified, X will stand for a general metric space and S, T, etc., are arbitrary subsets of X.

Part A: True or False Questions

For each of the following statements, determine if it is true or false. If it is true, prove it. If it is false, give a counterexample or provide proper justification.

1. The circle $S^1 := \{(x, y) \in \mathbb{R}^2 : x^2 + y^2 = 1\} \subset \mathbb{R}^2$ is compact.

   Answer: True.

   Proof. It is closed and bounded in $\mathbb{R}^2$, hence compact.

2. The ellipse $\left\{(x, y) \in \mathbb{R}^2 : \frac{x^2}{a^2} + \frac{y^2}{b^2} = 1\right\} \subset \mathbb{R}^2$ is compact.

   Answer: True.

   Proof. It is closed and bounded in $\mathbb{R}^2$, hence compact.

3. The parabola $\{(x, y) \in \mathbb{R}^2 : y^2 = x\} \subset \mathbb{R}^2$ is compact.

   Answer: False.

   Justification. Although it is closed in $\mathbb{R}^2$, since it is unbounded, it is noncompact.

4. The hyperbola $\left\{(x, y) \in \mathbb{R}^2 : \frac{x^2}{a^2} - \frac{y^2}{b^2} = 1\right\} \subset \mathbb{R}^2$ is compact.

   Answer: False.

   Justification. Although it is closed in $\mathbb{R}^2$, since it is unbounded, it is noncompact.

5. The upper half circle $S := \left\{(x, y) \in \mathbb{R}^2 : x^2 + y^2 = 1 \, , y > 0\right\} \subset \mathbb{R}^2$ is compact.

   Answer: False.

   Justification. Although it is bounded, since it is not closed in $\mathbb{R}^2$, it is noncompact.

6. $S := [0, \infty) \subset \mathbb{R}$ is compact.

   Answer: False.

   Justification. Although S is closed in $\mathbb{R}$, since it is unbounded, it is noncompact.

   Alternatively, $\{(-1, n) : n \in \mathbb{N}\}$ is an open cover of S without finite subcover. Hence S is noncompact.

7. The cone $S := \{(x, y, z) : z \geq \sqrt{x^2 + y^2}\} \subset \mathbb{R}^3$ is compact.

   Answer: False.

   Justification. Although it is closed in $\mathbb{R}^3$, since it is not bounded, it is noncompact.

8. The cylinder $S := \{(x, y, z) : x^2 + y^2 = 1\} \subset \mathbb{R}^3$ is compact.

   Answer: False.

   Justification. Although it is closed in $\mathbb{R}^3$, since it is not bounded, it is noncompact.

9. $S := (0, 1] \subset X = (0, \infty)$ with Euclidean metric is closed and bounded, hence compact.

   Answer: The first half is True, the second half is False.

   Justification. It is clear that S is closed and bounded in X. However, Heine-Borel Theorem does not apply, as the metric space X under consideration is not the Euclidean space $\mathbb{R}$. In fact, $\{(0, 1 - \frac{1}{n}) : n \in \mathbb{N}\}$ is an open cover of S which does not admit any finite subcover. Hence S is not compact.

10. $S := \{\frac{1}{n} : n \in \mathbb{N}\} \cup \{0\} \subset \mathbb{Q}$ is closed and bounded in $\mathbb{Q}$, hence compact.

    Answer: True.

    Proof. It is not hard to see that S is closed and bounded in $\mathbb{Q}$. It is also compact. However, the compactness does not follow directly from the closedness and boundedness of $S \subset \mathbb{Q}$. Instead, it follows from the fact that S when considered as a subset of $\mathbb{R}$, is closed and bounded. Hence by Heine-Borel Theorem, S is compact in $\mathbb{R}$. Finally, by Theorem 1.3.7, S is compact in $\mathbb{Q}$.

The compactness of S can also be proved by first principle: Let $\{U_\lambda\}_{\lambda\in\Lambda}$ be any open cover of S in X. Then there exists $\lambda_0 \in \Lambda$ such that $0 \in U_{\lambda_0}$. Note that U_{λ_0} covers all but finitely many points of S. In fact, as U_{λ_0} is open in $\mathbb{Q}$, there exists an open set $V \subset \mathbb{R}$ such that $U_{\lambda_0} = V \cap \mathbb{Q}$. Since $V \subset \mathbb{R}$ is open neighborhood of 0, there exists $\delta > 0$ such that $(-\delta, \delta) \subset V$. Let $N \in \mathbb{N}$ be large enough such that $\frac{1}{N} < \delta$. Then $\frac{1}{n} \in (-\delta, \delta) \cap \mathbb{Q} \subset V \cap \mathbb{Q} = U_{\lambda_0}$ for all $n \geq N$ and so the only points of S that are not covered by U_{λ_0} are $\{\frac{1}{n} : n = 1, \dots, N-1\}$. For each such n, there exists $\lambda_n \in \Lambda$ such that $\frac{1}{n} \in U_{\lambda_n}$. Hence $\{U_{\lambda_n} : n = 0, 1, 2, \dots, N-1\}$ is a finite subcover of the open cover $\{U_\lambda\}_{\lambda\in\Lambda}$ and so S is compact.

11. $S := \mathbb{Q} \cap [0, 1] \subset \mathbb{R}$ is compact.

Answer: False.

Justification. S is not closed in $\mathbb{R}$. In fact, $(\mathbb{R} \setminus \mathbb{Q}) \cap [0, 1]$ does not have any interior point, hence $(\mathbb{R} \setminus \mathbb{Q}) \cap [0, 1]$ is not open in $[0, 1]$ and so $S = [0, 1] \setminus \big((\mathbb{R} \setminus \mathbb{Q}) \cap [0, 1]\big)$ is not closed in $[0, 1]$, thus noncompact.

Alternatively, pick any irrational point $a \in S$. Then

$$\left\{\left(-\frac{1}{n}, a - \frac{1}{n}\right), \left(a - \frac{1}{n}, 1 + \frac{1}{n}\right)\right\}_{n\in\mathbb{N}}$$

is an open cover of S in $\mathbb{R}$ which has no finite subcover. Here, we used the natural convention that $(\alpha, \beta) := \phi$ for $\alpha \geq \beta$.

12. Let $a, b \in \mathbb{R} \setminus \mathbb{Q}$, $a < b$, and $S := (a, b) \cap \mathbb{Q}$.
    (a) When considered as a subset of the metric space $\mathbb{R}$ with the usual Euclidean metric, S is compact.
    (b) Consider S as a subset of the metric space $\mathbb{Q}$ with the usual Euclidean metric.

(i) S is bounded.
(ii) S is open.
(iii) S is closed.
(iv) S is compact.

Answer:

(a) False.

Justification. As the closure of S in $\mathbb{R}$ is $\overline{S} = [a, b] \supsetneqq S$, S is not closed in $\mathbb{R}$, hence not compact.

(b) (i) True.

Proof. It is obvious.

(ii) True.

Proof. Since (a, b) is open in $\mathbb{R}$, $S = (a, b) \cap \mathbb{Q}$ is open in $\mathbb{Q}$.

(iii) True.

Proof. Since $a,\ b \in \mathbb{R} \setminus \mathbb{Q}$, we have $S = (a, b) \cap \mathbb{Q} = [a, b] \cap \mathbb{Q}$. In particular, since $[a, b]$ is closed in $\mathbb{R}$, $S = \mathbb{Q} \cap [a, b]$ is closed in $\mathbb{Q}$.

(iv) False.

Justification. For any $n \in \mathbb{N}$ with $n > \frac{1}{b-a}$, let $U_n := \left(a, b - \frac{1}{n}\right) \cap \mathbb{Q}$. The argument used in Part (ii) shows that U_n is open in $\mathbb{Q}$. The sets $\{U_n : n \in \mathbb{N}\}$ form an open cover of S which obviously has no finite subcover. Hence $S \subset \mathbb{Q}$ is not compact.

Alternatively, by (a), S is not compact in $\mathbb{R}$. Hence by Theorem 1.3.7, it is not compact in $\mathbb{Q}$.

Part B: Problems

1. Without the closedness condition, determine whether Cantor Intersection Theorem is still valid.

Answer: Without the closedness condition, Cantor Intersection Theorem is no longer valid.

Example: Consider $Q_n := (0, \frac{1}{n}) \subset \mathbb{R}$, $n \in \mathbb{N}$. Then each Q_n is bounded but not closed in $\mathbb{R}$, and $\bigcap_{n=1}^{\infty} Q_n = \phi$.

2. Without the boundedness condition, determine whether Cantor Intersection Theorem is still valid.

   Answer: Without the boundedness condition, Cantor Intersection Theorem is no longer valid.

   Example: Consider $Q_n := [n, \infty) \subset \mathbb{R}$, $n \in \mathbb{N}$. Then each Q_n is closed but not bounded in $\mathbb{R}$, and $\bigcap_{n=1}^{\infty} Q_n = \phi$.

3. Determine whether Cantor Intersection Theorem is still valid for any general metric spaces.

   Answer: Cantor Intersection Theorem is in general not valid for a general metric space.

   Example: Consider $X := (0,1) \subset \mathbb{R}$. Let $Q_n := \left[\frac{n}{n+1}, 1\right)$ for $n \in \mathbb{N}$. Each Q_n is closed in $(0,1)$ and is obviously bounded, but $\bigcap_{n=1}^{\infty} Q_n = \phi$.

4. Let $I_0 := [0,1]$, $I_1 :=$ the residue of I_0 after the middle open $\frac{1}{3}$ of $[0,1]$ is removed, and inductively, for any $n \in \mathbb{N}$, I_{n+1} be the residue of I_n after the middle open $\frac{1}{3}$ of each of its subintervals is removed. To be more precise,

$$
\begin{aligned}
I_0 &:= [0,1] \\
I_1 &:= I_0 \setminus \left(\frac{1}{3}, \frac{2}{3}\right) = \left[0, \frac{1}{3}\right] \cup \left[\frac{2}{3}, 1\right] \\
I_2 &:= I_1 \setminus \left\{\left(\frac{1}{9}, \frac{2}{9}\right) \cup \left(\frac{7}{9}, \frac{8}{9}\right)\right\} \\
&= \left[0, \frac{1}{9}\right] \cup \left[\frac{2}{9}, \frac{1}{3}\right] \cup \left[\frac{2}{3}, \frac{7}{9}\right] \cup \left[\frac{8}{9}, 1\right] \\
&\vdots
\end{aligned}
$$

or, by abusing notations in the obvious way, we have

$$I_{n+1} := \frac{I_n}{3} \cup \left\{ \frac{2}{3} + \frac{I_n}{3} \right\}, \qquad n = 0, 1, 2 \ldots .$$

Define $C := \bigcap_{n=0}^{\infty} I_n$. Classically, X is known as the *Cantor Set*.

(a) Show that $C \neq \phi$.
(b) Find $\overline{C}$.
(c) Find C°.
(d) Find C'.
(e) Find ∂C.
(f) Determine if C is compact.

Solution:

(a) As $0 \in I_n$ for all n, we have $0 \in C$ and so $C \neq \phi$.
(b) Each I_n is closed in $\mathbb{R}$. Being the intersection of I_n's, C is closed and so $\overline{C} = C$.
(c) First observe that at step $n \in \mathbb{N}$, I_n consists of 2^n closed intervals, each of which has length $\frac{1}{3^n}$. So we can denote by I_n^k, $k = 1, \ldots, 2^n$, the subintervals which form I_n, each of which has length $\frac{1}{3^n}$. For any $x \in C$ and any $\delta > 0$, let $N \in \mathbb{N}$ be large enough such that $\frac{1}{3^N} < \frac{\delta}{2}$. Then as $C := \bigcap_{n=0}^{\infty} I_n$, we have $x \in I_N$ and so $x \in I_N^k$ for some $k = 1, \ldots, 2^N$. Note that when I_{N+1} is formed, the middle open $\frac{1}{3}$-interval of I_N^k which is a subset of $(x - \delta, x + \delta)$ would be removed. As a result, $(x - \delta, x + \delta) \not\subset C$. Since this is true for every $\delta > 0$, we conclude that $x \notin C^{\circ}$ and so $C^{\circ} = \phi$.
(d) Observe that every point in $x \in C \setminus \{0, 1\}$ is a boundary point of some open $\frac{1}{3}$-interval which has been removed when C is being constructed. In particular, every open interval centered at x would contain infinitely many open $\frac{1}{3}$-intervals which have been removed and hence in turn contains infinitely points of C. Hence $x \in C'$. Similarly, $0,\ 1 \in C'$ and so $C \subset C'$. On the other hand, by (b), C is closed and so $C' \subset C$. Thus $C' = C$.

(e) Since $C = \overline{C} = C^\circ \cup \partial C$ and $C^\circ = \phi$, we have $\partial C = C$.

(f) As C is both closed and bounded in $\mathbb{R}$, by Heine-Borel Theorem, C is compact.

5. Consider the metric space $(\mathbb{R}, d)$, where $d(x,y) := \dfrac{|x-y|}{1+|x-y|}$.

   (a) If $d(x,y) < r < \frac{1}{2}$, show that $|x-y| < \sqrt{r}$.

   (b) Show that $\mathbb{N}$ is closed and bounded in $(\mathbb{R}, d)$.

   (c) Determine whether $\mathbb{N}$ is compact.

   Solution:

   (a) Elementary.

   (b) Since $d(x,y) < 1$ for any x, $y \in \mathbb{R}$, $\mathbb{R}$ is bounded and so in particular, $\mathbb{N}$ is bounded. On the other hand, consider $x \in \mathbb{R} \setminus \mathbb{N}$. If we denote by $B(x,r)$ and $B_d(x,r)$ the open balls at x with radius $r > 0$ with respect to the Euclidean metric and the metric d, respectively, then there exists $\delta_0 > 0$ such that $B(x,\delta_0) \cap \mathbb{N} = \phi$. Observe that by (a), if $\delta < \frac{1}{\sqrt{2}}$, we have $B_d(x,\delta^2) \subset B(x,\delta)$. Hence for any $\delta < \min\{\frac{1}{\sqrt{2}}, \delta_0\}$, we have $B_d(x,\delta^2) \cap \mathbb{N} \subset B(x,\delta) \cap \mathbb{N} \subset B(x,\delta_0) \cap \mathbb{N} = \phi$. Thus $\mathbb{R} \setminus \mathbb{N}$ is open and so $\mathbb{N}$ is closed in $(\mathbb{R}, d)$.

   (c) Although $\mathbb{N}$ is closed and bounded in $(\mathbb{R}, d)$, Heine-Borel Theorem does not apply here, as the Theorem is valid only when the Euclidean metric is used. In fact, under the metric d defined here, $\mathbb{N}$ is actually not compact. To see this, we observe that for any $n \in \mathbb{N}$, $B_d(n, \frac{1}{2}) \cap \mathbb{N} = \{n\}$. Hence the open cover $\{B_d(n, \frac{1}{2}) : n \in \mathbb{N}\}$ does not have any finite subcover. Thus $\mathbb{N}$ is not compact.

6. Determine whether Cantor Intersection Theorem is valid for any compact metric spaces. That is, if X is a compact metric space, is it always true that every decreasing nest of closed nonempty subsets of X (note that subsets of X are automatically bounded) has nonempty intersection.

Answer: Cantor Intersection Theorem is still valid for arbitrary compact spaces.

Proof. Let X be a compact metric space. Let $\{Q_n : n \in \mathbb{N}\}$ be a decreasing nest of nonempty closed subsets of X. Assume that $\bigcap_{n=1}^{\infty} Q_n = \phi$. For any $n \in \mathbb{N}$, let $U_n := Q_1 \setminus Q_n$. Then

$$Q_1 = Q_1 \setminus \phi = Q_1 \setminus \left(\bigcap_{n=1}^{\infty} Q_n \right) = \bigcup_{n=1}^{\infty} (Q_1 \setminus Q_n) = \bigcup_{n=1}^{\infty} U_n \,.$$

Hence $\{U_n : n \in \mathbb{N}\}$ is an open cover (in Q_1) for Q_1. Being a closed subset of the compact space X, Q_1 is compact. Thus one can find a finite subcover $\left\{U_{n_j} : j = 1, \ldots, J\right\}$ of $\{U_n : n \in \mathbb{N}\}$ for Q_1. Let $m := \min\{n_j : j = 1, \ldots, J\}$. Then U_m is the largest one among all the U_{n_j}'s and

$$Q_1 = \bigcup_{j=1}^{J} U_{n_j} = U_m = Q_1 \setminus Q_m \,,$$

which forces $Q_m = \phi$, contradicting to the assumption that $Q_n \neq \phi$ for all n. Thus $\bigcap_{n=1}^{\infty} Q_n \neq \phi$.

7. Determine whether it is true that for every bounded subset S of $\mathbb{R}^n$, we can always find a smallest compact subset of $\mathbb{R}^n$ which contains S.

Answer: True.

Proof. By Heine-Borel Theorem, a subset $S \subset \mathbb{R}^n$ is compact if and only if it is closed and bounded. Also, the closure $\overline{S}$ of S is the smallest closed subset of $\mathbb{R}^n$ which contains S. So the problem boils down to showing that $\overline{S}$ of a bounded set $S \subset \mathbb{R}^n$ is also bounded. To see this, let $S \subset \mathbb{R}^n$ be bounded. If $S = \phi$, there is nothing to prove. So assume $S \neq \phi$. Fix $x_0 \in S$. As S is bounded, $S \subset B(x_0, R)$ for some $R > 0$. If $\overline{S}$ is unbounded, then there exists at least one point $x_1 \in \overline{S}$, such that $d(x_0, x_1) > R + 2$. So $B(x_1, 1) \cap S = \phi$. This contradicts with $x_1 \in \overline{S}$. Therefore, $\overline{S}$ is bounded in X.

8. Let $A \subset X$ be closed, with $A^\circ = \phi$. Show that for any nonempty open set $U \subset X$, there exists a nonempty open set V in X such that $\overline{V} \subset U$ and $\overline{V} \cap A = \phi$.

   Proof. As $A^\circ = \phi$ and $U \neq \phi$ is open, $A \not\supset U$. So there exists $y \in U \setminus A$. As U is open, $B(y, \varepsilon_1) \subset U$ for some $\varepsilon_1 > 0$. On the other hand, as $y \notin A$ and A is closed, $B(y, \varepsilon_2) \cap A = \phi$ for some $\varepsilon_2 > 0$. Take $\varepsilon := \min\{\varepsilon_1/2, \varepsilon_2/2\}$ and $V := B(y, \epsilon)$. Clearly, V is open, $V \neq \phi$, $\overline{V} \subset B(y, \varepsilon_1) \subset U$, and $\overline{V} \cap A \subset B(y, \varepsilon_2) \cap A = \phi$.

9. X is said to be a *Baire space* if the union of countably many closed subsets of X, each of which has empty interior, must have empty interior. That is, given any countable collection $\{A_n\}_{n\in\mathbb{N}}$ of closed subsets of X with $A_n^\circ = \phi$ for every n, we have $\left(\bigcup_{n\in\mathbb{N}} A_n\right)^\circ = \phi$.
   Show that every compact space X is a Baire space.

   Proof. Let $\{A_n\}_{n\in\mathbb{N}}$ be a countable collection of closed subsets of X with $A_n^\circ = \phi$ for every n. Let $V_0 \subset X$ be any nonempty open subset. As $A_1^\circ = \phi$, $A_1 \not\supset V_0$. By Exercise 1.4, Part B, Problem #8, there is a nonempty open set V_1 in X such that $\overline{V_1} \subset V_0$ and $\overline{V_1} \cap A_1 = \phi$. Inductively, we have a sequence of nonempty open sets $\{V_n\}_{n\in\mathbb{N}}$ satisfying

$$\overline{V_n} \subset V_{n-1} \qquad \text{and} \qquad \overline{V_n} \cap A_n = \phi \quad \text{for all } \ n \in \mathbb{N}\ .$$

   In particular, we have a sequence of nonempty closed sets

$$\cdots \subset \overline{V_n} \subset \overline{V_{n-1}} \subset \cdots \subset \overline{V_2} \subset \overline{V_1}$$

   with the finite intersection property. Since X is compact, by Exercise 1.4, Part B, Problem #6 or Exercise 1.3, Part B, Problem #12, we have a point

$$x \in \bigcap_{n\in\mathbb{N}} \overline{V_n} \neq \phi\ .$$

In particular, $x \in \overline{V_1} \subset V_0$. Furthermore, since $x \in \overline{V_n}$ and $\overline{V_n} \cap A_n = \phi$ for all n, it follows that $x \notin A_n$ for any n and so $x \notin \bigcup_{n\in\mathbb{N}} A_n$. To sum up, we have an $x \in V_0 \setminus \bigcup_{n\in\mathbb{N}} A_n$, hence $\bigcup_{n\in\mathbb{N}} A_n \not\supset V_0$. Since $V_0 \subset X$ is any arbitrary nonempty open subset of X, we conclude that $\left(\bigcup_{n\in\mathbb{N}} A_n\right)^\circ = \phi$ and thus X is a Baire space.

10. Let X be compact and $\{U_n\}_{n\in\mathbb{N}}$ be a countable collection of open sets in X, each of which is dense in X. Show that $\bigcap_{n\in\mathbb{N}} U_n$ is also dense in X.

Proof. Observe first that $S \subset X$ is dense in X if and only if $(X \setminus S)^\circ = \phi$. In fact,

$$\begin{aligned}
&\overline{S} = X \\
&\Leftrightarrow \forall\, x \in X,\ \forall\, \varepsilon > 0,\ B(x,\varepsilon) \cap S \neq \phi \\
&\Leftrightarrow \forall\, x \in X,\ \forall\, \varepsilon > 0,\ \exists\, z \in B(x,\varepsilon) \cap S \\
&\Leftrightarrow \forall\, x \in X,\ \forall\, \varepsilon > 0,\ \exists\, z \in B(x,\varepsilon) \text{ and } z \notin X \setminus S \\
&\Leftrightarrow \forall\, x \in X,\ \forall\, \varepsilon > 0,\ X \setminus S \not\supset B(x,\varepsilon) \\
&\Leftrightarrow (X \setminus S)^\circ = \phi\,.
\end{aligned}$$

Now if $\{U_n\}$ is a countable collection of open sets each of which is dense in X, then by the observation above, $X \setminus U_n$ is a countable collection of closed sets, each of which has empty interior. By Exercise 1.4, Part B, Problem #9, being compact, X is Baire and so in particular, $\left(\bigcup_{n\in\mathbb{N}}(X \setminus U_n)\right)^\circ = \phi$. By the observation above again,

$$\bigcap_{n\in\mathbb{N}} U_n = X \setminus \left(X \setminus \bigcap_{n\in\mathbb{N}} U_n\right) = X \setminus \left(\bigcup_{n\in\mathbb{N}} (X \setminus U_n)\right)$$

is dense in X.

11. Show that $[0,1] \setminus \mathbb{Q}$ is dense in $[0,1]$.

Proof. $\Big\{[0,1]\backslash\{r\} : r \in \mathbb{Q}\Big\}$ is a countable collection of open sets in $[0,1]$, each of which is dense in the compact space $[0,1]$. By Exercise 1.4, Part B, Problem #10,

$$[0,1] \setminus \mathbb{Q} = \bigcap_{r \in \mathbb{Q}} \Big([0,1] \setminus \{r\}\Big)$$

is dense in $[0,1]$.

12. A metric space X is said to be *locally compact* if every $x \in X$ has a compact neighborhood in X. That is, for any $x \in X$, there exists a compact subset $C \subset X$ which contains an open neighborhood U of x in X. Show that the following statements are equivalent:
    (a) X is locally compact.
    (b) For any $x \in X$, there exists an open neighborhood U of x in X with $\overline{U}$ compact.
    (c) For any $x \in X$, there exists $r > 0$ such that $\overline{B}(x,r)$ is compact.

    Proof.
    (a)$\Rightarrow$(b): By definition of local compactness, for any $x \in X$, there exists a compact set C and an open set U with $x \in U \subset C$. Hence we have $x \in U \subset \overline{U} \subset \overline{C} = C$. Note that being a closed subset of the compact set C, $\overline{U}$ is compact.
    (b)$\Rightarrow$(c): By assumption, there exists an open neighborhood U of x in X with $\overline{U}$ compact. Now as U is open, there exists $r > 0$ such that $B(x,r) \subset U$. Note that $\overline{B}(x,r) \subset \overline{U}$ and so $\overline{B}(x,r)$ is compact.
    (c)$\Rightarrow$(a): By assumption, there exists $r > 0$ such that $\overline{B}(x,r)$ is compact. As $B(x,r) \subset \overline{B}(x,r)$ is trivially an open neighborhood of x, (a) follows.

13. Show that
    (a) $\mathbb{R}$ is locally compact.
    (b) Every discrete metric space is locally compact.

Proof.

(a) It is obvious by Exercise 1.4, Part B, Problem #12, as by Heine-Borel Theorem, $\overline{B}(x, r)$ is compact for every $x \in \mathbb{R}$ and every $r > 0$.

(b) It is obvious since every singleton in a discrete metric space is open and compact.

14. (a) Show that every compact space is locally compact.
(b) Determine whether the converse of (a) is true or false.

Solution:

(a) If X is compact, then X is trivially open and compact in X, hence X itself is a compact neighborhood of every point in X and so X is locally compact.

(b) The converse of (a) is in general not true.

Example 1:

By Exercise 1.4, Part B, Problem #13, $\mathbb{R}$ is locally compact but not compact.

Example 2:

Let X be a discrete metric space with infinitely many elements. Then X is not compact but by Exercise 1.4, Part B, Problem #13, X is locally compact.

15. Show that every locally compact metric space is a Baire Space.

Proof. The proof is similar to that of Exercise 1.4, Part B, Problem #9. Let $\{A_n\}_{n\in\mathbb{N}}$ be a countable collection of closed subsets of X with $A_n^\circ = \phi$ for every n. Let $V_0 \subset X$ be any nonempty open subset. As $A_1^\circ = \phi$, $A_1 \not\supset V_0$. By Exercise 1.4, Part B, Problem #8, there is a nonempty open set V_1 in X such that $\overline{V_1} \subset V_0$ and $\overline{V_1} \cap A_1 = \phi$. By the local compactness of X, by choosing a "smaller" subset, we may simply assume that $\overline{V_1}$ is compact.

In fact, pick an arbitrary point $z \in V_1$. By local compactness, we can find an open neighborhood U of z in X such that $\overline{U}$ is compact. Take $W := V_1 \cap U$. Then $\overline{W} = \overline{V_1 \cap U} \subset \overline{V_1} \subset V_0$, $\overline{W} \cap A_1 \subset$

$\overline{V_1} \cap A_1 = \phi$, and $\overline{W} = \overline{V_1 \cap U} \subset \overline{U}$ is compact. So by replacing V_1 by W if necessary, we may assume that $\overline{V_1}$ is compact.

Inductively, we have a sequence of nonempty open sets $\{V_n\}_{n\in\mathbb{N}}$ satisfying

$$\overline{V_n} \subset V_{n-1} \qquad \text{and} \qquad \overline{V_n} \cap A_n = \phi \quad \text{for all} \quad n \in \mathbb{N} \, .$$

In particular, we have a decreasing nest of nonempty closed sets

$$\cdots \subset \overline{V_n} \subset \overline{V_{n-1}} \subset \cdots \subset \overline{V_2} \subset \overline{V_1}$$

in the compact set $\overline{V_1}$. By Exercise 1.4, Part B, Problem #6, we have a point

$$x \in \bigcap_{n\in\mathbb{N}} \overline{V_n} \neq \phi \, .$$

In particular, $x \in \overline{V_1} \subset V_0$. Furthermore, since $x \in \overline{V_n}$ and $\overline{V_n} \cap A_n = \phi$ for all n, it follows that $x \notin A_n$ for any n and so $x \notin \bigcup_{n\in\mathbb{N}} A_n$.

To sum up, we have an $x \in V_0 \setminus \bigcup_{n\in\mathbb{N}} A_n$, hence $\bigcup_{n\in\mathbb{N}} A_n \not\supset V_0$. Since $V_0 \subset X$ is any arbitrary nonempty open subset of X, we conclude that $\left(\bigcup_{n\in\mathbb{N}} A_n\right)^\circ = \phi$ and thus X is a Baire space.

Chapter 2

Limits and Continuity

Chapter 1 deals with basic topological properties of a single metric space. In this chapter, we shall treat relations of two or more metric spaces. Naturally, by a "relation" between two metric spaces we mean a function, or a mapping, from one metric space to another. In particular, the concepts of limits and continuity in the context of Euclidean spaces will be carried forward to the context of general metric spaces, with the upshot being the study of the "equivalence problem", namely, the classification of metric spaces into "equivalent classes". Same as the case for advanced calculus, the basic language here will be the "epsilon-delta" ($\varepsilon - \delta$) language. The idea is that in the setting of abstract metric spaces, we have the concept of metrics which leads to the concept of "closeness" among points. Hence the concepts of limits and continuity can easily be mastered once the epsilon-delta language is transformed into the intuition of closeness among points. In particular, for any function $f : X \to Y$ of a metric space X into a metric space Y, the limit of $f(x)$ as x approaches $a \in X$ is the point in Y for which $f(x)$ is getting arbitrarily close to, when x is getting arbitrarily close to a. Similarly, f is continuous at $a \in X$ if points near a are mapped to points near $f(a) \in Y$ by f, and two metric spaces are "equivalent" if one can be "continuously deformed into" another, with the process reversible. To certain extent, these intuitive definitions of the concepts of limits and continuity are more important than those using rigorous epsilon-delta language.

2.1 Convergence in a Metric Space

Definition 2.1.1. *A sequence $\{x_n\}_{n\in\mathbb{N}}$ in X is said to be convergent and converge to $a \in X$ if for every $\varepsilon > 0$, there exists $N \in \mathbb{N}$ such* *that*

$$d(x_n, a) < \varepsilon \quad \textit{whenever} \quad n \geq N \ ,$$

that is,

$$x_n \in B(a, \varepsilon) \quad \textit{whenever} \quad n \geq N \ .$$

In this case we write $\lim_{n\to\infty} x_n = a$, *or* $\{x_n\} \to a$ *as* $n \to \infty$, *or* $x_n \to a$ *as* $n \to \infty$, *or even just* $x_n \to a$ *in case no confusion may arise. If* $\{x_n\}_{n\in\mathbb{N}}$ *is not convergent in* X, *it is said to be divergent.*

Remarks.

(i) Note that $\{x_n\} \to a$ in X as $n \to \infty$ if and only if $\{d(x_n, a)\} \to 0$ in $\mathbb{R}$ as $n \to \infty$. Hence the convergence of a sequence of points $\{x_n\}_{n\in\mathbb{N}}$ in X to a is equivalent to the convergence of the corresponding sequence of real numbers $\{d(x_n, a)\}_{n\in\mathbb{N}}$ in $\mathbb{R}$ to 0.

Proof.

$$\begin{aligned} & \{x_n\} \to a \text{ as } n \to \infty \\ \iff & \text{for any } \varepsilon > 0, \text{ there exists } N \in \mathbb{N} \text{ s.t.} \\ & \qquad |d(x_n, a) - 0| = d(x_n, a) < \varepsilon \text{ for all } n \geq N \\ \iff & \{d(x_n, a)\} \to 0 \text{ as } n \to \infty \ . \end{aligned}$$

(ii) If a sequence $\{x_n\}_{n\in\mathbb{N}}$ in $S \subset X$ converges to a point $a \in S$, then as a sequence in X, $\{x_n\}_{n\in\mathbb{N}}$ also converges to a. However, in general, the converse need not be true. That is, if a sequence $\{x_n\}_{n\in\mathbb{N}}$ in $S \subset X$ converges in X to $a \in X$, then as a sequence in S, $\{x_n\}_{n\in\mathbb{N}}$ need not be convergent.

Proof. The first statement is straight forward, as the metrics in S and X are the same. For the second statement, it suffices to provide a counterexample. This will be done after the following Theorem.

Theorem 2.1.2. *A sequence $\{x_n\}_{n\in\mathbb{N}}$ in X can converge to at most one point in X, called the limit of $\{x_n\}_{n\in\mathbb{N}}$.*

Proof. It is obvious by triangle inequality. In fact, if $\{x_n\} \to a$ and $\{x_n\} \to b$ as $n \to \infty$ for some a, $b \in X$, then for any $\varepsilon > 0$, there exist N_1, $N_2 \in \mathbb{N}$ such that

$$d(x_n, a) < \frac{\varepsilon}{2} \qquad \text{for all } \ n \geq N_1 \ ,$$
$$d(x_n, b) < \frac{\varepsilon}{2} \qquad \text{for all } \ n \geq N_2 \ .$$

Take $N := \max\{N_1, N_2\}$. Then by triangle inequality, we have

$$d(a, b) \leq d(x_n, a) + d(x_n, b) < \frac{\varepsilon}{2} + \frac{\varepsilon}{2} = \varepsilon \quad \text{for all } n \geq N \ .$$

Since $\varepsilon > 0$ is arbitrary, we conclude that $d(a, b) = 0$ or $a = b$. $\square$

Example 2.1.3. Let $X := \mathbb{R}$. The sequence $\left\{\frac{1}{n}\right\}_{n\in\mathbb{N}}$ in X clearly converges to $0 \in X$, but when considered as a sequence in the subspace $S := (0, \infty) \subset X$, it is not convergent. In fact, if it converged to a point $a \in S$, then $\left\{\frac{1}{n}\right\}_{n\in\mathbb{N}}$ as a sequence in X would converge to two distinct points 0 and a, which is impossible by Theorem 2.1.2. This Example also completes the proof of (ii) of the previous Remark.

Theorem 2.1.4. *Let $\{x_n\}_{n\in\mathbb{N}}$ be a sequence in X. If $\{x_n\} \to a \in X$ as $n \to \infty$, then*

(a) *$\{x_n : n \in \mathbb{N}\}$ is bounded in X,*

(b) *$a \in \overline{\{x_n : n \in \mathbb{N}\}}$,*

(c) *If $\#\{x_n : n \in \mathbb{N}\} = \infty$, $a \in \{x_n : n \in \mathbb{N}\}'$.*
Here, $\#(S)$ stands for the cardinality of (or intuitively, the number of elements in) the set $S \subset X$.

Proof.

(a) By the definition of limits, outside a ball centered at a with finite radius there can be at most finitely many points of $\{x_n : n \in \mathbb{N}\}$. Hence the whole set $\{x_n : n \in \mathbb{N}\}$ is lying in a ball centered at a with a large enough radius.

(b) It is clear by definition of limits. In fact, for any $r > 0$, there exists $N \in \mathbb{N}$ such that $x_n \in B(a, r)$ for all $n \geq N$. In particular, $B(a, r) \cap \{x_n : n \in \mathbb{N}\} \neq \phi$ and so $a \in \overline{\{x_n : n \in \mathbb{N}\}}$.

(c) Similar to (b) above, for any $r > 0$, there exists $N \in \mathbb{N}$ such that $x_n \in B(a, r)$ for all $n \geq N$. Since $\#\{x_n : n \in \mathbb{N}\} = \infty$, $\#\big(B(a, r) \cap \{x_n : n \in \mathbb{N}\}\big) = \infty$ and so $a \in \{x_n : n \in \mathbb{N}\}'$. □

A very useful characterization of adherent points and accumulation points is the following.

Theorem 2.1.5. *Let $S \subset X$ and $a \in X$. Then*

(a) *$a \in \overline{S}$ if and only if there is a sequence $\{x_n\}_{n\in\mathbb{N}}$ in S such that $\{x_n\} \to a$ as $n \to \infty$;*

(b) *$a \in S'$ if and only if there is an infinite sequence of distinct points $\{x_n\}_{n\in\mathbb{N}}$ in $S \setminus \{a\}$ such that $\{x_n\} \to a$ as $n \to \infty$.*

Proof.

(a) ($\Rightarrow$) If $a \in \overline{S}$, then for every $n \in \mathbb{N}$, there exists $x_n \in S$ with $d(x_n, a) < \frac{1}{n}$. Clearly, this sequence $\{x_n\}_{n\in\mathbb{N}}$ constructed in S has the desired property that $\{x_n\} \to a$ as $n \to \infty$.

($\Leftarrow$) If there is a sequence $\{x_n\}_{n\in\mathbb{N}}$ in S converging to a, then for any $r > 0$, there exists $N \in \mathbb{N}$ such that $d(x_n, a) < r$ for all $n \geq N$. Hence $B(a, r) \cap \{x_n : n \in \mathbb{N}\} \neq \phi$ and so $a \in \overline{S}$.

(b) ($\Rightarrow$) If $a \in S'$, then near a there are infinitely many points of S. Hence for every $n \in \mathbb{N}$, there exists $x_n \in S \setminus \{x_1, \cdots, x_{n-1}\}$ with $0 < d(x_n, a) < \frac{1}{n}$. By construction, $\{x_n\}_{n\in\mathbb{N}}$ is an infinite sequence of distinct points in $S \setminus \{a\}$ with $\{x_n\} \to a$ as $n \to \infty$.

($\Leftarrow$) If there is an infinite sequence of distinct points $\{x_n\}_{n\in\mathbb{N}}$

in $S \setminus \{a\}$ converging to a, then for any $r > 0$, there exists $N \in \mathbb{N}$ such that $0 < d(x_n, a) < r$ for all $n \geq N$. Hence $B(a, r) \cap \{x_n : n \in \mathbb{N}\} \setminus \{a\} \neq \phi$ and so $a \in S'$. □

With Theorem 2.1.5, we have the following very useful necessary and sufficient condition of closedness of a subset $S \subset X$.

Theorem 2.1.6. *Let $S \subset X$. Then the following are equivalent:*

(a) *S is closed.*

(b) *If $\{x_n\}_{n\in\mathbb{N}}$ is a sequence in S with $\{x_n\} \to a$ as $n \to \infty$, then $a \in S$.*

Proof. It is rather straightforward by Theorem 2.1.5.
(a) ⇒ *(b)*: If S is closed and $\{x_n\}_{n\in\mathbb{N}}$ is a sequence in S with $\{x_n\} \to a$ as $n \to \infty$, then by (a) of Theorem 2.1.5, $a \in \overline{S} = S$.
(b) ⇒ *(a)*: Take any $a \in \overline{S}$. By (a) of Theorem 2.1.5, there is a sequence $\{x_n\}_{n\in\mathbb{N}}$ in S such that $\{x_n\} \to a$ as $n \to \infty$. But then by assumption (b), we have $a \in S$. Thus $\overline{S} = S$ and so S is closed. □

To end this section, we have the following trivial result.

Theorem 2.1.7. *Let $\{x_n\}_{n\in\mathbb{N}}$ be a sequence in X. Then $\{x_n\} \to a$ as $n \to \infty$ if and only if $\{x_{n_k}\} \to a$ as $k \to \infty$ for every subsequence $\{x_{n_k}\}_{k\in\mathbb{N}}$ of $\{x_n\}_{n\in\mathbb{N}}$.*

Proof. Obvious. □

Exercise 2.1

Part A: True or False Questions

For each of the following statements, determine if it is true or false. If it is true, prove it. If it is false, give a counterexample or provide proper justification.

1. The sequence $\left\{x_n := \frac{1}{n}\right\}_{n\in\mathbb{N}}$ in $\mathbb{R}$ is convergent.

   Answer: True.

   Proof. For any $\varepsilon > 0$, let $N \in \mathbb{N}$ be large enough such that $N > \frac{1}{\varepsilon}$. Then for any $n \geq N$,

   $$d(x_n, 0) = |x_n - 0| = |x_n| = \frac{1}{n} \leq \frac{1}{N} < \varepsilon\,.$$

   Hence $\{d(x_n, 0)\} \to 0$ as $n \to \infty$ and so $\{x_n\}_{n\in\mathbb{N}} \to 0$ as $n \to \infty$.

2. The sequence $\left\{x_n := (-1)^n + \frac{1}{n}\right\}_{n\in\mathbb{N}}$ in $\mathbb{R}$ is convergent.

   Answer: False.

   Justification. Intuitively, observe that when n is even, $x_n = 1 + \frac{1}{n}$ and so as n increases, $\{x_n\}_{n\in\mathbb{N},\text{even}}$ should be decreasing to 1. On the other hand, when n is odd, $x_n = -1 + \frac{1}{n}$ and so as n increases, $\{x_n\}_{n\in\mathbb{N},\text{odd}}$ should be decreasing to -1. So that means there does not exist a real number for which all x_n's will be close to when n is large. Hence the sequence $\{x_n\}_{n\in\mathbb{N}}$ does not converge to any real number. Actually this intuition can be made rigorous by applying Theorem 2.1.7. In fact, observe that the subsequence $\{x_{2n}\}_{n\in\mathbb{N}} = \left\{1 + \frac{1}{2n}\right\}_{n\in\mathbb{N}} \to 1$ as $n \to \infty$, while another subsequence $\{x_{2n+1}\}_{n\in\mathbb{N}} = \left\{-1 + \frac{1}{2n+1}\right\}_{n\in\mathbb{N}} \to -1$ as $n \to \infty$. By Theorem 2.1.7, the sequence $\{x_n\}_{n\in\mathbb{N}}$ does not converge.

   Alternatively, it will be constructive to show the divergence of the sequence $\{x_n\}_{n\in\mathbb{N}}$ by using definition.

   Recall that $\{x_n\}_{n\in\mathbb{N}} \to a$ as $n \to \infty$ if and only if for any $\varepsilon > 0$, there exists $N \in \mathbb{N}$ such that $d(x_n, a) < \varepsilon$ whenever $n \geq N$. Hence

$\{x_n\}_{n\in\mathbb{N}} \not\to a$ if and only if there exists $\varepsilon > 0$ such that for any $N \in \mathbb{N}$, we can find $n \geq N$ with $d(x_n, a) \geq \varepsilon$.

Back to our problem. We will show that for any real number $a \in \mathbb{R}$ and any integer $N \in \mathbb{N}$, we can find $n \geq N$ such that $d(x_n, a) \geq \frac{1}{2}$:

Case 1: $a \leq 0$. Then for any even number $n \geq N$,

$$d(x_n, a) = |x_n - a| = \left|1 + \frac{1}{n} - a\right| = 1 + \frac{1}{n} + |a| > \frac{1}{2} .$$

Case 2: $a > 0$. Then for any odd number $n \geq N$,

$$d(x_n, a) = |x_n - a| = \left|-1 + \frac{1}{n} - a\right| = a + 1 - \frac{1}{n} > \frac{1}{2} .$$

Hence $\{x_n\} \not\to a$ as $n \to \infty$.

3. (Compare with Theorem 2.1.7)

   A sequence $\{x_n\}_{n\in\mathbb{N}}$ in X is convergent if and only if every subsequence $\{x_{n_k}\}_{k\in\mathbb{N}}$ of $\{x_n\}_{n\in\mathbb{N}}$ is convergent.

   Answer: True.

   Proof. In view of Theorem 2.1.7, it suffices to show that if all subsequences $\{x_{n_k}\}_{k\in\mathbb{N}}$ of $\{x_n\}_{n\in\mathbb{N}}$ are convergent, then they converge to the same limit. So assume that we have two subsequences $\{x_{n_k}\}_{k\in\mathbb{N}}$ and $\{x_{m_k}\}_{k\in\mathbb{N}}$ of $\{x_n\}_{n\in\mathbb{N}}$, with

$$\lim_{k\to\infty} x_{n_k} = a , \qquad \lim_{k\to\infty} x_{m_k} = b .$$

   Then by assumption, a third subsequence $\{x_{\ell_k}\}_{k\in\mathbb{N}}$ of $\{x_n\}_{n\in\mathbb{N}}$ defined by

$$x_{\ell_k} := \begin{cases} x_{n_k} & \text{if } k \text{ is odd} \\ x_{m_k} & \text{if } k \text{ is even} \end{cases}$$

   must also be convergent. However, since

$$\begin{aligned} x_{\ell_{2k+1}} &= x_{n_{2k+1}} \longrightarrow a \quad \text{as } k \to \infty , \\ x_{\ell_{2k}} &= x_{m_{2k}} \longrightarrow b \quad \text{as } k \to \infty , \end{aligned}$$

   we must have $a = b$.

4. A convergent sequence $\{x_n\}_{n\in\mathbb{N}}$ in $\mathbb{R}$ with discrete metric is *eventually* a constant sequence, that is, there exists $N \in \mathbb{N}$ such that $x_n = x_N$ for all $n \geq N$.

   Answer: True.

   Proof. As $\{x_n\}_{n\in\mathbb{N}}$ is convergent, it is Cauchy and so there exists $N \in \mathbb{N}$ such that $d(x_n, x_N) < \frac{1}{2}$ for all $n \geq N$. However, as d is the discrete metric, it only has values 0 and 1 and so this forces $d(x_n, x_N) = 0$ for all $n \geq N$ and so $x_n = x_N$ for all $n \geq N$.

5. A convergent sequence $\{x_n\}_{n\in\mathbb{N}}$ in $\mathbb{Z}$ with the metric induced by the Euclidean metric on $\mathbb{R}$ is *eventually* a constant sequence, that is, there exists $N \in \mathbb{N}$ such that $x_n = x_N$ for all $n \geq N$.

   Answer: True.

   Proof. As $\{x_n\}_{n\in\mathbb{N}}$ is convergent, it is Cauchy and so there exists $N \in \mathbb{N}$ such that $d(x_n, x_N) < \frac{1}{2}$ for all $n \geq N$. However, as x_N and $x_n \in \mathbb{Z}$ for all n, we have $d(x_n, x_N) \in \mathbb{N} \cup \{0\}$. This forces $d(x_n, x_N) = 0$ for all $n \geq N$ and so $x_n = x_N$ for all $n \geq N$.

6. A convergent sequence $\{x_n\}_{n\in\mathbb{N}}$ in a metric space $X = X_1 \cup X_2$ with $X_1 \cap X_2 = \phi$ such that there are infinitely many distinct x_n's in each X_i, $i = 1, 2$, is eventually a constant sequence, that is, there exists $N \in \mathbb{N}$ such that $x_n = x_N$ for all $n \geq N$.

   Answer: False.

   Example: Consider $X := X_1 \cup X_2$, where $X_1 := (-\infty, 0)$ and $X_2 := [0, \infty)$. Then $X_1 \cap X_2 = \phi$. Let $x_n := \frac{(-1)^n}{n}$, $n \in \mathbb{N}$. Then $\{x_n\} \to 0 \in X$ as $n \to \infty$, and there are infinitely many distinct x_n's in X_1 and in X_2. But the sequence $\{x_n\}_{n\in\mathbb{N}}$ is clearly not eventually constant.

7. If X is compact, then every sequence in X has a convergent subsequence.

Answer: True.

Proof. Let $\{x_k\}_{k\in\mathbb{N}}$ be any infinite sequence in X.

If $\{x_k : k \in \mathbb{N}\} \subset X$ is a finite set, then there is at least one $K \in \mathbb{N}$ s.t. $x_k = x_K$ for infinitely many $k \in \mathbb{N}$. That is, there is a subsequence $\{x_{k_j}\}_{j\in\mathbb{N}}$ of $\{x_k\}_{k\in\mathbb{N}}$ with $x_{k_j} = x_K$ for all $j \in \mathbb{N}$. In particular, $\{x_{k_j}\}_{j\in\mathbb{N}}$ is a convergent subsequence of $\{x_k\}$.

If $\{x_k : k \in \mathbb{N}\} \subset X$ is infinite, by the Bolzano-Weierstrass Property of the compact space X, $\{x_k : k \in \mathbb{N}\}$ has an accumulation point $p \in X$. In particular, for any $j \in \mathbb{N}$, there exists $k_j \in \mathbb{N}$ s.t. $d(x_{k_j}, p) < \frac{1}{j}$. Without loss of generality, we may assume that $k_{j+1} > k_j$ for all $j \in \mathbb{N}$. Therefore, $\{x_{k_j}\}_{j\in\mathbb{N}}$ is a subsequence of $\{x_k\}_{k\in\mathbb{N}}$ which is convergent (to p).

8. Every bounded sequence in $\mathbb{R}^n$ has a convergent subsequence.

   Answer: True.

   Proof. Let $\{x_k\}_{k\in\mathbb{N}}$ be a bounded sequence in $\mathbb{R}^n$.

   If $\{x_k : k \in \mathbb{N}\} \subset \mathbb{R}^n$ is a finite set, then there is at least one $K \in \mathbb{N}$ s.t. $x_k = x_K$ for infinitely many $k \in \mathbb{N}$. That is, there is a subsequence $\{k_j\}_{j\in\mathbb{N}}$ of $\{k\}_{k\in\mathbb{N}}$ with $x_{k_j} = x_K$ for all $j \in \mathbb{N}$. In particular, $\{x_{k_j}\}_{j\in\mathbb{N}}$ is a convergent subsequence of $\{x_k\}$.

   If $\{x_k : k \in \mathbb{N}\} \subset X$ is infinite, by Bolzano-Weierstrass Theorem, $\{x_k : k \in \mathbb{N}\}$ has an accumulation point $p \in \mathbb{R}^n$. In particular, for any $j \in \mathbb{N}$, there exists $k_j \in \mathbb{N}$ s.t. $d(x_{k_j}, p) < \frac{1}{j}$. Without loss of generality, we may assume that $k_{j+1} > k_j$ for all $j \in \mathbb{N}$. Therefore, $\{x_{k_j}\}_{j\in\mathbb{N}}$ is a subsequence of $\{x_k\}_{k\in\mathbb{N}}$ which is convergent (to p).

9. Let $\{x_n\}_{n\in\mathbb{N}}$ be a sequence in X. If every *convergent* subsequence of $\{x_n\}_{n\in\mathbb{N}}$ converges to the same point $x \in X$, then the entire sequence $\{x_n\}_{n\in\mathbb{N}}$ is convergent to x.

   Answer: False.

   Example: Let $X := \mathbb{R}$ and $\{x_n\}_{n\in\mathbb{N}}$ be the sequence defined by

$$x_n := \begin{cases} 1 & \text{if } n = 2k \text{ for some } k \in \mathbb{N}\,, \\ k & \text{if } n = 2k-1 \text{ for some } k \in \mathbb{N}\,. \end{cases}$$

It is easy to see that every convergent subsequence of $\{x_n\}_{n\in\mathbb{N}}$ converges to 1, but $\{x_n\}_{n\in\mathbb{N}}$ itself does not converge.

10. Let $\{x_n\}_{n\in\mathbb{N}}$ be a sequence in a *compact* space X. If every *convergent* subsequence of $\{x_n\}_{n\in\mathbb{N}}$ converges to the same point $x \in X$, then the entire sequence $\{x_n\}_{n\in\mathbb{N}}$ is convergent to x.

    Answer: True.

    Suppose the contrary, then there exists $\varepsilon > 0$ such that there are infinitely many elements of $\{x_n : n \in \mathbb{N}\}$ not in $B(x, \varepsilon)$. By Exercise 2.1, Part A, Problem #7, there is a subsequence $\{x_{n_k}\}_{k\in\mathbb{N}}$ of $\{x_n\}_{n\in\mathbb{N}}$ such that $x_{n_k} \notin B(x, \varepsilon)$ for all k and $\{x_{n_k}\}_{k\in\mathbb{N}}$ converges, say to some $x^* \in X$. Clearly $x^* \neq x$, which is a contradiction.

11. Let X be a metric space and let d_1, d_2 be two metrics on X. Let $\{x_n\}_{n\in\mathbb{N}}$ be a sequence in X. Then $\{x_n\}$ is convergent in (X, d_1) if and only if it is convergent in (X, d_2).

    Answer: False.

    Example: Let $X = \mathbb{R}$, d_1 be the Euclidean metric, and d_2 be the discrete metric. Then the sequence $\left\{\frac{1}{n}\right\}_{n\in\mathbb{N}}$ is convergent in (X, d_1) but not in (X, d_2).

Part B: Problems

1. For each of the following sequences $\{x_n\}_{n\in\mathbb{N}}$, find the derived set $\{x_n : n \in \mathbb{N}\}'$ and the closure $\overline{\{x_n : n \in \mathbb{N}\}}$ of the *subset* $\{x_n : n \in \mathbb{N}\} \subset X$.

    (a) $x_n := (1 + \frac{1}{n})^n$;

    (b) $x_n := (-1)^n(1 + \frac{1}{n})$;

    (c) $x_n := n^{\sin(\frac{n\pi}{2})}$;

    (d) $x_n := \sin(\frac{n\pi}{3})$;

    (e) $x_n := \begin{cases} 2^{1/n} & \text{if } n \text{ is odd,} \\ 1 & \text{if } n \text{ is even.} \end{cases}$

Answer:

(a) $\{x_n : n \in \mathbb{N}\}' = \{e\}$; $\overline{\{x_n : n \in \mathbb{N}\}} = \{x_n : n \in \mathbb{N}\} \cup \{e\}$.
(b) $\{x_n : n \in \mathbb{N}\}' = \{-1, 1\}$; $\overline{\{x_n : n \in \mathbb{N}\}} = \{-1, 1\}$.
(c) $\{x_n : n \in \mathbb{N}\}' = \{0\}$; $\overline{\{x_n : n \in \mathbb{N}\}} = \{x_n : n \in \mathbb{N}\} \cup \{0\}$.
(d) $\{x_n : n \in \mathbb{N}\}' = \phi$; $\overline{\{x_n : n \in \mathbb{N}\}} = \{x_n : n \in \mathbb{N}\}$.
(e) $\{x_n : n \in \mathbb{N}\}' = \{1\}$; $\overline{\{x_n : n \in \mathbb{N}\}} = \{x_n : n \in \mathbb{N}\} \cup \{1\}$.

2. Construct a metric on $\mathbb{R}$ for which the sequence $\{\frac{1}{n}\}_{n\in\mathbb{N}}$ converges to a point other than 0.

Solution: It is clear that if we can construct one such metric, we can construct many. One such examples is as follows: Let $f : \mathbb{R} \to \mathbb{R}$ be the bijection given by

$$x \in \mathbb{R} \mapsto f(x) := \begin{cases} 1 & \text{if } x = 0 \\ 0 & \text{if } x = 1 \\ x & \text{otherwise} \end{cases},$$

and define

$$d(x, y) := |f(x) - f(y)| \qquad \text{for all } x, y \in \mathbb{R} .$$

It is not hard to show that d such defined is a metric on $\mathbb{R}$. Note that under this metric, $\{\frac{1}{n}\}_{n\in\mathbb{N}} \to 1$ as $n \to \infty$.

3. If $\{x_n\}_{n\in\mathbb{N}}$ and $\{y_n\}_{n\in\mathbb{N}}$ are sequences in X, $\{x_n\} \to x$, and $\{y_n\} \to y$ as $n \to \infty$, show that $\{d(x_n, y_n)\} \to d(x, y)$ as $n \to \infty$.

Proof. For any $\varepsilon > 0$, there exists $N_1, N_2 \in \mathbb{N}$ such that $d(x_n, x) < \varepsilon$ for all $n \geq N_1$ and $d(y_n, y) < \varepsilon$ for all $n \geq N_2$. Take $N := \max\{N_1, N_2\}$. Then for any $n \geq N$, we have, by triangle inequality,

$$\begin{aligned} &|d(x_n, y_n) - d(x, y)| \\ &\leq |d(x_n, y_n) - d(x_n, y)| + |d(x_n, y) - d(x, y)| \\ &\leq d(y_n, y) + d(x_n, x) < 2\varepsilon . \end{aligned}$$

Thus $\{d(x_n, y_n)\} \to d(x, y)$ as $n \to \infty$.

4. Let $X := C[0,1]$ with metric $d_1(f,g) := \displaystyle\int_0^1 \big|f(x) - g(x)\big|\, dx$. Determine whether the sequence $\{f_n\}_{n\in\mathbb{N}}$ in (X, d_1) defined by $f_n(x) := x^n$ for $x \in [0,1]$ is convergent.

   Answer: Yes, it is convergent in (X, d_1).

   Proof. Observe that for any $f,\, g \in X$, $d_1(f,g)$ is nothing but the area between the graphs of the functions f and g. Observe also that for any fixed $x \in [0,1]$,

$$f_n(x) = x^n \to f(x) := \begin{cases} 0 & \text{if } 0 \le x < 1 \\ 1 & \text{if } x = 1\,. \end{cases}$$

   So the function f such defined may look like a possible candidate for the limit of the sequence $\{f_n\}_{n\in\mathbb{N}}$. However, obviously, $f \notin X$, as it is discontinuous at $x = 1$. So it cannot be the limit. However, as $d(f,0)$ = the area between the graphs of f and the zero function which is obviously 0, so the zero function which is obviously in X is a good candidate for the limit of the sequence $\{f_n\}_{n\in\mathbb{N}}$. Indeed, we have

$$d_1(f_n, 0) = \int_0^1 \big|f_n(x) - 0\big|\, dx = \int_0^1 x_n\, dx = \frac{1}{n+1} \to 0$$

   as $n \to \infty$ and so we conclude that $\{f_n\}_{n\in\mathbb{N}} \to 0$ in (X, d_1).

5. Let $X := C[0,1]$ with metric

$$d_\infty(f,g) := \sup\{|f(x) - g(x)| : x \in [0,1]\}\,.$$

   Determine whether the sequence $\{f_n\}_{n\in\mathbb{N}}$ in (X, d_∞) defined by $f_n(x) := x^n$ for $x \in [0,1]$ is convergent.

   Answer: No, it is not convergent in (X, d_∞).

   Proof. As observed in Exercise 2.1, Part B, Problem #4,

$$f_n(x) = x^n \to f(x) := \begin{cases} 0 & \text{if } 0 \le x < 1 \\ 1 & \text{if } x = 1\,. \end{cases}$$

So if $\{f_n\}_{n\in\mathbb{N}} \to g \in X$ in d_∞, then for any $\varepsilon > 0$, there exists $N \in \mathbb{N}$ such that for every $x \in [0, 1]$,

$$|f_n(x) - g(x)| \leq d_\infty(f_n, g) < \varepsilon \quad \text{for all } n \geq N .$$

In particular, for all $x \in [0, 1)$, we have

$$|g(x)| = |f(x) - g(x)| = \lim_{n\to\infty} |f_n(x) - g(x)| \leq \varepsilon .$$

As $\varepsilon > 0$ is arbitrary, this shows $g(x) = 0$ for all $x \in [0, 1)$. Now the only possible function which is continuous on $[0, 1]$ and $= 0$ on $[0, 1)$ is the constant zero function. So the problem boils down to determining whether the sequence $\{f_n\}_{n\in\mathbb{N}}$ converges to the zero function 0 in d_∞. Now as

$$\begin{aligned} d_\infty(f_n, 0) &= \sup\{|f(x) - 0| : x \in [0, 1]\} \\ &= \sup\{x^n : x \in [0, 1]\} = 1 \quad \text{for all } n \in \mathbb{N} , \end{aligned}$$

we see that $\{d_\infty(f_n, 0)\}_{n\in\mathbb{N}} \not\to 0$ as $n \to \infty$. Thus $\{f_n\}_{n\in\mathbb{N}}$ does not converge in d_∞.

6. A subset $A \subset X$ is said to be *sequentially compact* if every sequence in A has a subsequence which is convergent in A. Prove or disprove the following statements:
   (a) Every sequentially compact set is closed and bounded.
   (b) Every closed and bounded subset of $\mathbb{R}^n$ is sequentially compact.
   (c) Every closed and bounded subset of a general metric space X is sequentially compact.

Answer:

(a) True.

Proof. Let $A \subset X$ be sequentially compact. The closedness of A follows from the definition of sequential compactness. In fact, for any $a \in \overline{A}$, there is a sequence $\{x_n\}_{n\in\mathbb{N}}$ such that

$\{x_n\} \to a$ as $n \to \infty$. Now since A is sequentially compact, there is a subsequence $\{x_{n_k}\}_{k\in\mathbb{N}}$ of $\{x_n\}_{n\in\mathbb{N}}$ which is convergent to a point $b \in A$. Since all subsequences of $\{x_n\}_{n\in\mathbb{N}}$ must converge to a, this forces $a = b \in A$ and so $\overline{A} = A$.

On the other hand, assume that A is not bounded. Fix $a \in A$. There is a sequence $\{x_n\}_{n\in\mathbb{N}}$ in A satisfying $d(x_n, a) > n$ for all $n \in \mathbb{N}$. Since A is sequentially compact, a subsequence $\{x_{n_k}\}_{k\in\mathbb{N}}$ of $\{x_n\}_{n\in\mathbb{N}}$ must be convergent in A. By construction, $\{x_{n_k}\}_{k\in\mathbb{N}}$ is unbounded. But this is impossible as every convergent sequence must be bounded. Hence A must be bounded.

(b) True.

Proof. Let $A \subset \mathbb{R}^n$ be closed and bounded. Let $\{x_k\}_{k\in\mathbb{N}}$ be any infinite sequence in A.

If $\{x_k : k \in \mathbb{N}\} \subset \mathbb{R}^n$ is finite, then there is at least one $K \in \mathbb{N}$ s.t. $x_k = x_K$ for infinitely many $k \in \mathbb{N}$. That is, there is a subsequence $\{x_{k_j}\}_{j\in\mathbb{N}}$ of $\{x_k\}_{k\in\mathbb{N}}$ with $x_{k_j} = x_K$ for all $j \in \mathbb{N}$. In particular, $\{x_{k_j}\}_{j\in\mathbb{N}}$ is a convergent subsequence of $\{x_k\}_{k\in\mathbb{N}}$.

If $\{x_k : k \in \mathbb{N}\} \subset \mathbb{R}^n$ is infinite, by Bolzano-Weierstrass Theorem, $\{x_k : k \in \mathbb{N}\}$ has an accumulation point $p \in \mathbb{R}^n$. In particular, for any $j \in \mathbb{N}$, there exists $k_j \in \mathbb{N}$ s.t. $d(x_{k_j}, p) < \frac{1}{j}$. Without loss of generality, we may assume that $k_{j+1} > k_j$ for all $j \in \mathbb{N}$. Therefore, $\{x_{k_j}\}_{j\in\mathbb{N}}$ is a subsequence of $\{x_k\}_{k\in\mathbb{N}}$ which is convergent (to p) and so A is sequentially compact.

(c) False.

Example: Consider $X := C[0,1]$ with

$$d_\infty(f, g) := \sup\{|f(x) - g(x)| : x \in [0,1]\}$$

for all $f, g \in X$ and

$$A := \overline{B}(0,1) = \{f \in X : d_\infty(f, 0) \leq 1\} \subset X\ .$$

Then trivially, A is bounded. On the other hand, suppose $\{f_n\}_{n\in\mathbb{N}}$ is a sequence in A with $\{f_n\} \to f \in X$ as $n \to \infty$.

So for any $\varepsilon > 0$, there exists $N \in \mathbb{N}$ s.t. $d_\infty(f_n, f) < \varepsilon$ for all $n \geq N$. Hence

$$d_\infty(f, 0) \leq d_\infty(f, f_n) + d_\infty(f_n, 0) < \varepsilon + 1 .$$

Since $\varepsilon > 0$ is arbitrary, this forces $d_\infty(f, 0) \leq 1$ and so $f \in \overline{B}(0, 1) = A$. By Theorem 2.1.6, A is closed X. However A is not sequentially compact. To see this, consider the sequence $\{f_n\}_{n\in\mathbb{N}}$ in X defined by

$$f_n(x) := \exp(-nx^2) \quad \text{for all } x \in [0, 1] .$$

Clearly, $f_n \in A$ for all $n \in \mathbb{N}$ and so $\{f_n\}_{n\in\mathbb{N}}$ is a sequence in A. Suppose $\{f_n\}_{n\in\mathbb{N}}$ has a subsequence $\{f_{n_j}\}_{j\in\mathbb{N}}$ which is convergent to some $g \in A$. Then there exists $J \in \mathbb{N}$ s.t.

$$\sup\{|f_{n_j}(x) - g(x)| : x \in [0, 1]\} = d(f_{n_j}, g) < \frac{1}{3}$$

for all $j \geq J$. In particular, for any fixed $x \in (0, 1]$, we have

$$|g(x)| \leq |f_{n_j}(x) - g(x)| + |f_{n_j}(x)| < \frac{1}{3} + \exp(-n_j x^2)$$

for all $j \geq J$. Letting $j \to \infty$ we have $|g(x)| \leq \frac{1}{3}$. Since this is true for all $x \in (0, 1]$ and g is continuous at 0, this forces $|g(0)| \leq \frac{1}{3}$ and so

$$d(f_{n_j}, g) \geq |f_{n_j}(0) - g(0)| \geq \frac{2}{3} \quad \text{for all } j \geq J ,$$

which is absurd.

7. A metric space X is sequentially compact if and only if it has the Bolzano-Weierstrass Property.

   Proof. ($\Rightarrow$): Let X be sequentially compact and $S \subset X$ be an infinite subset of X. Then we can choose an infinite sequence of distinct points $\{x_n\}_{n\in\mathbb{N}}$ in S. Since X is sequentially compact, $\{x_n\}_{n\in\mathbb{N}}$

has a convergent subsequence $\{x_{n_k}\}_{k\in\mathbb{N}}$, converging to some point say $p \in X$. Then we must have $p \in S'$.
In fact, for any $r > 0$, there exists $N \in \mathbb{N}$ such that $d(x_{n_k}, p) < r$ for all $k \geq N$, or equivalently, $x_{n_k} \in B(p, r)$ for all $k \geq N$. That means near p we have infinitely many x_{n_k}'s which are points in S and are all distinct. Hence $p \in S'$.
($\Leftarrow$): Suppose X has the Bolzano-Weierstrass Property. Let $\{x_n\}_{n\in\mathbb{N}}$ be any sequence in X. If the underlying set $S := \{x_n : n \in \mathbb{N}\} \subset X$ is finite, then at least one of the x_n's, say, x_K, is repeated infinitely many times by the sequence $\{x_n\}_{n\in\mathbb{N}}$. That is to say that the sequence $\{x_n\}_{n\in\mathbb{N}}$ has a constant subsequence $\{x_K, x_K, x_K, \dots\}$ which is obviously convergent (to x_K).
If S is infinite, then by Bolzano-Weierstrass Property, there exists $p \in S'$. By definition of accumulation points, there exists $n_1 \in \mathbb{N}$ such that $x_{n_1} \in B(p, 1) \setminus \{p\}$. Since near p there are infinitely many x_n's, there exists $n_2 > n_1$ such that $x_{n_2} \in B(p, \frac{1}{2}) \setminus \{p\}$. Inductively, there exists $n_k > n_{k-1}$ such that $x_{n_k} \in B(p, \frac{1}{k}) \setminus \{p\}$ for any $k = 2, 3, 4, \dots$. Hence in particular, $\{x_{n_k}\}_{k\in\mathbb{N}}$ is a subsequence of $\{x_n\}_{n\in\mathbb{N}}$ which is convergent (to p). Thus X is sequentially compact.

8. Compactness implies sequential compactness.

   Proof. It is obvious from Theorem 1.3.6 and Exercise 2.1, Part B, Problem #7.

2.2 Complete Metric Spaces

By elementary calculus, we know that $\mathbb{R}$ is complete in the sense that every Cauchy sequence in $\mathbb{R}$ is actually convergent in $\mathbb{R}$. In this section, we shall extend the concepts of Cauchyness and completeness to the context of general metric spaces. It turns out, as expected, that $\mathbb{R}^n$ is complete for every $n \in \mathbb{N}$. Furthermore, as we can imagine, being a very strong condition, compactness implies completeness.

Definition 2.2.1. *A sequence $\{x_n\}_{n\in\mathbb{N}}$ in X is said to be a Cauchy sequence if for every $\varepsilon > 0$, there exists $N \in \mathbb{N}$ such that*

$$d(x_n, x_m) < \varepsilon \quad \textit{whenever}\, n,\ m \geq N\ .$$

Theorem 2.2.2. *Every convergent sequence is Cauchy.*

Proof. It is obvious by triangle inequality. In fact, if $\{x_n\} \to a \in X$, then for every $\varepsilon > 0$, there exists $N \in \mathbb{N}$ such that

$$d(x_n, a) < \frac{\varepsilon}{2} \quad \text{for all } n \geq N\ .$$

Hence

$$d(x_n, x_m) \leq d(x_n, a) + d(x_m, a) < \varepsilon \quad \text{for all } n,\ m \geq N$$

and so $\{x_n\}_{n\in\mathbb{N}}$ is Cauchy. □

Example 2.2.3. In general, a Cauchy sequence may not converge. For instance, take the sequence $\left\{\frac{1}{n}\right\}_{n\in\mathbb{N}}$ in $X := (0, 1]$. Clearly $\left\{\frac{1}{n}\right\}_{n\in\mathbb{N}}$ is Cauchy but it is not convergent in X.

Definition 2.2.4. *A metric space X is said to be complete if every Cauchy sequence in X converges in X. A subset $S \subset X$ is said to be complete if, when considered as a metric space itself, (S, d) is complete.*

Example 2.2.5. $\mathbb{Q}$ is *not* complete under the Euclidean metric induced from $\mathbb{R}$. In fact, since $\mathbb{Q}$ is dense in $\mathbb{R}$, it is easy to construct a Cauchy sequence in $\mathbb{Q}$ which converges to a point in $\mathbb{R} \setminus \mathbb{Q}$. For example, fix $a \in \mathbb{R} \setminus \mathbb{Q}$. Then $x_n := \frac{\lfloor 10^n a \rfloor}{10^n}$, $n \in \mathbb{N}$, is a Cauchy sequence in $\mathbb{Q}$ which is convergent in $\mathbb{R}$ (to $a \in \mathbb{R} \setminus \mathbb{Q}$) but not convergent in $\mathbb{Q}$.

Theorem 2.2.6. *$\mathbb{R}^k$ is complete for every $k \in \mathbb{N}$.*

Proof. Let $\{x_n\}_{n\in\mathbb{N}}$ be a Cauchy sequence in $\mathbb{R}^k$. We split our consideration into two cases according to the cardinality of the set $\{x_n : n \in \mathbb{N}\} \subset \mathbb{R}^k$.

Case 1. $\{x_n : n \in \mathbb{N}\} \subset \mathbb{R}^k$ is a finite set.

In this case, at least one point in $\{x_n : n \in \mathbb{N}\}$ has to be repeated infinitely many times by the sequence $\{x_n\}_{n\in\mathbb{N}}$. However, as $\{x_n\}_{n\in\mathbb{N}}$ is Cauchy, when n is large, the x_n's will get arbitrarily close to each other. This could not happen if two or more points of $\{x_n : n \in \mathbb{N}\}$ are repeated infinitely many times by the sequence $\{x_n\}_{n\in\mathbb{N}}$. Hence exactly one point of $\{x_n : n \in \mathbb{N}\}$ is repeated infinitely many times by the sequence $\{x_n\}_{n\in\mathbb{N}}$ and so the sequence $\{x_n\}_{n\in\mathbb{N}}$ converges to this particular point.

Case 2. $\{x_n : n \in \mathbb{N}\} \subset \mathbb{R}^k$ is an infinite set.

Similar to the proof of (a) of Theorem 2.1.4, we see easily that $\{x_n : n \in \mathbb{N}\}$ is bounded. Thus by Bolzano-Weierstrass Theorem, the bounded infinite set $\{x_n : n \in \mathbb{N}\}$ has an accumulation point $a \in \mathbb{R}^k$. By the Cauchyness of $\{x_n\}_{n\in\mathbb{N}}$, for any $\varepsilon > 0$, there exists $N \in \mathbb{N}$ such that $d(x_n, x_m) < \varepsilon/2$ whenever $n, m \geq N$. As $a \in \{x_n : n \in \mathbb{N}\}'$, there exists $m \geq N$ such that $x_m \in B(a, \frac{\varepsilon}{2})$. Hence we have

$$d(x_n, a) \leq d(x_n, x_m) + d(x_m, a) < \frac{\varepsilon}{2} + \frac{\varepsilon}{2} = \varepsilon \quad \text{whenever } n \geq N\,.$$

Thus $\{x_n\} \to a$ as $n \to \infty$ and so $\mathbb{R}^n$ is complete. □

Theorem 2.2.7. *Every compact subset S of a metric space X is complete.*

Proof. Let $\{x_n\}_{n\in\mathbb{N}} \subset S$ be a Cauchy sequence. Similar to the proof of Theorem 2.2.6, we split our consideration into two cases according to the cardinality of the set $\{x_n : n \in \mathbb{N}\}$.

Case 1. $\{x_n : n \in \mathbb{N}\} \subset S$ is a finite set.

In this case, at least one point in $\{x_n : n \in \mathbb{N}\}$ has to be repeated infinitely many times by the sequence $\{x_n\}_{n\in\mathbb{N}}$. However, as $\{x_n\}_{n\in\mathbb{N}}$ is Cauchy, when n is large, the x_n's will get arbitrarily close to each other. This could not happen if two or more points of $\{x_n : n \in \mathbb{N}\}$ are repeated infinitely many times by the sequence $\{x_n\}_{n\in\mathbb{N}}$. Hence exactly one point of $\{x_n : n \in \mathbb{N}\}$ is repeated infinitely many times by the sequence $\{x_n\}_{n\in\mathbb{N}}$ and so the sequence $\{x_n\}_{n\in\mathbb{N}}$ converges to this particular point.

Case 2. $\{x_n : n \in \mathbb{N}\} \subset S$ is an infinite set.

Since S is compact, By Theorem 1.3.6, $\{x_n\}_{n\in\mathbb{N}}$ has an accumulation point $a \in S$. By the Cauchyness of $\{x_n\}_{n\in\mathbb{N}}$, for any $\varepsilon > 0$, there exists $N \in \mathbb{N}$ such that $d(x_n, x_m) < \varepsilon/2$ whenever n, $m \geq N$. As $a \in \{x_n : n \in \mathbb{N}\}'$, there exists $m \geq N$ such that $x_m \in B\big(a, \frac{\varepsilon}{2}\big)$. Hence we have

$$d(x_n, a) \leq d(x_n, x_m) + d(x_m, a) < \frac{\varepsilon}{2} + \frac{\varepsilon}{2} = \varepsilon \quad \text{whenever } n \geq N\,.$$

Thus $\{x_n\} \to a$ as $n \to \infty$ and so S is complete. □

Example 2.2.8. We show that the abstract metric space $\big(C[-1,1], d\big)$, where $d(f,g) := \int_{-1}^{1} |f(x) - g(x)|\, dx$, is *not* complete.

Proof. For every $n \in \mathbb{N}$, define

$$f_n(x) := \begin{cases} -1 & \text{if } x \in \Big[-1, -\dfrac{1}{n}\Big] \\ nx & \text{if } x \in \Big[-\dfrac{1}{n}, \dfrac{1}{n}\Big] \\ 1 & \text{if } x \in \Big[\dfrac{1}{n}, 1\Big]\,. \end{cases}$$

Then $\{f_n\}_{n\in\mathbb{N}}$ is a Cauchy sequence in $(C[-1,1], d)$ which is not convergent.

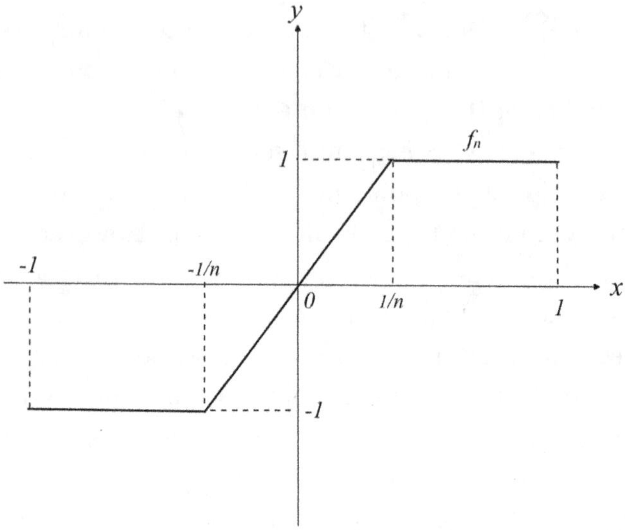

In fact, suppose $\{f_n\} \to g \in C[-1,1]$ in the metric d as $n \to \infty$. For any $x_0 \in (0,1]$, if $g(x_0) \neq 1$, there exists a nondegenerate closed interval I with $x_0 \in I \subset (0,1]$ such that

$$|g(x) - 1| > \frac{|g(x_0) - 1|}{2} > 0 \quad \text{for all } x \in I\ .$$

Note that we have

$$f_n(x) = 1 \quad \text{for all } x \in I, \text{ whenever } n \gg\ .$$

Hence

$$\begin{aligned} d(f_n, g) &= \int_{[-1,1]} |g(x) - f_n(x)|\, dx \\ &\geq \int_I |g(x) - f_n(x)|\, dx \\ &= \int_I |g(x) - 1|\, dx \\ &> \frac{|g(x_0) - 1|}{2} \cdot \ell(I) \quad \text{for all } n \gg\ , \end{aligned}$$

where $\ell(I) :=$ the length of I. Since I is nondegenerate, we have $\ell(I) > 0$. That means $\{f_n\} \not\to g$ in d as $n \to \infty$, which is a contradication. Therefore, we must have $g \equiv 1$ on $(0, 1]$. Similarly, we have $g \equiv -1$ on $[-1, 0)$. But there is no $g \in C[-1, 1]$ satisfying

$$g(x) = \begin{cases} 1 & \text{for } x \in (0, 1] \\ -1 & \text{for } x \in [-1, 0) \ . \end{cases}$$

Thus $\{f_n\}_{n\in\mathbb{N}}$ is not convergent in $(C[-1, 1], d)$ and so $(C[-1, 1], d)$ is not complete.

Exercise 2.2

Part A: True or False Questions

For each of the following statements, determine if it is true or false. If it is true, prove it. If it is false, give a counterexample or provide proper justification.

1. $(0, 1) \subset \mathbb{R}$ is complete.

   Answer: False.

   Justification. It is evident that $\{\frac{1}{n}\}_{n=2}^{\infty}$ is a Cauchy sequence in $(0, 1)$ which is not convergent there.

   Alternatively, this follows immediately from Exercise 2.2, Part A, Problem #8 below.

2. Let $X := C[0, 1]$ with metric
$$d(f, g) := \sup\{|f(x) - g(x)| : x \in [0, 1]\} .$$
   The sequence $\{f_n\}_{n\in\mathbb{N}}$ in (X, d) given by $f_n(x) := x^n$, $x \in [0, 1]$, is a Cauchy sequence.

   Answer: False.

   Justification. This follows from the observation that
$$\begin{aligned} d\big(f_{2n}, f_n\big) &= \sup\{|f_{2n}(x) - f_n(x)| : x \in [0, 1]\} \\ &\geq |f_{2n}(2^{-1/n}) - f_n(2^{-1/n})| \\ &= |(2^{-1/n})^{2n} - (2^{-1/n})^n| = \frac{1}{4} \quad \text{for all } n \in \mathbb{N} . \end{aligned}$$
   Hence $\{f_n\}_{n\in\mathbb{N}}$ is not Cauchy.

3. Let $X := C[0, 1]$ with metric
$$d(f, g) := \sup\{|f(x) - g(x)| : x \in [0, 1]\} .$$
   The sequence $\{f_n\}_{n\in\mathbb{N}}$ in (X, d) given by
$$f_n(x) := \begin{cases} nx & \text{if } 0 \leq x \leq 1/n \\ 1 & \text{if } 1/n < x \leq 1 \end{cases}$$
   is a Cauchy sequence.

Answer: False.

Justification. This follows from the observation that

$$\begin{aligned} d(f_{2n}, f_n) &= \sup\{|f_{2n}(x) - f_n(x)| : x \in [0,1]\} \\ &\geq \left| f_{2n}\left(\frac{1}{2n}\right) - f_n\left(\frac{1}{2n}\right) \right| \\ &= 1 - \frac{n}{2n} = \frac{1}{2} \qquad \text{for all } n \in \mathbb{N}\,. \end{aligned}$$

Hence $\{f_n\}_{n\in\mathbb{N}}$ is not Cauchy.

4. Let $\{x_n\}_{n\in\mathbb{N}}$ be a sequence in X and $E_m := \{x_n : n \geq m\} \subset X$. Then $\{x_n\}_{n\in\mathbb{N}}$ is Cauchy if and only if $\lim_{m\to\infty} d(E_m) = 0$.

   Answer: True.

   Proof. As $E_{m+1} \subset E_m$ for all $m \in \mathbb{N}$, we have

$$\begin{aligned} &\lim_{m\to\infty} d(E_m) = 0 \\ \iff &\text{ for all } \varepsilon > 0, \text{there exists } N \in \mathbb{N} \text{ } s.t. \text{ } d(E_N) < \varepsilon \\ \iff &\text{ for all } \varepsilon > 0, \text{there exists } N \in \mathbb{N} \text{ } s.t. \\ &\qquad d(x_n, x_m) < \varepsilon \text{ for all } n, m \geq N \\ \iff &\ \{x_n\}_{n\in\mathbb{N}} \text{ is Cauchy}\,. \end{aligned}$$

5. Let $\{x_n\}_{n\in\mathbb{N}}$ be a Cauchy sequence in X. Then for any $n, m \in \mathbb{N}$ large enough, we have

$$d(x_{n+1}, x_{m+1}) \leq d(x_n, x_m)\,.$$

   Answer: False.

   Example: Let $X = \mathbb{R}$ with the Euclidean metric. Define

$$x_n := \begin{cases} 0 & \text{if } n \text{ is odd} \\ \frac{1}{n} & \text{if } n \text{ is even.} \end{cases}$$

Then $\{x_n\} \to 0$ and hence in particular, it is Cauchy. However, whenever n, $m \in \mathbb{N}$ are odd and distinct, no matter how large they are, we have

$$d(x_{n+1}, x_{m+1}) = \left| \frac{1}{n+1} - \frac{1}{m+1} \right| > 0 = d(x_n, x_m) .$$

6. Every Cauchy sequence is bounded.

   Answer: True.

   Proof. The proof is similar to that of Theorem 2.1.4 (a). In fact, let $\{x_n\}_{n\in\mathbb{N}}$ be a Cauchy sequence in X. Then there exists and $N \in \mathbb{N}$ such that $d(x_m, x_N) < 1$, or equivalently, $x_m \in B(x_N, 1)$ for all $m \geq N$. Outside the ball $B(x_N, 1)$ there can be at most finitely many points of $\{x_n : n \in \mathbb{N}\}$. Hence the whole set $\{x_n : n \in \mathbb{N}\}$ is lying in a ball centered at x_N with finite radius.

7. The "term-wise product" of Cauchy sequences in $\mathbb{R}$ is Cauchy, that is, if $\{x_n\}_{n\in\mathbb{N}}$ and $\{y_n\}_{n\in\mathbb{N}}$ are Cauchy sequences in $\mathbb{R}$, then so is $\{x_n y_n\}_{n\in\mathbb{N}}$.

   Answer: True.

   Proof. It should be evident from the inequality

$$\begin{aligned} |x_n y_n - x_m y_m| &= |x_n y_n - x_n y_m + x_n y_m - x_m y_m| \\ &\leq |x_n - x_m|\,|y_m| + |y_n - y_m|\,|x_m| . \end{aligned}$$

In fact, by Exercise 2.2, Part A, Problem #6, there exists $K_1, K_2 > 0$ such that $|x_n| \leq K_1$ and $|y_n| \leq K_2$ for all $n \in \mathbb{N}$. Let $K := \max\{K_1, K_2\}$. Then we have $|x_n|, |y_n| \leq K$ for all $n \geq N$. On the other hand, by the definition Cauchyness, for any $\varepsilon > 0$, there exist N_1, $N_2 \in \mathbb{N}$ such that

$$\begin{aligned} |x_n - x_m| &< \frac{\varepsilon}{2K} \qquad \text{for all } n, m \geq N_1 \\ |y_n - y_m| &< \frac{\varepsilon}{2K} \qquad \text{for all } n, m \geq N_2 . \end{aligned}$$

Let $N := \max\{N_1, N_2\}$. Then for any n, $m \geq N$, we have

$$|x_n y_n - x_m y_m| < \varepsilon \ .$$

Thus $\{x_n y_n\}_{n\in\mathbb{N}}$ is Cauchy.

8. Every complete subset of a metric space is closed.

   Answer: True.

   Proof. Let $S \subset X$ be complete. For any $a \in \overline{S}$, there exists a sequence $\{x_n\}_{n\in\mathbb{N}}$ in S which is convergent to a. Being a convergent sequence, $\{x_n\}_{n\in\mathbb{N}}$ is Cauchy. Since S is complete, $\{x_n\}_{n\in\mathbb{N}}$ must be convergent in S. That means $a = \lim_{n\to\infty} x_n \in S$. Hence S is closed.

9. If $S \subset X$ is open, then S is not complete.

   Answer: False.

   Note. By Exercise 2.2, Part A, Problem #8, $S \subset X$ is complete implies S is closed. That is equivalent to saying that if $S \subset X$ is not closed, then it is not complete. It *does not* mean that if $S \subset X$ is open then it is not complete. The point is, "not closed" is not equivalent to "open".

   Example: Let $X := \mathbb{R}$ and $S := X$. Then S is open in X and it is complete.

10. Every closed subset of a metric space is complete.

    Answer: False.

    Example: Let $X := (0,\infty) \times \mathbb{R} = \{(x, y) \in \mathbb{R}^2 : x > 0\}$ and $S := (0,\infty) \times \{0\} = \{(x, 0) : x > 0\} \subset X$. Then it is clear that S is a closed subset of X. Furthermore, $\{(\frac{1}{n}, 0)\}_{n\in\mathbb{N}}$ is a Cauchy sequence in S which is not convergent in X. Hence S is not complete.

11. Every closed and bounded subset of a metric space is complete.

    Answer: False.

Example: Let $X := \mathbb{Q}$ and $S := [\sqrt{2}, \sqrt{3}] \cap \mathbb{Q} \subset \mathbb{Q}$. Then S is closed and bounded in X but it is not complete. In fact, it is easy to construct a sequence $\{x_n\}_{n\in\mathbb{N}}$ in S which converges to $\sqrt{2}$ in $\mathbb{R}$. Hence it is a Cauchy sequence in S which is not convergent in S.

12. Every closed subset of a complete metric space is complete.

Answer: True.

Proof. Let X be complete and $Y \subset X$ be closed. Let $\{x_n\}_{n\in\mathbb{N}}$ be a Cauchy sequence in Y. As the metric in Y is induced by that in X, when considered as a sequence in X, $\{x_n\}_{n\in\mathbb{N}}$ is also Cauchy. Since X is complete, $\{x_n\}_{n\in\mathbb{N}}$ converges to some $x \in X$ as $n \to \infty$. Note that $x \in \overline{Y} = Y$. Therefore, $\{x_n\}_{n\in\mathbb{N}} \to x \in Y$ as $n \to \infty$ and so Y is complete.

13. The metric space $X := (0, 1]$ equipped with metric

$$d(x,y) := \left|\frac{1}{x} - \frac{1}{y}\right| , \qquad x, y \in X ,$$

is a complete metric space.

Answer: True.

Proof. Let $\{x_n\}_{n\in\mathbb{N}}$ be a Cauchy sequence in (X, d). By definition, for any $\varepsilon > 0$, there exists $N \in \mathbb{N}$ such that

$$\left|\frac{1}{x_n} - \frac{1}{x_m}\right| = d(x_n, x_m) < \varepsilon \quad \text{for all } n, m \geq N .$$

In particular, $\left\{\frac{1}{x_n}\right\}_{n\in\mathbb{N}}$ is a Cauchy sequence in $[1, \infty)$ with the usual Euclidean metric, which is, by Exercise 2.2, Part A, Problem #12, a complete metric space. Hence $\left\{\frac{1}{x_n}\right\}_{n\in\mathbb{N}} \to$ some $y \in [1, \infty)$ in the usual Euclidean metric, as $n \to \infty$. But then that also means

$$d\left(x_n, \frac{1}{y}\right) = \left|\frac{1}{x_n} - y\right| \to 0 \qquad \text{as } n \to \infty ,$$

that is, $\{x_n\}_{n\in\mathbb{N}} \to \frac{1}{y}$ in d as $n \to \infty$.

14. The metric space $\mathbb{N}$ equipped with metric $d(m,n) := \left|\frac{1}{m} - \frac{1}{n}\right|$ is a complete metric space.

 Answer: False.

 Justification. For any $n \in \mathbb{N}$, let $x_n := n$. It is not hard to see that $\{x_n\}_{n\in\mathbb{N}}$ is a Cauchy sequence in $(\mathbb{N}, d)$. However, for any $n_0 \in \mathbb{N}$,

$$d(x_n, x_{n_0}) = d(n, n_0) = \left|\frac{1}{n} - \frac{1}{n_0}\right| \to \frac{1}{n_0} \neq 0 \qquad \text{as } n \to \infty\,.$$

 Thus $\{x_n\}_{n\in\mathbb{N}}$ is not convergent and so $(\mathbb{N}, d)$ is not complete.

15. The metric space $\mathbb{N}$ equipped with metric

$$d(m,n) := \left|e^{-m} - e^{-n}\right| \qquad m, n \in \mathbb{N}$$

 is a complete metric space.

 Answer: False.

 Justification. For any $n \in \mathbb{N}$, let $x_n := n$. For any $\varepsilon > 0$, let $N \in \mathbb{N}$ be such that $e^{-N} < \varepsilon$. Then

$$\begin{aligned} d(x_m, x_n) = d(m,n) &= \left|e^{-m} - e^{-n}\right| \\ &\leq e^{-m} + e^{-n} \leq 2e^{-N} < 2\varepsilon \quad \text{for all } m, n \geq N \end{aligned}$$

 and so $\{x_n\}_{n\in\mathbb{N}}$ is Cauchy in $(\mathbb{N}, d)$. However, for any $n_0 \in \mathbb{N}$,

$$d(x_n, n_0) = d(n, n_0) = \left|e^{-n} - e^{-n_0}\right| \to e^{-n_0} \neq 0 \quad \text{as } n \to \infty\,.$$

 Thus $\{x_n\}_{n\in\mathbb{N}}$ is not convergent and so $(\mathbb{N}, d)$ is not complete.

Part B: Problems

1. Let (X, d_1) be a metric space and $d_2(x,y) := \min\{1, d_1(x,y)\}$ for any x, $y \in X$. It is easily verified that d_2 is a well-defined metric on X. Let $\{x_n\}_{n\in\mathbb{N}}$ be a sequence in X. Show that it is Cauchy in d_1 if and only if it is Cauchy in d_2.

Proof. ($\Rightarrow$): Suppose $\{x_n\}_{n\in\mathbb{N}}$ is Cauchy in d_1. Then for any $1 > \varepsilon > 0$, there exists $N \in \mathbb{N}$ such that $d_1(x_n, x_m) < \varepsilon$ for any n, $m \geq N$. It follows that

$$d_2(x_n, x_m) = \min\{1, d_1(x_n, x_m)\} < \varepsilon \quad \text{for all } n, m \geq N$$

and so $\{x_n\}_{n\in\mathbb{N}}$ is Cauchy in d_2.
($\Leftarrow$): Suppose $\{x_n\}_{n\in\mathbb{N}}$ is Cauchy in d_2. Then for any $1 > \varepsilon > 0$, there exists $N \in \mathbb{N}$ such that

$$\min\{1, d_1(x_n, x_m)\} = d_2(x_n, x_m) < \varepsilon \quad \text{for all } n, m \geq N\ .$$

It follows that

$$d_1(x_n, x_m) < \varepsilon \qquad \text{for all } n, m \geq N$$

and so $\{x_n\}_{n\in\mathbb{N}}$ is Cauchy in d_1.

2. Recall that a subset $S \subset X$ is said to be totally bounded (Exercise 1.3, Part B, Problem #10) if for every $\varepsilon > 0$, there exist $k > 0$ and $x_1, \ldots, x_k \in S$ such that $\{B(x_i, \varepsilon)\}_{i=1}^k$ is an open cover of S. Show that the underlying set of every Cauchy sequence $\{x_n\}_{n\in\mathbb{N}}$ in X is totally bounded, that is, the set $S := \{x_n : n \in \mathbb{N}\} \subset X$ is totally bounded. In particular, S is bounded.

   Proof. Let $\varepsilon > 0$ be given. Since $\{x_n\}_{n\in\mathbb{N}}$ is Cauchy, there exists $N \in \mathbb{N}$ such that $d(x_n, x_m) < \varepsilon$ for all n, $m \geq N$. In particular, $x_n \in B(x_N, \varepsilon)$ for all $n \geq N$. Hence $\{B(x_1, \varepsilon), \ldots , B(x_N, \varepsilon)\}$ is an open cover of S. Therefore, S is totally bounded. The last statement is obvious.

3. Show that every infinite sequence of distinct points $\{x_n\}_{n\in\mathbb{N}}$ in a totally bounded subset $S \subset X$ contains a non-constant Cauchy subsequence.

Proof. Since S is totally bounded, it can be decomposed into finitely many subsets each of which has diameter less than 1. One of these finitely many subsets would contain infinitely many x_n's. Call it S_1. Pick an $n_1 \in \mathbb{N}$ such that $x_{n_1} \in S_1$. Since $S_1 \subset S$ and S is totally bounded, S_1 is also totally bounded. So S_1 can be decomposed into finitely many subsets each of which has diameter less than $\frac{1}{2}$. One of these subsets, say S_2, contains infinitely many x_n's. Hence there exists $n_2 \in \mathbb{N}$, $n_2 > n_1$, such that $x_{n_2} \in S_2$. Inductively, we have a decreasing nest of subsets

$$S \supset S_1 \supset S_2 \supset S_3 \supset \cdots$$

with $d(S_j) < \frac{1}{j}$ and a subsequence $\{x_{n_j}\}_{j=1}^{\infty}$ of $\{x_n\}_{n=1}^{\infty}$ such that $x_{n_j} \in S_j$ for every $j \in \mathbb{N}$. Observe that $\{x_{n_j}\}_{j=1}^{\infty}$ is Cauchy. In fact, for any $\varepsilon > 0$, let $N \in \mathbb{N}$ be large enough such that $\frac{1}{N} < \varepsilon$. Then for any p, $q \geq N$, we have $x_{n_p} \in S_p \subset S_N$, $x_{n_q} \in S_q \subset S_N$, and so

$$d(x_{n_p}, x_{n_q}) \leq d(S_N) < \frac{1}{N} < \varepsilon .$$

4. For any sequence $\{x_n\}_{n\in\mathbb{N}}$ in X, write $S_n := \{x_k : k \geq n\} \subset X$. Clearly, $\{S_n\}_{n\in\mathbb{N}}$ is a decreasing nest of subsets of X. Show that a sequence $\{x_n\}_{n\in\mathbb{N}}$ in X is Cauchy if and only if $d(S_n) \to 0$ as $n \to \infty$.

   Proof. ($\Rightarrow$): Suppose $\{x_n\}_{n\in\mathbb{N}}$ is Cauchy. For any $\varepsilon > 0$, there exists $N \in \mathbb{N}$ such that $d(x_n, x_m) < \varepsilon$ for all $n, m \geq N$. Hence $d(S_n) \leq \varepsilon$ for all $n \geq N$. That is, $d(S_n) \to 0$ as $n \to \infty$.
   ($\Leftarrow$): Suppose $d(S_n) \to 0$ as $n \to \infty$. For any $\varepsilon > 0$, there exists $N \in \mathbb{N}$ such that $d(S_n) < \varepsilon$ for all $n \geq N$. Hence for any $n \geq m \geq N$, $d(x_n, x_m) \leq d(S_m) < \varepsilon$ and so $\{x_n\}_{n\in\mathbb{N}}$ is Cauchy.

5. For any sequences $\{x_n\}_{n\in\mathbb{N}}$, $\{y_n\}_{n\in\mathbb{N}}$ in X, if $\{x_n\}_{n\in\mathbb{N}}$ is Cauchy and $\lim_{n\to\infty} d(x_n, y_n) = 0$, show that $\{y_n\}_{n\in\mathbb{N}}$ is also Cauchy.

Proof. For any $\varepsilon > 0$, there exists $N_1 \in \mathbb{N}$ such that $d(x_n, y_n) < \varepsilon$ for all $n \geq N_1$. On the other hand, there exists $N_2 \in \mathbb{N}$ such that $d(x_n, x_m) < \varepsilon$ for all $n, m \geq N_2$. Take $N := \max\{N_1, N_2\}$. Then by triangle inequality,

$$d(y_n, y_m) \leq d(y_n, x_n) + d(x_n, x_m) + d(x_m, y_m) < 3\varepsilon$$

for any $n, m \geq N$. Hence $\{y_n\}_{n\in\mathbb{N}}$ is Cauchy.

6. For any sequences $\{x_n\}_{n\in\mathbb{N}}$, $\{y_n\}_{n\in\mathbb{N}}$ in X, if $\lim_{n\to\infty} x_n = a$ and $\lim_{n\to\infty} d(x_n, y_n) = 0$, show that $\lim_{n\to\infty} y_n = a$.

   Proof. For any $\varepsilon > 0$, there exists $N_1 \in \mathbb{N}$ such that $d(x_n, y_n) < \varepsilon$ for all $n \geq N_1$. On the other hand, there exists $N_2 \in \mathbb{N}$ such that $d(x_n, a) < \varepsilon$ for all $n \geq N_2$. Take $N := \max\{N_1, N_2\}$. Then by triangle inequality,

$$d(y_n, a) \leq d(y_n, x_n) + d(x_n, a) < 2\varepsilon$$

   for any $n \geq N$. Hence $\{y_n\} \to a$ as $n \to \infty$.

7. Consider $X := C[0, 1]$ with metrics

$$d_\infty(f, g) := \sup_{x\in[0,1]} |f(x) - g(x)| \quad \text{and}$$

$$d_1(f, g) := \int_0^1 |f(x) - g(x)|\, dx\ .$$

   For each $n \in \mathbb{N}$, define

$$f_n(x) := \begin{cases} 2n - 2n^2 x & \text{if } 0 \leq x \leq 1/n \\ 0 & \text{if } 1/n < x \leq 1\ . \end{cases}$$

   Discuss whether the sequence $\{f_n\}_{n\in\mathbb{N}}$ is Cauchy in (X, d_∞) and in (X, d_1).

Solution: $\{f_n\}_{n\in\mathbb{N}}$ is not Cauchy in either (X, d_∞) or (X, d_1).
First observe that

$$d_\infty(f_n, 0) = \sup_{x\in[0,\frac{1}{n}]} |2n - 2n^2 x| = 2n \quad \text{for all } n \in \mathbb{N}$$

and so the sequence $\{f_n\}_{n\in\mathbb{N}}$ is not bounced, hence in particular, not Cauchy.
On the other hand, for any n, $k \in \mathbb{N}$, elementary calculation shows

$$d_1(f_{n+k}, f_n) = \frac{2k}{2n+k} \,.$$

In particular, $d_1(f_{2n}, f_n) = \frac{2}{3}$ for any $n \in \mathbb{N}$ and so $\{f_n\}_{n\in\mathbb{N}}$ is not Cauchy.

8. Suppose $\{x_n\}_{n\in\mathbb{N}}$ is a sequence in X such that the underlying set $\{x_n : n \in \mathbb{N}\} \subset X$ is finite. Show that if $\{x_n\}_{n\in\mathbb{N}}$ is Cauchy, then it must be convergent.

Proof. Under the given assumption, exactly one element of the finite set $\{x_n : n \in \mathbb{N}\}$ is assumed by the sequence $\{x_n\}_{n\in\mathbb{N}}$ infinitely many times. Hence this particular element is the limit of the sequence $\{x_n\}_{n\in\mathbb{N}}$.

In more details, first observe that if the sequence $\{x_n\}_{n\in\mathbb{N}}$ is a constant sequence, then there is nothing to prove. So assume that it is not a constant sequence. Let

$$r := \inf\{d(x_n, x_m) : n, m \in \mathbb{N}\}.$$

Then as there can be at most finitely many different values of the x_n's, we have $r > 0$. Since the sequence $\{x_n\}_{n\in\mathbb{N}}$ is Cauchy, there exists $N \in \mathbb{N}$ such that

$$d(x_n, x_m) < \frac{r}{2} \leq r \qquad \text{for all } n, m \geq N \,.$$

This is possible only if $x_n = x_m$ for all n, $m \geq N$. That is, $x_n = x_N$ for all $n \geq N$ and hence $\{x_n\} \to x_N$ as $n \to \infty$.

9. If a Cauchy sequence $\{x_n\}_{n\in\mathbb{N}}$ in X has a convergent subsequence, then $\{x_n\}_{n\in\mathbb{N}}$ is itself convergent.

   Proof. Let $\{x_{n_k}\}_{k\in\mathbb{N}}$ be a convergent subsequence of $\{x_n\}_{n\in\mathbb{N}}$, with limit $x \in X$. Then for any $\varepsilon > 0$, there exists $N_1 \in \mathbb{N}$ such that

   $$d(x_{n_k}, x) < \frac{\varepsilon}{2} \qquad \text{for all } k \geq N_1.$$

   On the other hand, since $\{x_n\}_{n\in\mathbb{N}}$ is Cauchy, there exists $N_2 \in \mathbb{N}$, such that

   $$d(x_m, x_n) < \frac{\varepsilon}{2} \qquad \text{for all } m, n \geq N_2.$$

   Let $K := \max\{N_1, N_2\}$. Then for all $n \geq K$, on the one hand we have $n \geq N_2$ and $n_K \geq K \geq N_2$, hence

   $$d(x_{n_K}, x_n) < \frac{\varepsilon}{2}\ ;$$

   and on the other hand we have $n_K \geq K \geq N_1$, hence

   $$d(x_{n_K}, x) < \frac{\varepsilon}{2}\ .$$

   Combining, we conclude that

   $$d(x_n, x) \leq d(x_n, x_{n_K}) + d(x_{n_K}, x) < \varepsilon\ ,$$

   which implies that $\{x_n\}_{n\in\mathbb{N}}$ is convergent and converges to x.

10. Show that every sequentially compact metric space is totally bounded.

    Proof. It would be more convenient to use contra-positive arguments. So assume that X is not totally bounded. Then by definition, there exists $\varepsilon > 0$ such that X cannot be covered by finitely many open balls of radius ε. Pick $x_1 \in X$. Clearly $B(x_1, \varepsilon) \neq X$, or else X is covered by one ε ball. So there exists $x_2 \in X \setminus B(x_1, \varepsilon)$. Hence we have $d(x_1, x_2) \geq \varepsilon$. Next, $B(x_1, \varepsilon) \cup B(x_2, \varepsilon) \neq X$, or else X is covered by two ε balls. So there exists $x_3 \in X \setminus \big(B(x_1, \varepsilon) \cup B(x_2, \varepsilon)\big)$.

Hence we have $d(x_3, x_i) \geq \varepsilon$ for $i = 1, 2$, and $B(x_1, \varepsilon) \cup B(x_2, \varepsilon) \cup B(x_3, \varepsilon) \neq X$. Inductively, we can construct a sequence $\{x_n\}_{n \in \mathbb{N}}$ in X such that

$$x_{n+1} \notin \bigcup_{i=1}^{n} B(x_i, \varepsilon) \qquad \text{for all } n \in \mathbb{N}$$

and so

$$d(x_{n+1}, x_i) \geq \varepsilon \qquad \text{for all } i = 1, \dots, n \text{ and all } n \in \mathbb{N} \,.$$

In particular, we have $d(x_n, x_m) \geq \varepsilon$ for all $n \neq m$. So $\{x_n\}_{n \in \mathbb{N}}$ is not Cauchy and so it cannot contain any convergent subsequence. Hence X is not sequentially compact. Hence the assertion.

11. Show that every sequentially compact metric space is complete.

    Proof. Let X be sequentially compact and $\{x_n\}_{n \in \mathbb{N}}$ be a Cauchy sequence in X. So for any $\varepsilon > 0$, there exists $N \in \mathbb{N}$ such that $d(x_n, x_m) < \varepsilon$ for all n, $m \geq N$. Since X is sequentially compact, the Cauchy sequence $\{x_n\}_{n \in \mathbb{N}}$ has a convergent subsequence $\{x_{n_k}\}_{k \in \mathbb{N}}$, which converges to, say, $p \in X$. So for any $j > k \geq N$, we have $n_j > n_k \geq k \geq N$ and so

    $$d(x_k, p) \leq d(x_k, x_{n_j}) + d(x_{n_j}, p) < \varepsilon + d(x_{n_j}, p) \,.$$

    Letting $j \to \infty$, we have

    $$d(x_k, p) \leq \varepsilon \qquad \text{for all } k \geq N$$

    and so $\{x_k\} \to p$ as $k \to \infty$. Hence completeness.

12. Show that every totally bounded complete metric space is sequentially compact.

    Proof. Let X be totally bounded and complete, and $\{x_n\}_{n \in \mathbb{N}}$ be any sequence in X. If $\{x_n : n \in \mathbb{N}\}$ is finite, we have seen before that it must have a constant subsequence and so the result is trivial. Hence we only need to consider the case where $\{x_n : n \in \mathbb{N}\}$ is an infinite set. For this, we simply assume that all x_n's are distinct. Since X is totally bounded, finitely many open balls of radius 1 cover X. One of these open balls, say $B(y_0, 1)$, would cover infinitely many x_n's. So if we write $\Lambda_0 := \{n \in \mathbb{N} : x_n \in B(y_0, 1)\} \subset \mathbb{N}$, we have $\#(\Lambda_0) = \infty$. Similarly, finitely many open balls of radius $\frac{1}{2}$ cover X. One of these balls, say $B(y_1, \frac{1}{2})$, would cover infinitely many x_n's with $n \in \Lambda_0$. Write $\Lambda_1 := \{n \in \Lambda_0 : x_n \in B(y_1, \frac{1}{2})\} \subset \Lambda_0$. Then we have $\#(\Lambda_1) = \infty$. Inductively, for every $k \in \mathbb{N}$, we have an open ball $B(y_k, \frac{1}{2^k})$ containing infinitely many x_n's with $n \in \Lambda_{k-1}$. Write $\Lambda_k := \{n \in \Lambda_{k-1} : x_n \in B(y_k, \frac{1}{2^k})\} \subset \Lambda_{k-1}$. Then $\#(\Lambda_k) = \infty$. In other words, we have constructed a decreasing nest of index sets Λ_k, each of which is infinite. So we can pick a strictly increasing sequence of natural numbers $\{n_k\}_{k\in\mathbb{N}}$ such that $n_k \in \Lambda_k$ for each $k \in \mathbb{N}$. Observe that the corresponding sequence $\{x_{n_k}\}_{k\in\mathbb{N}}$ is a Cauchy subsequence of $\{x_n\}_{n\in\mathbb{N}}$. In fact, for any $k, p \in \mathbb{N}$, we have $n_k \in \Lambda_k$ and $n_{k+p} \subset \Lambda_{k+p} \subset \Lambda_k$ and so $x_{n_k}, x_{n_{k+p}} \in B(y_k, \frac{1}{2^k})$. Therefore,

$$
\begin{aligned}
d(x_{n_k}, x_{n_{k+p}}) &\leq d(x_{n_k}, y_k) + d(y_k, x_{n_{k+p}}) \\
&< \frac{1}{2^k} + \frac{1}{2^k} = \frac{1}{2^{k-1}} ,
\end{aligned}
$$

from which we see that the subsequence $\{x_{n_k}\}_{k\in\mathbb{N}}$ of $\{x_n\}_{n\in\mathbb{N}}$ is Cauchy. Since X is complete, $\{x_{n_k}\}_{k\in\mathbb{N}}$ is convergent. Hence X is sequentially compact.

13. Recall that an open cover $\mathcal{U}$ of X is said to have a Lebesgue number $\lambda > 0$ if every subset of X of diameter less than λ is contained in some element in $\mathcal{U}$. Show that if X is a sequentially compact metric space, then every open cover of X has a Lebesgue number. (Compare with Exercise 1.3, Part B, Problem #11.)

Proof. Suppose X is sequentially compact. Assume to the contrary that there is an open cover $\mathcal{U} = \{U_\alpha\}_{\alpha\in\Lambda}$ of X, where Λ is an arbitrary index set, without a Lebesgue number. Then for any $n \in \mathbb{N}$, there exists $S_n \subset X$ with $d(S_n) < \frac{1}{n}$ and $S_n \not\subset U_\alpha$ for any $\alpha \in \Lambda$. Pick a representative element $x_n \in S_n$ for each $n \in \mathbb{N}$. Then as a sequence in the sequentially compact space X, $\{x_n\}_{n\in\mathbb{N}}$ has a convergent subsequence $\{x_{n_k}\}_{k\in\mathbb{N}}$, converging to, say, $p \in X$. As $\mathcal{U}$ covers X, there exists $\alpha_0 \in \Lambda$ such that $p \in U_{\alpha_0}$. Since U_{α_0} is open, there exists $r > 0$ such that $p \in B(p,r) \subset U_{\alpha_0}$. As $\{x_{n_k}\} \to p$, there exists $N \in \mathbb{N}$ such that

$$d(x_{n_k}, p) < \frac{r}{2} \qquad \text{for all } k \geq N\,.$$

By choosing a larger N if necessary, we may simply assume that $\frac{1}{N} < \frac{r}{2}$. Observe that

$$x_{n_N} \in S_{n_N} \cap B\left(p, \frac{r}{2}\right)$$

and so

$$S_{n_N} \cap B\left(p, \frac{r}{2}\right) \neq \phi\,.$$

On the other hand, since $d(S_{n_N}) < \frac{1}{n_N} \leq \frac{1}{N} < \frac{r}{2}$, we have

$$S_{n_N} \subset B(p,r) \subset U_{\alpha_0}\,,$$

which contradicts to the assumption that $S_{n_N} \not\subset U_\alpha$ for any $\alpha \in \Lambda$.

14. Sequential compactness implies compactness.

Proof. Let X be sequentially compact, and $\mathcal{U} = \{U_\alpha\}_{\alpha\in\Lambda}$ be any open cover of X. By Exercise 2.2, Part B, Problem #13, $\mathcal{U}$ has a Lebesgue number $\lambda > 0$. Let $\varepsilon := \frac{\lambda}{3}$. By Exercise 2.2, Part B, Problem #10, X is totally bounded and so there exists $x_1, \ldots, x_n \in X$ such that $X = \bigcup_{i=1}^{n} B(x_i, \varepsilon)$. For each $i = 1, \ldots, n$, we have

$$d\left(B(x_i, \varepsilon)\right) \leq 2\varepsilon = \frac{2\lambda}{3} < \lambda$$

and so by the definition of λ, there exists $U_{\alpha_i} \in \mathcal{U}$ such that $B(x_i, \varepsilon) \subset U_{\alpha_i}$. Therefore,

$$X = \bigcup_{i=1}^{n} B(x_i, \varepsilon) \subset \bigcup_{i=1}^{n} U_{\alpha_i} .$$

Hence $\{U_{\alpha_i}\}_{i=1}^{n}$ is a finite subcover of $\mathcal{U}$ and so X is compact.

15. A closed subspace of a complete metric space is compact if and only if it is totally bounded.

Proof. Let $S \subset X$ be a closed subset of a complete metric space X. By Exercise 2.2, Part A, Problem #12 , S is complete. If S is compact, by Exercise 1.3, Part B, Problem #10, it is totally bounded. Conversely, if S is totally bounded, by Exercise 2.2, Part B, Problem #12, S is sequentially compact, which in turn is compact by Exercise 2.2, Part B, Problem #14.

16. Let $\mathcal{B}(X, \mathbb{R})$ be the set of all bounded functions functions of $X \subset \mathbb{R}$ to $\mathbb{R}$. Define a metric ρ on $\mathcal{B}(X, \mathbb{R})$ by

$$\rho(f, g) := \sup\{|f(x) - g(x)| : x \in X\} .$$

Prove that $(\mathcal{B}(X, \mathbb{R}), \rho)$ is complete.

Proof. The fact that ρ is a well-defined metric is elementary and is left as an exercise. If $\{f_n\}_{n \in \mathbb{N}}$ is Cauchy in $(\mathcal{B}(X, \mathbb{R}), \rho)$, then for each $x \in X$, we have

$$|f_n(x) - f_m(x)| \leq \sup\{|f_n(x) - f_m(x)| : x \in X\} = \rho(f_n, f_m) ,$$

which means that $\{f_n(x)\}_{n \in \mathbb{N}}$ is a Cauchy sequence in $\mathbb{R}$ which is complete. Hence $\{f_n(x)\}_{n \in \mathbb{N}} \to$ some real number which obviously depends on the choice of $x \in X$. Call this limit $f(x)$. Do this for every $x \in X$ and we have defined a function $f : X \to \mathbb{R}$ by $f(x) := \lim_{n \to \infty} f_n(x)$. Therefore, the problem boils down to showing (i) $f \in \mathcal{B}(X, \mathbb{R})$ and (ii) $\{f_n\}_{n \in \mathbb{N}} \to f$ in ρ.

(i) $f \in \mathcal{B}(X, \mathbb{R})$:
As $\{f_n\}_{n\in\mathbb{N}}$ is Cauchy in $(\mathcal{B}(X, \mathbb{R}), \rho)$, there exists $N \in \mathbb{N}$ such that $\rho(f_n, f_N) < 1$ for all $n \geq N$. Since $f_N \in \mathcal{B}(X, \mathbb{R})$, there exists $M > 0$ such that $|f_N| \leq M$ on X. Hence for any $x \in X$,

$$|f_n(x)| \leq |f_n(x) - f_N(x)| + |f_N(x)| < 1 + M$$

whenever $n \geq N$ and so

$$|f(x)| = \lim_{n\to\infty} |f_n(x)| \leq 1 + M \; .$$

Hence $|f| \leq 1 + M$ on X and so $f \in \mathcal{B}(X, \mathbb{R})$.

(ii) $\{f_n\}_{n\in\mathbb{N}} \to f$ *in* ρ, *equivalently,* $\rho(f_n, f) \to 0$:
Since $\{f_n\}_{n\in\mathbb{N}}$ is Cauchy in ρ, for any $\varepsilon > 0$, there exists $N \in \mathbb{N}$ large enough such that $\rho(f_n, f_m) < \frac{\varepsilon}{2}$ for all n, $m \geq N$. For each $x \in X$, since $\lim_{m\to\infty} f_m(x) = f(x)$, we can choose $m > N$ large enough such that $|f_m(x) - f(x)| < \frac{\varepsilon}{2}$. Hence

$$\begin{aligned}|f_n(x) - f(x)| &\leq |f_n(x) - f_m(x)| + |f_m(x) - f(x)| \\ &\leq \rho(f_n, f_m) + |f_m(x) - f(x)| < \varepsilon\end{aligned}$$

for all $n \geq N$ and so $\rho(f_n, f) = \sup_{x\in X} |f_n(x) - f(x)| \leq \varepsilon$ for all $n \geq N$. Since $\varepsilon > 0$ is arbitrary, we conclude that $\{\rho(f_n, f)\} \to 0$ as $n \to \infty$. Hence $\{f_n\}_{n\in\mathbb{N}} \to f$ in ρ as $n \to \infty$.

17. Let M_{22} be the space of all 2×2 matrices over $\mathbb{R}$, equipped with the metric

$$d(A, B) := \max_{1\leq i,j\leq 2} |a_{ij} - b_{ij}|$$

for all $A = (a_{ij})$, $B = (b_{ij}) \in M_{22}$. Show that (M_{22}, d) is complete.

Proof. It is elementary to verify that d is a well-defined metric on M_{22}. Let $\{A_n = (a_{ij}^n)\}_{n\in\mathbb{N}}$ be a Cauchy sequence in (M_{22}, d). By definition, for any $\varepsilon > 0$, there exists $N \in \mathbb{N}$ such that

$$\max_{1\le i,j\le 2} |a_{ij}^n - a_{ij}^m| = d(A_n, A_m) < \varepsilon \quad \text{whenever } n,\ m \ge N\ .$$

It follows that for any fixed pair (i, j),

$$|a_{ij}^n - a_{ij}^m| < \epsilon \quad \text{whenever} \quad n, m \ge N.$$

Hence $\{a_{ij}^n\}_{n\in\mathbb{N}}$ is a Cauchy sequence in $\mathbb{R}$, which must be convergent. Let $a_{ij} := \lim_{n\to\infty} a_{ij}^n$. Then $A := (a_{ij})$ is the limit of A_n in d. In fact, for any $\varepsilon > 0$, for any $i,\ j = 1, 2$, there exists $N_{ij} \in \mathbb{N}$ such that

$$|a_{ij}^n - a_{ij}| < \varepsilon \quad \text{whenever } n \ge N_{ij}\ .$$

Take $N := \max_{1\le i,j\le 2} N_{ij}$. Then

$$d(A_n, A) = \max_{1\le i,j\le 2} |a_{ij}^n - a_{ij}| < \varepsilon \quad \text{whenever } n \ge N\ .$$

18. Determine whether the metric space $C([-1, 1], d)$ with metric

$$d(f, g) := \int_{-1}^{1} \big|f(x) - g(x)\big|\, dx \qquad \text{for any } f,\ g \in C[-1, 1]$$

is complete.

Answer: It is not complete.

Justification. Observe first that for any f, $g \in C[-1, 1]$, $d(f, g) =$ the area of the region bounded by the graphs of f and g. Consider the sequence of functions

$$f_n(x) := \begin{cases} -1 & -1 \le x \le -1/n, \\ nx & -1/n \le x \le 1/n, \\ 1 & 1/n \le x \le 1\ . \end{cases}$$

It is clear that $f_n \in C[-1,1]$ for all $n \in \mathbb{N}$. Furthermore, it is easy to verify that $d(f_n, f_m) = \left|\frac{1}{n} - \frac{1}{m}\right|$. For any $\varepsilon > 0$, let $K \in \mathbb{N}$ be large enough such that $\frac{1}{K} < \varepsilon$. Then

$$d(f_n, f_m) = \left|\frac{1}{n} - \frac{1}{m}\right| = \left|\frac{m-n}{mn}\right| < \frac{1}{n} \leq \frac{1}{N} < \varepsilon$$

for any $m > n \geq K$ and so $\{f_n\}_{n\in\mathbb{N}}$ is a Cauchy sequence in $C([-1,1], d)$.

To show that $\{f_n\}_{n\in\mathbb{N}}$ is not convergent, we use contra-positive argument. Suppose that $\{f_n\} \to f$ in $C([-1,1], d)$ as $n \to \infty$. By intuition, we should have $f \equiv 1$ on $(0,1]$. This can easily be justified. In fact, suppose $f(x_0) \neq 1$ for some $x_0 \in (0,1]$. Denote by $\alpha := |f(x_0) - 1| > 0$. By the continuity of f, there exists a closed non-degenerate sub-interval $[a,b] \subset (0,1]$ containing x_0 such that $|f(x) - 1| > \alpha/2$ for all $x \in [a,b]$. However, as $f_n \equiv 1$ on $[a,b]$ whenever $n \geq \frac{1}{a}$, we have

$$\begin{aligned} d(f_n, f) &= \int_{-1}^{1} |f_n(x) - f(x)|\, dx \geq \int_a^b |f_n(x) - f(x)|\, dx \\ &= \int_a^b |f(x) - 1|\, dx \geq \frac{\alpha}{2}(b-a) > 0 \quad \text{for all } n \geq \frac{1}{a}, \end{aligned}$$

which contradicts to the convergence of $\{f_n\}$ to f in $C([-1,1], d)$. Similarly, we also have $f \equiv -1$ on $[-1,0)$. However, no $f \in C[-1,1]$ can have such values.

19. Suppose $\{F_n\}_{n\in\mathbb{N}}$ is a decreasing nest of closed nonempty subsets of a complete metric space X, with $\lim_{n\to\infty} d(F_n) = 0$.
    (a) Show that $\bigcap_{n=1}^{\infty} F_n \neq \phi$.
    (b) Determine what $\bigcap_{n=1}^{\infty} F_n$ is.
    (c) If the F_n's are not all closed, would (a) still hold?
    (d) Without the condition $\lim_{n\to\infty} d(F_n) = 0$, would (a) still hold?

(e) If the condition $\lim_{n\to\infty} d(F_n) = 0$ is replaced by that all F_n's are bounded, would (a) still hold?

(f) If X fails to be complete, would (a) still hold?

Solution:

(a) Since $\{d(F_n)\}_{n\in\mathbb{N}} \to 0$ as $n \to \infty$, for any $\varepsilon > 0$, there exists $N \in \mathbb{N}$ such that $d(F_N) < \varepsilon$. For each $n \in \mathbb{N}$, pick $x_n \in F_n$. As $\{F_n\}_{n\in\mathbb{N}}$ is decreasing, for any $n, m \geq N$, we have $x_m \in F_m \subset F_N$ and $x_n \in F_n \subset F_N$. Hence

$$d(x_m, x_n) \leq d(F_N) < \varepsilon .$$

Thus $\{x_n\}_{n\in\mathbb{N}}$ is a Cauchy sequence in the complete metric space X and so $\{x_n\}_{n\in\mathbb{N}} \to$ some $x \in X$ as $n \to \infty$. In particular, for any $m \in \mathbb{N}$, $\{x_n\}_{n=m}^{\infty} \to x$ as $n \to \infty$ and so $x \in \overline{\{x_n : n \geq m\}} \subset \overline{F_m} = F_m$ for all $m \in \mathbb{N}$. Hence $x \in \bigcap_{n=1}^{\infty} F_n$.

(b) $\bigcap_{n=1}^{\infty} F_n$ is a singleton.
To see this, let $x, y \in \bigcap_{n=1}^{\infty} F_n$. For any $\varepsilon > 0$, there exists $N \in \mathbb{N}$ such that $d(F_N) < \varepsilon$. Since $x, y \in F_N$, it follows that $d(x, y) \leq d(F_N) < \varepsilon$. As $\varepsilon > 0$ is arbitrary, we have $d(x, y) = 0$, which means $x = y$. Hence the result.

(c) No.
Example: Consider $X := \mathbb{R}$ and $F_n := \left(0, \frac{1}{n}\right)$, $n \in \mathbb{N}$. Then $F_n \neq \phi$ for all $n \in \mathbb{N}$, $\{F_n\} \downarrow$ with $\lim_{n\to\infty} d(F_n) = 0$, but $\bigcap_{n=1}^{\infty} F_n = \phi$. Note that F_n's are not closed.

(d) No.
Example: Consider $X := \mathbb{R}$ and $F_n := [n, \infty)$, $n \in \mathbb{N}$. Then each F_n is closed and nonempty, $\{F_n\} \downarrow$, but $\bigcap_{n=1}^{\infty} F_n = \phi$. Note that in this case $\lim_{n\to\infty} d(F_n) \neq 0$.

(e) No.

Example: Consider $X := \mathbb{R}$ with discrete metric and $F_n := \left(0, \frac{1}{n}\right)$, $n \in \mathbb{N}$. Then each F_n is closed and nonempty, and $\{F_n\} \downarrow$. Furthermore, since $d(F_n) = 1$ for all $n \in \mathbb{N}$, all F_n's are bounded. But $\bigcap_{n=1}^{\infty} F_n = \phi$.

(f) No.

Example: Consider $X := [-1, 0) \cup (0, 1]$ with the usual Euclidean metric. Note that X is not complete. For any $n \in \mathbb{N}$, let $F_n := \left[-\frac{1}{n}, 0\right) \cup \left(0, \frac{1}{n}\right]$. Then each F_n is closed and nonempty, $\{F_n\} \downarrow$, with $\lim_{n\to\infty} d(F_n) = 0$. But $\bigcap_{n=1}^{\infty} F_n = \phi$.

20. Let X be a metric space. If $\bigcap_{n=1}^{\infty} F_n \neq \phi$ for any decreasing nest of closed subsets $\{F_n\}_{n\in\mathbb{N}}$ of X with $\lim_{n\to\infty} d(F_n) = 0$, show that X is complete.

Proof. Let $\{x_n\}_{n\in\mathbb{N}}$ be a Cauchy sequence in X. Denote by $E_m := \{x_n : n \geq m\}$. By Exercise 2.2, Part A, Problem #4, we have $\lim_{m\to\infty} d(E_m) = 0$. By Exercise 1.2, Part B, Problem #8, we have $\lim_{m\to\infty} d(\overline{E_m}) = 0$. Applying the given condition to the decreasing nest of closed subsets $\{\overline{E_m}\}$ of X, we have $\bigcap_{m=1}^{\infty} \overline{E_m} \neq \phi$. But then as $\lim_{m\to\infty} d(\overline{E_m}) = 0$, by Exercise 2.2, Part B, Problem #19, $\bigcap_{m=1}^{\infty} \overline{E_m}$ must be a singleton, say $= \{a\}$. In particular, we have $a \in \overline{E_m}$ for all $m \in \mathbb{N}$. Now again by $\lim_{m\to\infty} d(\overline{E_m}) = 0$, for any $\varepsilon > 0$, there exists $N_1 \in \mathbb{N}$ such that $d(x_m, a) \leq d(\overline{E}_m) < \varepsilon/2$ for all $m \geq N_1$. On the other hand, by the Cauchyness of $\{x_n\}_{n\in\mathbb{N}}$, there exists $N_2 \in \mathbb{N}$ such that $d(x_n, x_m) < \varepsilon/2$ for all $n, m \geq N_2$. Take $N := \max\{N_1, N_2\}$. Then for any $n, m \geq N$, we have

$$d(x_n, a) \leq d(x_n, x_m) + d(x_m, a) < \varepsilon .$$

Hence $\lim_{n\to\infty} x_n = a$ and so X is complete.

21. For any metric space X, show that every sequence in X has a Cauchy sub-sequence if and only if for every $r > 0$, there exist points $x_1, \ldots, x_n \in X$, such that $\bigcup_{j=1}^{n} B(x_j, r) = X$.

Proof. ($\Leftarrow$): Let $\{y_n\}_{n\in\mathbb{N}}$ be any sequence in X. By assumption, there exist points $x_1, \ldots, x_N \in X$, such that $\bigcup_{j=1}^{N} B(x_j, \frac{1}{2}) = X$. So there is $j \in \{1, \ldots, N\}$ such that $B(x_j, \frac{1}{2})$ contains infinitely many terms of the sequence $\{y_n\}_{n\in\mathbb{N}}$. Hence there is a subsequence $\{y_{1,n}\}_{n=1}^{\infty}$ of $\{y_n\}_{n=1}^{\infty}$ with

$$d(y_{1,p},\ y_{1,q}) \leq d(y_{1,p},\ x_j) + d(y_{1,q},\ x_j) < 1 \quad \text{for all } p, q \in \mathbb{N}\ .$$

Similarly, since X is a union of finitely many balls of radius $\frac{1}{4}$, we can pick one of these balls which contains infinitely many terms of $\{y_{1,n}\}$, and in turn we can find a subsequence $\{y_{2,n}\}_{n=1}^{\infty}$ of $\{y_{1,n}\}_{n=1}^{\infty}$ with

$$d(y_{2,p},\ y_{2,q}) < \frac{1}{2} \quad \text{for all } p, q \in \mathbb{N}\ .$$

Continuing in this process, inductively, for every natural number m, we can find a subsequence $\{y_{m+1,n}\}_{n=1}^{\infty}$ of $\{y_{m,n}\}_{n=1}^{\infty}$ with

$$d(y_{m+1,p},\ y_{m+1,q}) < \frac{1}{2^m} \quad \text{for all } p, q \in \mathbb{N}\ .$$

Hence in particular, for any $M \in \mathbb{N}$ and any $p,\ q > M$,

$$d(y_{p,p},\ y_{q,q}) < \frac{1}{2^M}\ .$$

That is, the "diagonal" subsequence $\{y_{k,k}\}_{k=1}^{\infty}$ of $\{y_n\}_{n=1}^{\infty}$ is Cauchy.
($\Rightarrow$): We will argue by contra-positive argument. Suppose there is $r > 0$ such that any finite union of balls of radius r cannot cover X. Choose $x_1 \in X$. By the assumption, $B(x_1, r) \subsetneqq X$. Hence there exists $x_2 \in X \setminus B(x_1, r)$. As $B(x_1, r) \cup B(x_2, r) \subsetneqq X$, there exists

$x_3 \in X \setminus \bigcup_{i=1}^{2} B(x_i, r)$. Inductively, we have a sequence $\{x_n\}_{n\in\mathbb{N}}$ in X such that

$$x_n \notin X \setminus \bigcup_{i=1}^{n-1} B(x_i, r) \quad \text{for all } n \ .$$

By construction, $d(x_n, x_m) \geq r$ for $m \neq n$ and so the sequence $\{x_n\}_{n=1}^{\infty}$ has no Cauchy subsequence.

2.3 Continuity and Homeomorphism

In this section, we will treat the central concepts of metric space Topology, namely, continuity and homeomorphism. The central problem in metric space Topology is the "equivalence problem", which is to determine when two metric spaces are in certain sense "equivalent" to each other. Actually this is not a specific problem in metric space Topology, but is the main theme of many branches of mathematics. In each branch of mathematics, there are basic objects to be studied, and the main theme would be to characterize when two basic objects are equivalent. Of course, we have different equivalence relations in different fields. For example, in Linear Algebra, the basic objects are vector spaces, and the equivalence of vector spaces is called *isomorphism.* In Abstract Algebra, the basic objects are groups, rings, and fields, and the equivalence is also called *isomorphism.* In Complex Analysis, the basic objects are regions, and the equivalence is *biholomorphism.* In Differential Geometry, the basic objects are manifolds, and the equivalence is *diffeomorphism.* In Topology, the basic objects are metric spaces, and the equivalence is called *homeomorphism.* A homeomorphism is simply a bijective continuous function whose inverse is also continuous.

As usual, before introducing the concept of continuity of a function between two metric spaces, it would be handy to first introduce the concept of limits. Again, our usual convention is that letters X, Y, Z, etc., will stand for general metric spaces.

Definition 2.3.1. *Let X, Y be metric spaces and $S \subset X$ be a subset. Let $f : S \to Y$ be a function. If $p \in X$ is an accumulation point of S, we write*

$$\lim_{x \to p} f(x) = b \quad \textit{or} \quad f(x) \to b \textit{ as } x \to p$$

if for any $\varepsilon > 0$, *there exists* $\delta > 0$ *such that*

$$d\big(f(x), b\big) < \varepsilon \quad \text{whenever } x \in S \setminus \{p\} \text{ and } d(x, p) < \delta \ ,$$

that is,

$$f(x) \in B(b, \varepsilon) \quad \text{whenever } x \in B(p, \delta) \cap S \setminus \{p\} \ ,$$

or

$$f\Big((B(p, \delta)) \cap S \setminus \{p\}\Big) \subset B(b, \varepsilon) \ .$$

Theorem 2.3.2. *Let* $p \in X$ *be an accumulation point of* $S \subset X$, $b \in Y$, *and* $f : S \to Y$ *be any function. Then*

$$\lim_{x \to p} f(x) = b$$

if and only if

$$\lim_{n \to \infty} f(x_n) = b$$

for every sequence $\{x_n\}_{n \in \mathbb{N}}$ *in* $S \setminus \{p\}$ *which converges to* p.

Proof. ($\Rightarrow$) As $\lim_{x \to p} f(x) = b$, by definition, for any $\varepsilon > 0$, there exists $\delta > 0$ such that

$$f(B(p, \delta) \cap S \setminus \{p\}) \subset B(b, \varepsilon) \ .$$

So if $\{x_n\}_{n \in \mathbb{N}}$ is a sequence in $S \setminus \{p\}$ such that $\{x_n\} \to p$ as $n \to \infty$, then there exists $N \in \mathbb{N}$ such that

$$x_n \in B(p, \delta) \cap S \setminus \{p\} \quad \text{whenever} \quad n \geq N$$

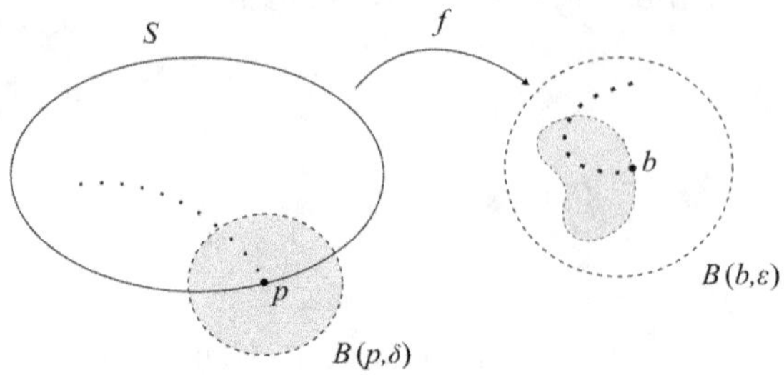

and so

$$f(x_n) \in f\Big(B(p,\delta) \cap S \setminus \{p\}\Big) \subset B(b,\varepsilon) \quad \text{whenever } n \geq N ,$$

that is, $\{f(x_n)\} \to b$ as $n \to \infty$.

($\Leftarrow$) Suppose $f(x) \not\to b$ as $x \to p$. Then there exists $\varepsilon > 0$ so that for each $n \in \mathbb{N}$ we can find $x_n \in S \setminus \{p\}$ with

$$0 < d(x_n, p) < \frac{1}{n} \quad \text{and} \quad d\big(f(x_n), b\big) \geq \varepsilon ,$$

that is, we have a sequence $\{x_n\}_{n\in\mathbb{N}}$ in $S \setminus \{p\}$ which converges to p but $\lim_{n\to\infty} f(x_n) \neq b$. □

Let $S \subset X$ be a subset and $f, g : S \to Y$. Operations in Y can easily induce (pointwise) operations on f, g. For instance, if $Y = \mathbb{C}$ or $\mathbb{R}$, then the elementary operations on Y like addition, multiplication, division, etc., would induce operations

$$f + g\ , \quad \lambda f \quad (\lambda \in \mathbb{C} \text{ or } \mathbb{R})\ , \quad fg\ , \quad \frac{f}{g} \quad (\text{if } g \neq 0)$$

pointwisely by

$$\begin{aligned}
(f+g)(x) &:= f(x) + g(x) \\
(\lambda f)(x) &:= \lambda f(x) \\
(f\,g)(x) &:= f(x)\,g(x) \\
\left(\frac{f}{g}\right)(x) &:= \frac{f(x)}{g(x)} \quad \text{if } g(x) \neq 0
\end{aligned}$$

for any $x \in S$. It is elementary to verify that all usual formulae for limits of real-valued functions defined on $\mathbb{R}$ are still valid in the context of metric spaces, that is,

$$\begin{aligned}
\lim_{x\to p}(f+g)(x) &= \lim_{x\to p} f(x) + \lim_{x\to p} g(x) ,\\
\lim_{x\to p} \lambda f(x) &= \lambda \lim_{x\to p} f(x) , \quad \lambda \in \mathbb{C} \text{ or } \mathbb{R} ,\\
\lim_{x\to p}(fg)(x) &= \lim_{x\to p} f(x) \lim_{x\to p} g(x) ,\\
\lim_{x\to p}\left(\frac{f}{g}\right)(x) &= \frac{\lim_{x\to p} f(x)}{\lim_{x\to p} g(x)} \quad \text{if } g \neq 0 \text{ near } p \text{ and } \lim_{x\to p} g(x) \neq 0 .
\end{aligned}$$

Similarly, if $Y = \mathbb{C}^n$ or $\mathbb{R}^n$, we may define

$$f+g , \quad \lambda f \ (\lambda \in \mathbb{C} \text{ or } \mathbb{R}) , \quad f\cdot g \quad \text{(inner product)} , \quad \text{etc.,}$$

pointwisely and again it is easily verified that

$$\begin{aligned}
\lim_{x\to p}(f+g)(x) &= \lim_{x\to p} f(x) + \lim_{x\to p} g(x) ,\\
\lim_{x\to p}(\lambda f)(x) &= \lambda \lim_{x\to p} f(x) , \quad \lambda \in \mathbb{C} \text{ or } \mathbb{R} ,\\
\lim_{x\to p}(f\cdot g)(x) &= \lim_{x\to p} f(x) \cdot \lim_{x\to p} g(x) .
\end{aligned}$$

Remark. Observe that $\lim_{x\to p} f(x) = b \iff \lim_{x\to p} d\big(f(x), b\big) = 0$.

Theorem 2.3.3. *Let $S \subset X$ be a subset, $p \in S'$, and $f : S \to \mathbb{R}^n$. Then $\lim_{x\to p} f(x) = b \Longrightarrow \lim_{x\to p} \big\|f(x)\big\| = \|b\|$.*

Proof. It is obvious by the triangle inequality. In fact, we have

$$0 \leq \Big| \|f(x)\| - \|b\| \Big| \leq \|f(x) - b\| .$$

If $\lim_{x\to p} f(x) = b$, by the preceding Remark, we have

$$\lim_{x\to p} \|f(x) - b\| = \lim_{x\to p} d\big(f(x), b\big) = 0$$

and so by Sandwich Theorem, we have

$$\lim_{x \to p} \Big| \|f(x)\| - \|b\| \Big| = 0 \,. \qquad \square$$

Remark. The converse of Theorem 2.3.3 is clearly not true. For example, consider $f : \mathbb{R} \to \mathbb{R}$ given by

$$f(x) := |x| \,.$$

Then $\lim_{x \to -1} \|f(x)\| = \lim_{x \to -1} |x| = 1 = \|-1\|$ but $\lim_{x \to -1} f(x) = 1 \neq -1$.

Definition 2.3.4. *Let $S \subset X$ be a subset. A function $f : S \to Y$ is said to be continuous at $p \in S$ if for every $\varepsilon > 0$ there exists $\delta > 0$ such that*

$$d\big(f(x), f(p)\big) < \varepsilon \quad \textit{whenever } x \in S \textit{ with } d(x,p) < \delta \,,$$

that is,

$$f(x) \in B(f(p), \varepsilon) \quad \textit{whenever } x \in B(p, \delta) \cap S \,,$$

or

$$f\big(B(p,\delta) \cap S\big) \subset B\big(f(p), \varepsilon\big) \,.$$

Note that if p is an accumulation point of S, then f is continuous at p if and only if

$$\lim_{\substack{x \in S \\ x \to p}} f(x) = f(p) \,,$$

and if p is an isolated point of S, that is, if $p \in S \setminus S'$, the condition is automatically satisfied and so f is automatically continuous at p by default.

f is said to be continuous on S if it is continuous at every $p \in S$.

Theorem 2.3.5. *Let $S \subset X$ be a subset, $p \in S$, and $f : S \to Y$ be a function. Then the following conditions are equivalent:*

(a) *f is continuous at p.*

(b) $\lim_{n\to\infty} f(x_n) = f(p)$ *whenever $\{x_n\}_{n\in\mathbb{N}}$ is a sequence in S with* $\lim_{n\to\infty}\{x_n\} = p$.

Proof. If $p \in S'$, the Theorem follows immediately from definition and Theorem 2.3.2. If $p \in S \setminus S'$, the assertion is trivial. □

Example 2.3.6. The Euclidean norm $\|\cdot\| : \mathbb{R}^n \to \mathbb{R}$ is a continuous function on $\mathbb{R}^n$.

Proof. This is a simple consequence of Theorem 2.3.3. To see this, let $f : \mathbb{R}^n \to \mathbb{R}^n$ be the identity function $f(x) := x$, $x \in \mathbb{R}^n$. Then for any $p \in \mathbb{R}^n$, it is obvious that $\lim_{x\to p} f(x) = \lim_{x\to p} x = p$ and so by Theorem 2.3.3, we have

$$\lim_{x\to p} \|x\| = \lim_{x\to p} \|f(x)\| = \|p\| \ ,$$

which means $\|\cdot\|$ is continuous at p. Since $p \in \mathbb{R}^n$ is arbitrary, $\|\cdot\|$ is continuous on $\mathbb{R}^n$.
Alternatively, the assertion can also be proved by direct verification from first principle. This will be left to the readers as an exercise.

Example 2.3.7. If (X, d) is a metric space, then the metric d is a function $d : X \times X \to \mathbb{R}$, and we can talk about the continuity of the function d once $X \times X$ is equipped with a metric. For example, it is easy to check that

$$\rho\big((x_1, y_1), (x_2, y_2)\big) := d(x_1, x_2) + d(y_1, y_2)$$

is a well-defined metric on $X \times X$. For any $\varepsilon > 0$ and any (x_0, y_0), $(x, y) \in X \times X$, since

$$\begin{aligned} d(x, y) &\le d(x, x_0) + d(x_0, y_0) + d(y_0, y) \ , \\ d(x_0, y_0) &\le d(x_0, x) + \ \ d(x, y) \ + d(y, y_0) \ , \end{aligned}$$

we have

$$|d(x,y) - d(x_0,y_0)| \leq d(x,x_0) + d(y,y_0) < 2\rho\big((x,y),(x_0,y_0)\big) .$$

Hence if we take $\delta := \varepsilon/2$, then

$$\rho\big((x,y),(x_0,y_0)\big) < \delta \Longrightarrow |d(x,y) - d(x_0,y_0)| < \varepsilon$$

and so $d : X \times X \to \mathbb{R}$ is continuous at (x_0, y_0). Since $(x_0, y_0) \in X \times X$ is arbitrary, d is continuous on $(X \times X, \rho)$.

Alternatively, suppose $\{(x_n, y_n)\}_{n\in\mathbb{N}}$ is a sequence in $X \times X$ with $\{(x_n, y_n)\} \to (x, y)$ in ρ as $n \to \infty$. To show the continuity of $d : (X \times X, \rho) \to \mathbb{R}$, it suffice to show

$$d(x_n, y_n) \to d(x,y) \quad \text{as } n \to \infty .$$

Now observe that

$$\begin{aligned} & (x_n, y_n) \to (x,y) \text{ in } \rho \text{ as } n \to \infty \\ \Longleftrightarrow \; & d(x_n, x) + d(y_n, y) = \rho\big((x_n,y_n),(x,y)\big) \to 0 \text{ as } n \to \infty . \end{aligned}$$

Since

$$\begin{aligned} d(x,y) &\leq d(x,x_n) + d(x_n,y_n) + d(y_n,y) \\ d(x_n,y_n) &\leq d(x_n,x) + \;\; d(x,y) \;\; + d(y,y_n) , \end{aligned}$$

we have

$$\Big|d(x_n,y_n) - d(x,y)\Big| \leq d(x,x_n) + d(y,y_n) \to 0 \text{ as } n \to \infty .$$

Hence $d(x_n, y_n) \to d(x, y)$ as $n \to \infty$ and so d is continuous.

Remark. Observe that similar arguments can show that $d : (X \times X, \rho') \to \mathbb{R}$ is also continuous, where the metric ρ' is defined by

$$\rho'\big((x_1,y_1),(x_2,y_2)\big) := \sqrt{d(x_1,x_2)^2 + d(y_1,y_2)^2} .$$

Theorem 2.3.8. *If $f : X \to Y$ is continuous at $p \in X$ and $g : Y \to Z$ is continuous at $f(p) \in Y$, then $g \circ f : X \to Z$ is continuous at p.*

Proof. For any $\varepsilon > 0$, since g is continuous at $f(p)$, there exists $\delta_1 > 0$ such that

$$g\big(B(f(p), \delta_1)\big) \subset B\big(g \circ f(p), \varepsilon\big) .$$

On the other hand, since f is continuous at p, there exists $\delta > 0$ such that

$$f\big(B(p, \delta)\big) \subset B\big(f(p), \delta_1\big) .$$

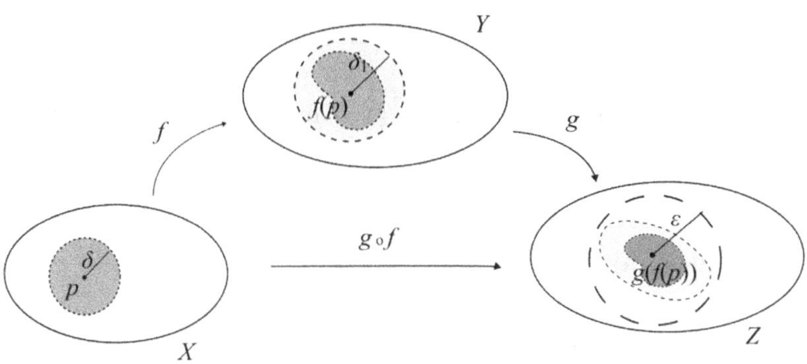

Hence

$$g \circ f\big(B(p, \delta)\big) \subset g\big(B(f(p), \delta_1\big) \subset B\big(g \circ f(p), \varepsilon\big) . \qquad \square$$

Theorem 2.3.9.

(a) *If f, $g : X \to \mathbb{C}$ (or $\mathbb{R}$) are continuous at p, then so are $f + g$, $f - g$, fg, and f/g (provided that $g(p) \neq 0$).*

(b) *If f, $g : X \to \mathbb{C}^n$ (or $\mathbb{R}^n$) are continuous at p, then so are $f + g$, λf, $f \cdot g$, and $\|f\|$ for any $\lambda \in \mathbb{C}$ (or $\mathbb{R}$).*

(c) *Let $f_i : X \to \mathbb{C}$ or $\mathbb{R}$ be functions on X, $i = 1, \ldots, n$. Then*

$$f := (f_1, \ldots, f_n) : X \to \mathbb{C}^n \text{ (or } \mathbb{R}^n\text{)}$$

is continuous at p if and only if f_i is continuous at p for each $i = 1, \ldots, n$.

Proof. It is elementary and will be left as an exercise. □

Remark. It is clear that the identity function of $\mathbb{C}$ to $\mathbb{C}$ (or $\mathbb{R}$ to $\mathbb{R}$) is continuous. Hence Theorem 2.3.9 gives rise to plenty of continuous functions, for example, polynomials in x, etc.

Remark. By now the readers should be fairly familiar with the interplay between open balls and open neighborhoods. Hence for the sake of simplicity, from this point on, sometimes we will talk about open balls centered at a point p and open neighborhoods of p interchangeably. So for example, $f : X \to Y$ is continuous at $p \in X$ if for every open neighborhood V of $f(p)$ in Y, there exists an open neighborhood U of p in X such that $f(U) \subset V$.

Same as for other properties, to verify continuity of a function by definition could be tedious and time-consuming, not to mention that such a process in general may not lead to any insightful picture. So as usual, we will try and establish handy necessary and/or sufficient conditions for continuity which could lead to useful characterization and/or properties of continuity.

Theorem 2.3.10. *$f : X \to Y$ is continuous on X if and only if the inverse image $f^{-1}(T)$ of any open set $T \subset Y$ is open in X.*

Proof. ($\Rightarrow$) Let $T \subset Y$ be open and $p \in f^{-1}(T)$. Write $q := f(p) \in T$. Since T is open in Y, there exists $\varepsilon > 0$ such that $B(q, \varepsilon) \subset T$. Since f is continuous at p, there exists $\delta > 0$ such that $f\big(B(p, \delta)\big) \subset B(q, \varepsilon)$. Hence

$$B(p, \delta) \subset f^{-1}\big(B(q, \varepsilon)\big) \subset f^{-1}(T)$$

and so $p \in f^{-1}(T)^\circ$. Hence $f^{-1}(T) = f^{-1}(T)^\circ$ is open.

($\Leftarrow$) Let $p \in X$ and $q := f(p)$. For any $\varepsilon > 0$, since $B(q, \varepsilon) \subset Y$ is open, $f^{-1}\big(B(q, \varepsilon)\big) \subset X$ is open. Since $f(p) = q$, we have $p \in f^{-1}\big(B(q, \varepsilon)\big)$ and so there exists $\delta > 0$ such that $p \in B(p, \delta) \subset f^{-1}\big(B(q, \varepsilon)\big)$ and thus $f\big(B(p, \delta)\big) \subset B(q, \varepsilon)$. Hence f is continuous at p. Since $p \in X$ is arbitrary, f is continuous on X. □

Theorem 2.3.11. *$f : X \to Y$ is continuous on X if and only if the inverse image $f^{-1}(T)$ of any closed set $T \subset Y$ is closed in X.*

Proof. This follows from Theorem 2.3.10 and the fact that

$$f^{-1}(Y \setminus T) = X \setminus f^{-1}(T) \ .$$

In fact,

$$\begin{aligned} f \text{ is continuous} \quad &\Leftrightarrow \quad f^{-1}(S) \subset X \text{ is open for any open } S \subset Y \\ &\Leftrightarrow \quad f^{-1}(Y \setminus T) \subset X \text{ is open for any closed } T \subset Y \\ &\qquad\quad \| \\ &\Leftrightarrow \quad X \setminus f^{-1}(T) \subset X \text{ is open for any closed } T \subset Y \ . \end{aligned}$$

□

Remark. In the proof of Theorem 2.3.11, we have made used of the fact that

$$f^{-1}(Y \setminus T) = X \setminus f^{-1}(T) \ .$$

However, for direct images of sets we only have

$$f(X) \setminus f(S) \subsetneqq f(X \setminus S)$$

and the quality holds only when f is $1-1$. Details will be left as an exercise.

Remark. At first sight, the two preceding Theorems on equivalence conditions for continuity may seem to be a little unpleasant, and it may look a lot more pleasant in case we have a result saying that continuous functions map open sets to open sets, and closed sets to closed sets, and vice versa. Unfortunately, it is not the case. In general, continuous functions may not map open sets to open sets. For example, the constant function $f : \mathbb{R} \to \mathbb{R}$ given by $f(x) := 0$ for all $x \in \mathbb{R}$ is clearly continuous. However, the open set $(0, 1) \subset \mathbb{R}$ is being mapped to $\{0\} \subset \mathbb{R}$ which is *not* open in $\mathbb{R}$. Similarly, continuous functions may not map closed sets to closed sets. For example, the identity function $f : (0, 1) \to \mathbb{R}$ given by $f(x) := x$ for all $x \in \mathbb{R}$ is clearly continuous. However, $(0, 1)$ is closed in $(0, 1)$ while $f((0, 1)) = (0, 1)$ is *not* closed in $\mathbb{R}$.

In general, a function which maps open sets to open sets is called an *open mapping*. A function which maps closed sets to closed sets is called a *closed mapping*. In general, open mappings may not be closed, closed mappings may not be open, open mappings may not be continuous, and closed mappings may not be continuous either. Interested readers are referred to Exercise 2.3, Part A, Problem #26, 28, 29, 30, 31, 32.

Having said that, however, openness, closedness, and continuity are not completely unrelated. For example, the following is a nice relation among them. Other examples can be found in Exercise 2.3, Part A, Problem #33 and Exercise 2.3, Part B, Problem #3.

Example 2.3.12. If $f : X \to Y$ is $1-1$, then $f^{-1} : f(X) \to X$ exists, and the following statements are equivalent:

(a) f is continuous;

(b) f^{-1} is open;

(c) f^{-1} is closed.

Proof. First of all, let us set the language straight. In general, for an arbitrary function $f : X \to Y$, $f^{-1}(S)$ stands for the pre-image of the set $S \subset Y$ under f, or more precisely,

$$f^{-1}(S) := \{x \in X : f(x) \in S\} .$$

In case $f : X \to Y$ is $1-1$, then $f : X \to f(X)$ is bijective and so there is an inverse function mapping $f(X)$ back onto X. As usual, it will be denoted by $f^{-1} : f(X) \to X$. So if $S \subset f(X)$ is a subset of $f(X)$, the inverse function f^{-1} will map the set S to a set $f^{-1}(S) \subset X$. So then $f^{-1}(S)$ is having two meanings. One, it is the pre-image of the set $S \subset f(X) \subset Y$ under f, and two, it is the image of the set $S \subset f(X)$ under the function f^{-1}. This may seem a bit confusing, but then the good news is, whichever meaning we choose, $f^{-1}(S)$ represents the same set and so there should indeed be no ambiguity.

(a)⇒(b): Let $S \subset f(X)$ be open. Then there is an open set $U \subset Y$ such that $S = U \cap f(X)$. Since f is continuous, by Theorem 2.3.10,

$$\begin{aligned} f^{-1}(S) &= f^{-1}(U \cap f(X)) = f^{-1}(U) \cap f^{-1}\big(f(X)\big) \\ &= f^{-1}(U) \cap X = f^{-1}(U) \subset X \qquad \text{is open .} \end{aligned}$$

Thus f^{-1} is an open mapping.

(b)⇒(c): Let $C \subset f(X)$ be closed. Then $f(X) \setminus C \subset f(X)$ is open. Since f^{-1} is an open mapping, $f^{-1}\big(f(X) \setminus C\big) \subset X$ is open and so

$$f^{-1}(C) = X \setminus f^{-1}\big(f(X) \setminus C\big)$$

is closed. Hence f^{-1} is a closed mapping.

(c)⇒(a): If $C \subset Y$ is closed, then $C \cap f(X) \subset f(X)$ is closed in $f(X)$. Since f^{-1} is a closed mapping, it maps closed sets in $f(X)$ to closed sets in X. In particular,

$$f^{-1}(C) = f^{-1}(C \cap f(X)) \subset X \qquad \text{is closed in } X \text{ .}$$

By Theorem 2.3.11, f is continuous. □

The interested readers are encouraged to establish the other implications (b)⇒(a), (c)⇒(b), and (a)⇒(c) of Example 2.3.12 directly. These will help achieving a better understanding of the relations among the 3 conditions.

Theorem 2.3.13. *Continuous functions map compact sets to compact sets.*

Proof. Let $f : X \to Y$ be continuous and $C \subset X$ be a compact set. Let $\mathcal{U} = \{U_\alpha\}_{\alpha\in\Lambda}$ be an open cover of $f(C)$ in Y. Since f is continuous, $f^{-1}(\mathcal{U}) := \{f^{-1}(U_\alpha)\}_{\alpha\in\Lambda}$ is an open cover of C. In fact,

$$
\begin{aligned}
& f(C) \subset \bigcup_{\alpha\in\Lambda} U_\alpha \\
\implies & C \subset f^{-1}(f(C)) \subset f^{-1}\left(\bigcup_{\alpha\in\Lambda} U_\alpha\right) = \bigcup_{\alpha\in\Lambda} f^{-1}(U_\alpha) \,.
\end{aligned}
$$

Since C is compact, the open cover $f^{-1}(\mathcal{U})$ of C has a finite sub-cover, say $\{f^{-1}(U_i)\}_{i=1}^n$. Then

$$
\begin{aligned}
& C \subset \bigcup_{i=1}^n f^{-1}(U_i) \\
\implies & f(C) \subset f\Big(\bigcup_{i=1}^n f^{-1}(U_i)\Big) = \bigcup_{i=1}^n f(f^{-1}(U_i)) \subset \bigcup_{i=1}^n U_i
\end{aligned}
$$

and thus $f(C)$ is compact. □

Remark. In the proof of Theorem 2.3.13, we have made used of the facts that

$$f(f^{-1}(T)) \subset T \quad \text{and} \quad f^{-1}(f(S)) \supset S \,.$$

Note that in general, equalities may not hold, so this is the most we could say (exercise).

Theorem 2.3.14. *If $f : X \to Y$ is continuous and $C \subset X$ is compact, then f is bounded on C.*

Proof. It is obvious from Theorems 2.3.13 and 1.3.5. □

Theorem 2.3.15. *If $f : X \to \mathbb{R}$ is continuous on a compact set $C \subset X$, then f attains its maximum and minimum in C, that is, there exist p, $q \in C$ such that*

$$f(p) = \sup f(C) = \max f(C) ,$$
$$f(q) = \inf f(C) = \min f(C) .$$

Proof. By Theorems 2.3.13 and 1.3.5, $f(C) \subset \mathbb{R}$ is compact and hence closed and bounded. Let $m = \inf f(C) \in \mathbb{R}$. By definition of infimum, we have $m \leq f(c)$ for all $c \in C$. Observe that m must be an adherent point of $f(C)$. In fact, if not, then there exists $r > 0$ such that $B(m, r) \cap f(C) = \phi$. This implies $m < m + \frac{r}{2} < f(c)$ for all $c \in C$, violating the assumption that $m = \inf f(C)$. Thus $m \in \overline{f(C)} = f(C)$ and so $m = f(q)$ for some $q \in C$. Similarly, $f(p) = \sup f(C)$ for some $p \in C$. □

Remark. By Theorem 2.3.15, every continuous function $f : \mathbb{R} \to \mathbb{R}$ maps closed and bounded intervals to closed and bounded intervals. In fact, for any $[a, b] \subset \mathbb{R}$, f attains its maximum M and minimum m in $[a, b]$. Thus $f([a, b]) \subset [m, M]$ with m, M being attained. By Intermediate Value Theorem, $f([a, b]) = [m, M]$.

Theorem 2.3.16. *Let $f : X \to Y$ be a function. If*

(a) *X is compact,*

(b) *f is $1 - 1$,*

(c) *f is continuous,*

then the inverse function $f^{-1} : f(X) \to X$ is also continuous.

Proof. The existence of the inverse function follows from (b). Now let $C \subset X$ be closed. By Theorem 1.3.8, C is compact and so by Theorem 2.3.13, $(f^{-1})^{-1}(C) = f(C) \subset Y$ is compact. In particular, it is closed in Y. However, since $f(C) \subset f(X)$, it follows that $(f^{-1})^{-1}(C) = f(C)$ is also closed in $f(X)$. Therefore, by Theorem 2.3.11, f^{-1} is continuous. □

Examples 2.3.17. (The compactness of X in Theorem 2.3.16 is essential.)

(i) Let $f : [1,2] \cup (3,4] \to \mathbb{R}$ be given by

$$f(x) = \begin{cases} x & x \in [1,2] \\ x-1 & x \in (3,4] \,. \end{cases}$$

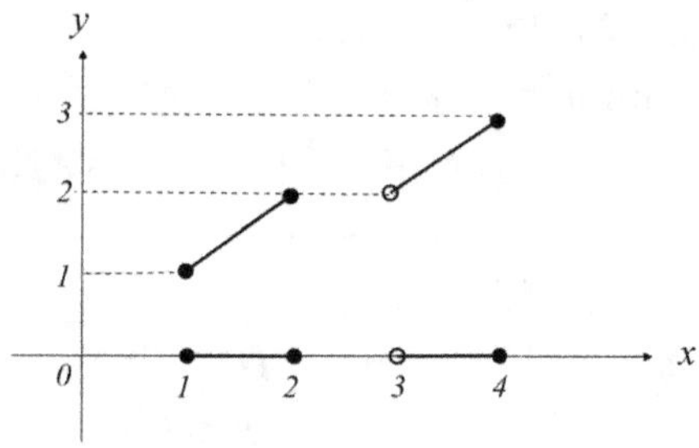

Then f is $1-1$ and continuous. However,

$$\lim_{y \to 2^-} f^{-1}(y) = 2 \neq 3 = \lim_{y \to 2^+} f^{-1}(y)$$

and so f^{-1} is not continuous at 2. This does not violate Theorem 2.3.16 as in this case the domain $[1,2] \cup (3,4]$ is not compact.

(ii) let $X = [0,1)$ with the usual Euclidean metric and consider

$$f : X \to \mathbb{C}$$

given by

$$f(x) = e^{2\pi i x} \,, \quad x \in [0,1) \,.$$

Clearly f is $1-1$ and continuous on X with $f(X) = S^1 \subset \mathbb{C}$.

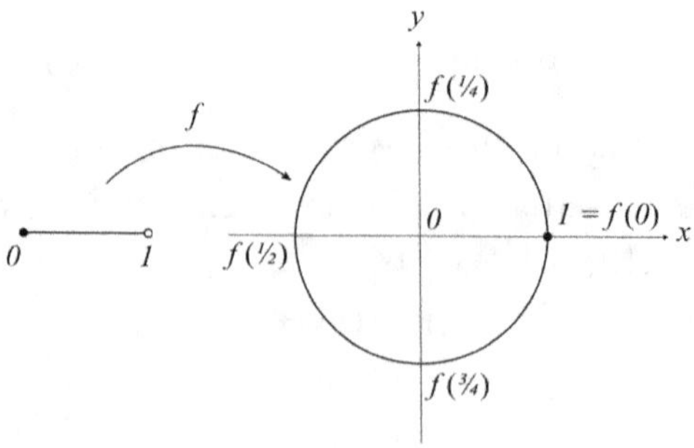

Hence in particular, $f^{-1} : S^1 \to X = [0,1)$ exists. However, f^{-1} is not continuous at the point $1 \in S^1 \subset \mathbb{C}$. This can be seen by either observing that images of points near $1 \in f(X) = S^1 \subset \mathbb{C}$ under f^{-1} are not necessarily close $0 \in X$, or noting that we have $\left\{f\left(1-\frac{1}{n}\right)\right\}_{n\in\mathbb{N}} \to f(0)$ as $n \to \infty$ but

$$\left\{f^{-1}\left(f\left(1-\frac{1}{n}\right)\right)\right\}_{n\in\mathbb{N}} = \left\{1-\frac{1}{n}\right\}_{n\in\mathbb{N}} \not\to 0 = f^{-1}(f(0)) .$$

Again, Theorem 2.3.16 is not violated, as in this case the domain X is not compact.

Definition 2.3.18. *A bijection $f : X \to Y$ which is continuous and with continuous inverse $f^{-1} : Y \to X$ is called a homeomorphism or a topological mapping of X onto Y. Any two metric spaces X, Y are said to be homeomorphic, denoted as $X \cong Y$, if there is a homeomorphism of X onto Y.*

Remark. Clearly, being homeomorphic is an equivalence condition. In fact, it is easy to see that

(i) the identity function $\iota : X \to X$ is a homeomorphism and hence $X \cong X$;

(ii) if $f : X \to Y$ is a homeomorphism, then so $f^{-1} : Y \to X$ is a homeomorphism;

(iii) if $f : X \to Y$ and $g : Y \to Z$ are homeomorphisms, then $g \circ f : X \to Z$ is also a homeomorphism.

Remark (*more important than customary*). Let $f : X \to Y$ be a continuous function and imagine that X, Y are elastic geometric objects. Then what f does to X is to move the points in X in a "continuous" manner, that is, to bend, twist, stretch, dilate, contract etc., but there is no tearing. The point is, the action of tearing is not continuous, because points that are originally close to each other will no longer be close to each other after the tearing. So intuitively, a homeomorphism of X onto Y is an action that transforms the space X in a "continuous manner" through bending, twisting, stretching, dilating, contracting, etc., but without tearing, onto the space Y in such a way that the process can be reversed. In short, we say that two metric spaces are homeomorphic if one can be "continuously deformed" to another, and back.

Examples 2.3.19.

(a) $(0,1) \cong (0,2)$.

(b) $[0,1) \cong [0,2)$.

(c) $[0,1] \cong [0,2]$.

(d) $(0,1) \cong \mathbb{R}$.

(e) $S^1 \setminus \{n\} \cong \mathbb{R}$, where n is the "north pole" $(0,1) \in S^1 \subset \mathbb{R}^2$.

(f) $S^2 \setminus \{n\} \cong \mathbb{R}^2$, where n is the "north pole" $(0,0,1) \in S^2 \subset \mathbb{R}^3$.

Proof.

(a) The function $f : (0,1) \to (0,2)$ given by $f(x) := 2x$, $x \in (0,1)$, is $1-1$, onto, continuous, with inverse $f^{-1}(y) = \frac{y}{2}$, $y \in (0,2)$, which is clearly also continuous, gives a homeomorphism $f : (0,1) \xrightarrow{\cong} (0,2)$.

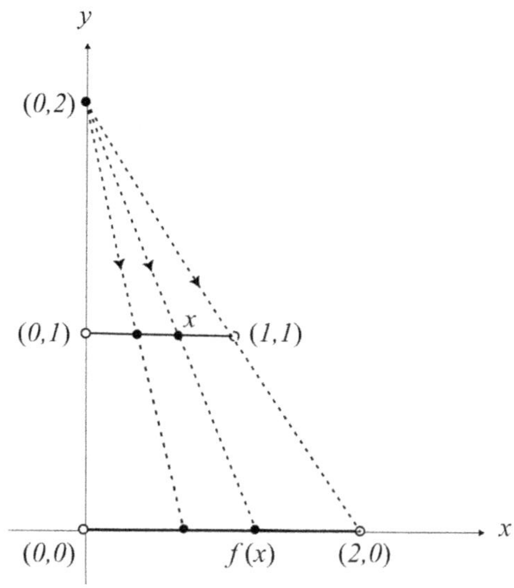

It is evident that other homeomorphisms of $(0, 1)$ onto $(0, 2)$ can easily be constructed. For example, the function $g : (0, 1) \to (0, 2)$ given by $g(x) := 2x^2$, $x \in (0, 1)$, is also $1 - 1$, onto, continuous, with continuous inverse $g^{-1}(y) = \sqrt{\frac{y}{2}}$, $y \in (0, 2)$. Hence it is also a homeomorphism.

(b) The same functions f, g as defined in (a) are homeomorphisms $[0, 1) \xrightarrow{\cong} [0, 2)$.

(c) The same function f as defined in (a) is a homeomorphism $[0, 1] \xrightarrow{\cong} [0, 2]$.

(d) It is easily seen that the function $f : (0, 1) \to (-\frac{\pi}{2}, \frac{\pi}{2})$ given by $f(x) := (x - \frac{1}{2})\pi$, $x \in (0, 1)$, gives a homeomorphism $(0, 1) \xrightarrow{\cong} (-\frac{\pi}{2}, \frac{\pi}{2})$. Meanwhile, the function $g : (-\frac{\pi}{2}, \frac{\pi}{2}) \to \mathbb{R}$ given by $g(x) := \tan x$, $x \in (-\frac{\pi}{2}, \frac{\pi}{2})$, gives a homeomorphism $(-\frac{\pi}{2}, \frac{\pi}{2}) \xrightarrow{\cong} \mathbb{R}$. Thus the composition $g \circ f : (0, 1) \xrightarrow{\cong} \mathbb{R}$ is a homeomorphism.

(e) We will prove this descriptively by our geometric insight. The interested readers could try to fill in the detailed computations with explicit formula.

In the xy-plane, lay the punctured unit circle $S^1 \setminus \{n\}$ onto the $\mathbb{R}_x$ axis. Then imagine that there is a light source emitting light from the north pole down through the punctured circle onto $\mathbb{R}_x$. In this way, we have a $1-1$, onto, continuous mapping from the punctured unit circle to the real line $\mathbb{R}_x$ with continuous inverse.

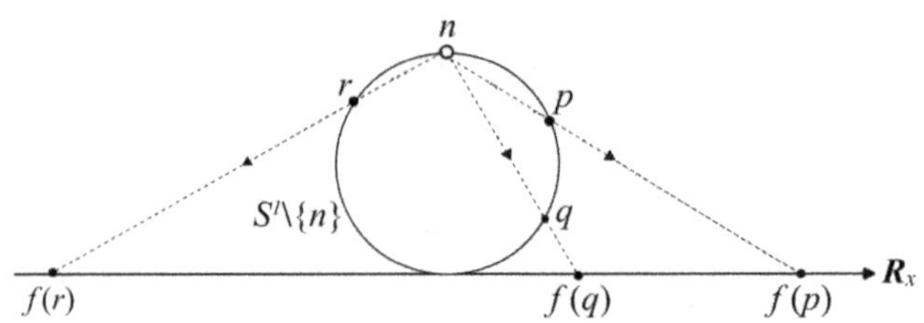

$f: S^1\setminus\{n\} \to \mathbf{R}_x$ *Stereographic Projection*

That is a homeomorphism of the punctured unit circle onto $\mathbb{R}$, called the *stereographic projection*.

(d) revisited: Based on the result of (e), another homeomorphism $(0,1) \xrightarrow{\cong} \mathbb{R}$ can be constructed geometrically by means of the stereographic projection as follows. First of all, by imagining that $(0,1)$ is an elastic piece of string of unit length, then it can be bent and stretched into a unit circle with the "north pole" removed. That means the line segment $(0,1)$ is homeomorphic to the punctured circle. By (e), the stereographic projection gives a homeomorphism of the punctured circle onto $\mathbb{R}$. Hence $(0,1)$ which is homeomorphic to the punctured circle is also homeomorphic to $\mathbb{R}$.

(f) Similar to (e), the stereographic projection of the punctured unit ball $S^2 \setminus \{n\} \subset \mathbb{R}^3$ onto the plane $\mathbb{R}^2$ is a homeomorphism. Details are left for the interested readers.

Example 2.3.20. Let $f: X \to Y$ be bijective. Then the following conditions are all equivalent:

(a) f is a homeomorphism.
(b) f and f^{-1} are continuous.
(c) f is continuous and open.
(d) f is continuous and closed.

Proof. (a)$\Leftrightarrow$(b) follows from the definition of homeomorphism. The equivalence of (b), (c), and (d) follows immediately from Example 2.3.12, as f^{-1} is continuous if and only if $f = (f^{-1})^{-1}$ is open if and only if $f = (f^{-1})^{-1}$ is closed. $\square$

Definition 2.3.21. *A property of a set which remains invariant under (or, is preserved by) topological mappings (or homeomorphisms) is called a topological property or a topological invariant.*

Example 2.3.22. Determine whether openness, closedness, compactness, boundedness, and the distance between two points are topological invariants.

Proof. By Example 2.3.20, openness and closedness are preserved by homeomorphisms, hence they are topological invariants.

By Theorem 2.3.13, compactness is preserved by continuous functions, hence by homeomorphisms and so it is also a topological invariant.

However, boundedness is not preserved by homeomorphisms. Example 2.3.19 (d) shows that $(0,1) \cong \mathbb{R}$, $(0,1)$ is bounded, but $\mathbb{R}$ is not. Hence boundedness is not a topological invariant.

Finally, the distance between two points is not preserved by homeomorphism. For example, from Example 2.3.19 (a), the function

$$\begin{aligned} f : (0,1) &\to (0,2) \\ x &\mapsto 2x \end{aligned}$$

is a homeomorphism. However, it is clear that

$$d(a,b) = |b-a| \neq 2|b-a| = d\big(f(a), f(b)\big)$$

for any $a \neq b$ in $(0,1)$. Hence the distance between two points is not a topological invariant.

Definition 2.3.23. *A mapping $f : X \to Y$ is said to be an isometry if it preserves the metric, or preserves the distance, that is,*

$$d_Y\big(f(p), f(q)\big) = d_X(p,q) \quad \textit{for all } p, q \in X \ .$$

Two metric spaces X and Y are said to be isometric if there is an isometry of X onto Y. Observe that surjective isometries are homeomorphisms.

Example 2.3.24. The infinite cylinder

$$C := \{(x,y,z) \in \mathbb{R}^3 : x^2 + y^2 = 1\} \subset \mathbb{R}^3$$

equipped with the metric d given by $d(p,q)$:= the shortest path among all paths on C joining the points p, $q \in C$. It is evident that (C,d) is a well-defined metric space. Let

$$S := C \setminus \{\text{a vertical straight line in } C\} \subset S \ .$$

Then it is intuitively clear that S is isometric to the open infinite paper strip $T := \{(x,y) \in \mathbb{R}^2 : |x| < \pi\}$. The interested readers are encouraged to try and work out the full details.

Exercise 2.3

Part A: True or False Questions

For each of the following statements, determine if it is true or false. If it is true, prove it. If it is false, give a counterexample or provide proper justification.

1. Let (X, d) be a metric space, $x \in X$ and $\phi \neq A \subset X$. Recall that the distance between x and A is defined by $d(x, A) := \inf_{a \in A} d(x, a)$. We can extend the notion of distance between a point and a set to distance between two nonempty subsets: if $A,\ B \subset X$, the distance between A and B is

$$d(A, B) := \inf\{d(a, b) : a \in A, b \in B\}\ .$$

If A and B are closed and $A \cap B = \phi$, then $d(A, B) > 0$.

Answer: False.
Example: In $X := \mathbb{R}^2$, take

$$A := \{(x, 0) : x \in \mathbb{R}\},\ B := \left\{\left(x, \frac{1}{x}\right) : x > 0\right\}\ .$$

Then $A,\ B$ are closed, $A \cap B = \phi$, but $d(A, B) = 0$.

2. Let $\pi_x,\ \pi_y : \mathbb{R}^2 \to \mathbb{R}$ be the projection of the plane $\mathbb{R}^2$ onto the x- and y- coordinate, respectively, i.e.,

$$\pi_x(x, y) := x \quad \text{and} \quad \pi_y(x, y) := y \quad \text{for all } (x, y) \in \mathbb{R}^2\ .$$

Then π_x and π_y are continuous on $\mathbb{R}^2$.

Answer: True.
Proof. Suffices to show the continuity of π_x, as that for π_y can be shown analogously. Fix $(a, b) \in \mathbb{R}^2$. For any $\varepsilon > 0$, pick $\delta := \varepsilon$. Then for any $(x, y) \in B_{\mathbb{R}^2}\big((a, b), \delta\big)$, we have

$$|x - a| \leq \sqrt{|x - a|^2 + |y - b|^2} < \delta = \varepsilon\ ,$$

that is, $\pi_x(a, b) = x \in B_{\mathbb{R}}(a, \varepsilon)$ and so π_x is continuous at (a, b). Since $(a, b) \in \mathbb{R}^2$ is arbitrary, π_x is continuous on $\mathbb{R}^2$.

3. Let $f : X \to \mathbb{R}$ be continuous. Then $\{x \in X : f(x) > 0\} \subset X$ is open.

   Answer: True.

   Proof. Fix $a \in \{x \in X : f(x) > 0\}$. Write $\varepsilon := f(a)$. Then by assumption, we have $\varepsilon > 0$. By the continuity of f, there exists $\delta > 0$ such that $f\big(B_X(a, \delta)\big) \subset B_{\mathbb{R}}\big(f(a), \varepsilon\big) = (0, 2\varepsilon) \subset (0, \infty)$. That is, $B_X(a, \delta) \subset \{x \in X : f(x) > 0\}$ and so $\{x \in X : f(x) > 0\}$ is open.

   Alternatively, since f is continuous and $(0, \infty) \subset \mathbb{R}$ is open, $\{x \in X : f(x) > 0\} = f^{-1}\big((0, \infty)\big) \subset X$ is open.

4. Let $f : X \to \mathbb{R}$ be continuous. Then $\{x \in X : f(x) \geq 0\} \subset X$ is closed.

   Answer: True.

   Proof. Similar to Exercise 2.3, Part A, Problem #3, it is easy to see that $\{x \in X : f(x) < 0\}$ is open. Hence, being the complement of an open set in X, $\{x \in X : f(x) \geq 0\}$ is closed.

5. If $f : \mathbb{R} \to \mathbb{R}$ be continuous, then

$$A := \{(x, y) \in \mathbb{R}^2 : yf(x) > 0\} \subset \mathbb{R}^2$$

   is open.

   Answer: True.

   Proof. Let $\pi_i : \mathbb{R}^2 \to \mathbb{R}$ be the projection of $\mathbb{R}^2$ onto its i-th coordinate, $i = 1, 2$. By Exercise 2.3, Part A, Problem #2, π_i is continuous for $i = 1, 2$. Define $F : \mathbb{R}^2 \to \mathbb{R}$ by

$$F(x, y) := y\, f(x) = (\pi_2(x, y)) \cdot (f \circ \pi_1(x, y))\ .$$

   Then we have $A = F^{-1}\big((0, \infty)\big)$. Since f, π_1 and π_2 are continuous, so is F. As $(0, \infty) \subset \mathbb{R}$ is open, $A = F^{-1}\big((0, \infty)\big) \subset \mathbb{R}^2$ is open.

6. If $f : X \to Y$ maps compact sets to compact sets, then f is continuous.

   Answer: False.

   Example: Let $X := \mathbb{R}$, $Y := \{0, 1\}$, and $f : X \to Y$ be given by

$$f(x) := \begin{cases} 0 & \text{if } x = 0\,, \\ 1 & \text{if } x \neq 0\,. \end{cases}$$

   Then clearly f sends every compact set to a compact set but f is not continuous.

7. Continuous functions map Cauchy sequences to Cauchy sequences. That is, if $\{x_n\}_{n\in\mathbb{N}}$ is a Cauchy sequence in X and $f : X \to Y$ is continuous, then $\{f(x_n)\}_{n\in\mathbb{N}}$ is a Cauchy sequence in Y.

   Answer: False.

   Example: Let $X = Y := \mathbb{R}^+ = (0, \infty)$, $x_n = 1/n$, $n \in \mathbb{N}$, and $f(x) := \frac{1}{x}$ for every $x \in \mathbb{R}^+$. Then $\{x_n\}_{n\in\mathbb{N}}$ is a Cauchy sequence in X and f is continuous on X. But $\{f(x_n)\}_{n\in\mathbb{N}} = \{n\}_{n\in\mathbb{N}}$ is not Cauchy.

8. If $f : X \to Y$ maps Cauchy sequences to Cauchy sequences, then f is continuous.

   Answer: True.

   Proof. We prove this by contra-positive arguments. Assume that f is discontinuous at some point $x \in X$. Then there exists $\varepsilon > 0$ such that for any $n \in \mathbb{N}$, there exists $s_n \in X$ with $0 < d(s_n, x) < \frac{1}{n}$ and $d(f(s_n), f(x)) \geq \varepsilon$. Define, for any $n \in \mathbb{N}$,

$$x_n := \begin{cases} s_n & \text{if } n \text{ is even} \\ x & \text{otherwise.} \end{cases}$$

   It is clear that the sequence $\{x_n\}_{n\in\mathbb{N}}$ is convergent to x, hence in particular is Cauchy, but $\{f(x_n)\}_{n\in\mathbb{N}}$ is not Cauchy.

9. All functions $f : X \to Y$, where X is any metric space and Y is a discrete metric space, are continuous.

   Answer: False.

   Example: Let $X := \mathbb{R}$ with Euclidean metric and $Y := \mathbb{R}$ with discrete metric. Consider the identity function $f : X \to Y, f(x) := x$ for all $x \in X$. As $\{0\} \subset Y$ is open but $f^{-1}(\{0\}) = \{0\} \subset \mathbb{R}$ is not open, f is not continuous.

10. All functions $f : X \to Y$, where X is a discrete metric space and Y is any metric space, are continuous.

    Answer: True.

    Proof. For any $x \in X$ and any $\varepsilon > 0$, we have

    $$f(B_X(x,1)) = f(\{x\}) = \{f(x)\} \subset B_Y(f(x), \varepsilon) .$$

    Hence f is continuous at x. As $x \in X$ is arbitrary, f is continuous on X.

    Alternatively, since all subsets of a discrete metric space are open, $f^{-1}(V) \subset X$ is open for any (open) subset $V \subset Y$. Hence f is continuous.

11. If $f : X \to Y$ is continuous, then $f(S^\circ) \subset f(S)^\circ$ for all $S \subset X$.

    Answer: False.

    Example: Let $X := \mathbb{R}$ with discrete metric and $Y := \mathbb{R}$ with Euclidean metric. Let $f : X \to Y$ be the identity mapping $f(x) := x$ for all $x \in X$. As X is discrete, by Exercise 2.3, Part A, Problem #10, f is continuous. Take $S := \mathbb{Q} \subset X$. Notice that $S^\circ = S$ and so $f(S^\circ) = f(S)$, but $f(S)^\circ = \mathbb{Q}^\circ$ (in $\mathbb{R}$ with the Euclidean metric) $= \phi$. Hence $f(S^\circ) \not\subset f(S)^\circ$.

12. If $f : X \to Y$ is such that $f(S^\circ) \subset f(S)^\circ$ for all $S \subset X$, then f is continuous.

    Answer: False.

Example: Let $X := \mathbb{R}$ and $Y := \{0, 1\}$ with discrete metric. Let $f : X \to Y$ be the function

$$f(x) := \begin{cases} 1 & x \geq 0 \\ 0 & x < 0 \,. \end{cases}$$

For any $S \subset X$, since $S^\circ \subset S$, we have $f(S^\circ) \subset f(S)$. But since Y is discrete, $f(S) = f(S)^\circ$ and so $f(S^\circ) \subset f(S)^\circ$. However, it is obvious that f is not continuous at 0.

13. If $f : X \to Y$ is continuous, then $f^{-1}(T^\circ) \subset f^{-1}(T)^\circ$ for all $T \subset Y$.

    Answer: True.

    Proof. Suppose f is continuous on X and $T \subset Y$. Since T° is an open subset of Y, $f^{-1}(T^\circ)$ is open in X. As $f^{-1}(T^\circ) \subset f^{-1}(T)$, we conclude that $f^{-1}(T^\circ) = \left(f^{-1}(T^\circ)\right)^\circ \subset f^{-1}(T)^\circ$.

14. If $f : X \to Y$ is such that $f^{-1}(T^\circ) \subset f^{-1}(T)^\circ$ for all $T \subset Y$, then f is continuous.

    Answer: True.

    Proof. Suppose $f^{-1}(T^\circ) \subset f^{-1}(T)^\circ$ for all $T \subset Y$. Let $V \subset Y$ be open. Then as $V^\circ = V$, we have $f^{-1}(V) = f^{-1}(V^\circ) \subset f^{-1}(V)^\circ$, which implies that $f^{-1}(V)^\circ = f^{-1}(V)$ and in particular, $f^{-1}(V)$ is open in X. Therefore, f is continuous on X.

15. If $f : X \to Y$ is continuous, then $f^{-1}(T^\circ) = f^{-1}(T)^\circ$ for all $T \subset Y$.

    Answer: Although by Exercise 2.3, Part A, Problem #13, we have $f^{-1}(T^\circ) \subset f^{-1}(T)^\circ$ for all $T \subset Y$, in general, we do not have the reverse inclusion relation. Hence the answer to this question is False.
    Example: Consider the function $f : [0, 1) \to S^1 \subset \mathbb{C}$ given by

    $$f(x) := e^{2\pi i x}\,, \qquad x \in [0, 1)\,.$$

It is obvious that f is continuous. Let

$$T := S^1 \cap \{\text{the closed First Quadrant}\} .$$

Then

$$T^\circ := S^1 \cap \{\text{the open First Quadrant}\}$$

and

$$f^{-1}(T)^\circ = \left[0, \frac{1}{4}\right]^\circ = \left[0, \frac{1}{4}\right) \not\subset \left(0, \frac{1}{4}\right) = f^{-1}(T^\circ) .$$

16. If $f : X \to Y$ is continuous, then $f(\overline{S}) \subset \overline{f(S)}$ for all $S \subset X$.

    Answer: True.
    Proof. Suppose f is continuous. Let $S \subset X$. Since $\overline{f(S)}$ is closed, its pre-image $f^{-1}(\overline{f(S)})$ is also closed. Since $S \subset f^{-1}(f(S)) \subset f^{-1}(\overline{f(S)})$, we have $\overline{S} \subset \overline{f^{-1}(\overline{f(S)})} = f^{-1}(\overline{f(S)})$ and so $f(\overline{S}) \subset \overline{f(S)}$.

17. If $f : X \to Y$ is a function such that $f(\overline{S}) \subset \overline{f(S)}$ for all $S \subset X$, then f is continuous.

    Answer: True.
    Proof. Suppose $f(\overline{S}) \subset \overline{f(S)}$ for all $S \subset X$. Then $\overline{S} \subset f^{-1}(\overline{f(S)})$ for all $S \subset X$. Let $T \subset Y$ be closed. Write $S := f^{-1}(T)$. Since $f(f^{-1}(T)) \subset T$, we have

$$\begin{aligned} \overline{f^{-1}(T)} = \overline{S} \subset f^{-1}(\overline{f(S)}) &= f^{-1}\left(\overline{f(f^{-1}(T))}\right) \\ &\subset f^{-1}\left(\overline{T}\right) = f^{-1}(T) . \end{aligned}$$

It follows that $f^{-1}(T)$ is closed and so f is continuous.

18. If $f : X \to Y$ is continuous, then $f(\overline{S}) = \overline{f(S)}$ for all $S \subset X$.

Answer: Although by Exercise 2.3, Part A, Problem #16, we have $f(\overline{S}) \subset \overline{f(S)}$ for all $S \subset X$, in general, we do not have the reverse inclusion relation. Hence the answer to this question is False.
Example: Consider the function $f : [0, 1) \to S^1 \subset \mathbb{C}$ given by

$$f(x) := e^{2\pi ix} , \qquad x \in [0, 1) .$$

It is obvious that f is continuous. Let $S := \left(\frac{3}{4}, 1\right) \subset [0, 1)$. Then $\overline{S} = \left[\frac{3}{4}, 1\right)$,

$$\begin{aligned} f(S) &= S^1 \cap \{\text{the open Fourth Quadrant}\} , \\ \overline{f(S)} &= S^1 \cap \{\text{the closed Fourth Quadrant}\} , \end{aligned}$$

while

$$f(\overline{S}) = (S^1 \cap \{\text{the open First Quadrant}\}) \cup \Big\{(0, -1)\Big\} .$$

Hence $f(\overline{S}) \not\supset \overline{f(S)}$.

19. Let $f : X \to Y$ be a continuous function. Then $\overline{f^{-1}(T)} \subset f^{-1}(\overline{T})$ for all $T \subset Y$.

Answer: True.
Proof. Suppose f is continuous. For any $T \subset Y$, $\overline{T}$ is closed and hence $f^{-1}(\overline{T})$ is closed. Clearly we have $f^{-1}(T) \subset f^{-1}(\overline{T})$. Since $\overline{f^{-1}(T)}$ is the smallest closed set containing in $f^{-1}(T)$, we conclude that $\overline{f^{-1}(T)} \subset f^{-1}(\overline{T})$.

20. If $f : X \to Y$ is a function such that $\overline{f^{-1}(T)} \subset f^{-1}(\overline{T})$ for all $T \subset Y$, then f is continuous.

Answer: True.
Proof. Suppose $\overline{f^{-1}(T)} \subset f^{-1}(\overline{T})$ for all $T \subset Y$. So when T is closed, we have

$$\overline{f^{-1}(T)} \subset f^{-1}(\overline{T}) = f^{-1}(T) \subset \overline{f^{-1}(T)} ,$$

which gives $\overline{f^{-1}(T)} = f^{-1}(T)$. Hence $f^{-1}(T)$ is closed whenever T is closed, which implies that f is continuous.

21. If $f : X \to Y$ is continuous, then $\overline{f^{-1}(T)} = f^{-1}(\overline{T})$ for all $T \subset Y$.

Answer: Although by Exercise 2.3, Part A, Problem #19, we have $\overline{f^{-1}(T)} \subset f^{-1}(\overline{T})$ for all $T \subset Y$, in general, we do not have the reverse inclusion relation. Hence the answer to this question is False.
Example: Consider the function $f : [0,1) \to S^1 \subset \mathbb{C}$ given by

$$f(x) := e^{2\pi ix} , \qquad x \in [0,1) .$$

It is obvious that f is continuous. Let

$$T := S^1 \cap \{\text{the open Fourth Quadrant}\} .$$

Then

$$\overline{T} := S^1 \cap \{\text{the closed Fourth Quadrant}\}$$

and $f^{-1}(T) = \left(\frac{3}{4}, 1\right)$, $\overline{f^{-1}(T)} = \left[\frac{3}{4}, 1\right)$, $f^{-1}(\overline{T}) = \left[\frac{3}{4}, 1\right) \cup \{0\}$. Hence $\overline{f^{-1}(T)} \not\supset f^{-1}(\overline{T})$.

22. Let $f : X \to Y$ be a continuous function. Then $f(S') \subset f(S)'$ for all $S \subset X$.

Answer: False.
Example: Let $f : \mathbb{R} \to \mathbb{R}$ be given by $f :\equiv 0$. Then f is continuous. Let $S := \mathbb{R}$. Then $f(S') = f(\mathbb{R}) = \{0\} \not\subset \phi = \{0\}' = f(S)'$.

23. Let $f : X \to Y$ be a continuous function. Then $f(S)' \subset f(S')$ for all $S \subset X$.

Answer: False.
Example: Let $X := \mathbb{R}$ with discrete metric and $Y := \mathbb{R}$ with Euclidean metric. Consider $f : X \to Y$ given by $f(x) := x$, $x \in X$. As X is discrete, it is obvious that f is continuous. Let $S := X$. Then $f(S)' = f(X)' = \mathbb{R}' = \mathbb{R} \not\subset \phi = f(\phi) = f(S')$.

24. Let $f : X \to Y$ be a continuous function. Then $f(\partial S) \subset \partial f(S)$ for all $S \subset X$.

    Answer: False.

    Example: Let $X := \mathbb{R}$, $Y := \{0\}$, and $f : X \to Y$ be the trivial function $f \equiv 0$. Then f is continuous. Let $S := \mathbb{Q} \subset X$. Then $f(\partial S) = f(\mathbb{R}) = \{0\} \not\subset \phi = \partial\{0\} = \partial f(S)$.

25. Let $f : X \to Y$ be a continuous function. Then $\partial f(S) \subset f(\partial S)$ for all $S \subset X$.

    Answer: False.

    Example: Let $X := (0, 1)$ and $Y := \mathbb{R}$, both with the Euclidean metric. Let $f : X \to Y$ be the identity function $f(x) := x$ for all $x \in X$. Clearly f is continuous. Let $S := X$. Then $\partial f(S) = \partial(0, 1) = \{0, 1\} \not\subset \phi = f(\phi) = f(\partial S)$.

26. A continuous function $f : \mathbb{R} \to \mathbb{R}$ must be either open or closed.

    Answer: False.

    Example: Let $f : \mathbb{R} \to \mathbb{R}$ be given by

    $$f(x) := \frac{x^2}{x^2 + 1}, \qquad x \in \mathbb{R}.$$

    Then $f(\mathbb{R}) = [0, 1) \subset \mathbb{R}$ and so in particular, it is neither open nor closed.

27. Any mapping from a metric space to a discrete metric space is necessarily open.

    Answer: True.

    Proof. This is clear since every subset of a discrete metric space is open. In fact, if Y is discrete and $f : X \to Y$ is a function, then every (open) subset $U \subset X$ is being mapped to $f(U) \subset Y$ which is open in Y.

28. An open mapping must be continuous.

    Answer: False.

    Example: Let $f : \mathbb{R} \to \{0, 1\} \subset \mathbb{R}$ be defined by

$$f(x) := \begin{cases} 0 & \text{if } x \geq 0 \\ 1 & \text{if } x < 0 \, . \end{cases}$$

    Clearly f is an open mapping but it is not continuous at 0.

29. A closed mapping must be continuous.

    Answer: False.

    Example: The function f in the Solution to Exercise 2.3, Part A, Problem #28 is a closed mapping but it is not continuous at 0.

    An alternative interesting counterexample for Problems #28 and #29:

    Let $S^1 = \{z \in \mathbb{C} : |z| = 1\}$ and $\arg(z) \in [0, 2\pi)$ denote the polar angle of $z \in S^1$. So we have a well-defined function *arg*: $S^1 \to [0, 2\pi)$. It is not difficult to see that *arg* is both open and closed but not continuous at the point $1 \in S^1$.

30. An open mapping must be closed.

    Answer: False.

    Example: Since open subsets of $(0, 1)$ are also open in $\mathbb{R}$, the identity function $f : (0, 1) \to \mathbb{R}$ is an open mapping. However, $(0, 1) \subset (0, 1)$ is closed, but $f\big((0, 1)\big) = (0, 1) \subset \mathbb{R}$ is not. Hence f is not a closed mapping.

31. A closed mapping must be open.

    Answer: False.

    Example: The zero function $f : \mathbb{R} \to \mathbb{R}$, $f :\equiv 0$, is clearly a closed mapping which is not open.

32. If $f : X \to Y$ is bijective and open, then f is continuous.

Answer: False.

Example: Consider $f : [0, \infty) \to \{-1\} \cup (0, \infty)$ given by

$$f(x) = \begin{cases} -1 & \text{if } x = 0 \ , \\ x & \text{if } x > 0 \ . \end{cases}$$

Then f is bijective, open, but it is not continuous at $x = 0$.

33. If $f : X \to Y$ is bijective and open, and Y is compact, then f is continuous.

    Answer: True.

    Prove. Since f is bijective, $f^{-1} : Y \to X$ exists. Since f is open, by Example 2.3.3, f^{-1} is a continuous function. For any closed subset $E \subset Y$, since Y is compact, E is also compact and so $f^{-1}(E) \subset X$ is compact and hence is closed in X. Hence f is continuous on X.

34. Let $\{A_n\}_{n=0}^{\infty}$ be a countable collection of closed subsets of a metric space X and $f : \bigcup_{n=0}^{\infty} A_n \to Y$. If $f\big|_{A_n}$ is continuous for each $n \in \mathbb{N}$, then f is continuous.

    Answer: False.

    Example: Let $X = Y := \mathbb{R}$, $A_0 := \{0\}$, $A_n := \left\{\frac{1}{n}\right\}$ for every $n \in \mathbb{N}$. Consider the function $f : \bigcup_{n=0}^{\infty} A_n \to Y$ defined by $f\big|_{A_0} := 1$ and $f\big|_{A_n} := 0$ for every $n \in \mathbb{N}$. Clearly $f\big|_{A_n}$ is continuous for each $n \in \mathbb{N}$, but f is not continuous at $0 \in \bigcup_{n=0}^{\infty} A_n$.

35. $f : X \to Y$ is continuous on X if and only if f is continuous on every compact subset of X.

    Answer: True.

    Proof. Necessity is clear. For the sufficiency, suppose that f is continuous on every compact subset of X. Fix $p \in X$. If p is an isolated point of X, then f is automatically continuous at p. Otherwise, p is an accumulation point of X. Hence there exists a sequence $\{x_n\}_{n\in\mathbb{N}}$ of distinct points in X with $\{x_n\} \to p$ as $n \to \infty$. Write $S := \{p\} \cup \{x_n : n \in \mathbb{N}\} \subset X$. Since every infinite sequence in S has an accumulation point (p to be precise) in S, S is sequentially compact. By Exercise 2.2, Part B, Problem #14, S is compact. By assumption, f is continuous at every point of S, hence in particular, continuous at p. Since p is arbitrary, f is continuous on X.

Part B: Problems

1. In each of the following situations, give an example of a function $f : S \to T$ such that it is continuous and onto, or else explain why there cannot be such an f:
   (a) $S = (0, 1)$, $T = (0, 1]$
   (b) $S = [0, 1] \cup [2, 3]$, $T = \{0, 1\}$
   (c) $S = [0, 1] \times [0, 1]$, $T = \mathbb{R}^2$

   Solution:
   (a) *Example*: Define $f(x) =: \begin{cases} 2x & x \in (0, \frac{1}{2}] \\ 1 & x \in (\frac{1}{2}, 1) \end{cases}$.
   It is clear that f is continuous on $(0, 1)$ and $f((0, 1)) = (0, 1]$.

   (b) *Example*: Define $f(x) := \begin{cases} 0 & x \in [0, 1] \\ 1 & x \in [2, 3] \end{cases}$.
   Note that the only open sets in T are $\phi, \{0\}, \{1\}, \{0, 1\}$. As
   $$f^{-1}(\phi) = \phi, \qquad f^{-1}(\{0\}) = [0, 1],$$
   $$f^{-1}(\{1\}) = [2, 3], \; f^{-1}(\{0, 1\}) = [0, 1] \cup [2, 3]$$
   are all open in S, we conclude that f is continuous on S.
   (c) Since S is compact, T is noncompact, and compactness should be preserved by continuous functions, there cannot be any continuous function mapping S onto T.

2. For any $A \subset X$, show that the function $f : X \to \mathbb{R}$ given by $f(x) := d(x, A)$ is continuous.

Proof. For any x, $y \in X$ and any $a \in A$, we have

$$d(x,a) \le d(x,y) + d(y,a)\ .$$

Thus

$$\begin{aligned} d(x,A) &= \inf_{a\in A} d(x,a) \le \inf_{a\in A}\ [d(x,y)+d(y,a)] \\ &= d(x,y) + \inf_{a\in A} d(y,a) \\ &= d(x,y) + d(y,A). \end{aligned}$$

Reversing x and y, we have $d(y,A) \le d(x,y) + d(x,A)$. Combining the two inequalities, we have

$$|f(x) - f(y)| = |d(x,A) - d(y,A)| \le d(x,y)$$

for all x, $y \in X$, which gives the continuity of f on X. In fact, for any $\varepsilon > 0$, let $\delta := \varepsilon$, then

$$|f(x) - f(y)| \le d(x,y) < \delta = \varepsilon$$

for all x, $y \in X$ with $d(x,y) < \delta$.

3. If $f : X \to Y$ is a bijective open mapping and Y is compact, show that f is continuous.

   Proof. Since f is bijective, $f^{-1} : Y \to X$ exists. By assumption, f is an open mapping and so for any open $S \subset X$, $(f^{-1})^{-1}(S) = f(S)$ is open in Y. Hence $f^{-1} : Y \to X$ is continuous. Let $T \subset Y$ be any closed subset of Y. Being a closed subset of the compact space Y, T is compact. By the continuity of f^{-1}, $f^{-1}(T)$ is compact and thus is closed. Therefore, f is continuous.

4. Let A, B be disjoint subsets of X.
   (a) If A is compact and B is closed, show that $d(A,B) > 0$.
   (b) Is (a) still valid if A, B are closed but not compact?

Proof.

(a) By Exercise 2.3, Part B, Problem #2, the function

$$\begin{aligned} d(\,\cdot\,, B) : X &\to \mathbb{R} \\ x &\mapsto d(x, B) \end{aligned}$$

is continuous. As A is compact, by Theorem 2.3.10, there exists $a_0 \in A$ such that

$$d(a_0, B) \le d(a, B) \qquad \text{for all } a \in A\,.$$

Since B is closed and $a_0 \notin B$, by Exercise 1.2, Part B, Problem #7,
we have $d(a_0, B) > 0$. Hence

$$d(a, b) \ge d(a, B) \ge d(a_0, B) > 0 \quad \text{for all } a \in A,\ b \in B\,.$$

Hence

$$d(A, B) = \inf_{a\in A, b\in B} d(a, b) \ge d(a_0, B) > 0\,.$$

(b) Without the compactness condition, (a) will no longer hold. *Example*: Consider $X := \mathbb{R}^2$,

$$\begin{aligned} A &:= \{(x, 0) : x \in \mathbb{R}\}\,, \\ B &:= \left\{\left(x\,, \frac{1}{x}\right) : x > 0\right\}\,. \end{aligned}$$

It is elementary to check that both A, B are closed in $\mathbb{R}^2$, $A \cap B = \phi$, and $d(A, B) = 0$.

5. For any $f,\ g \in C[a, b]$, define $h(x) := \min\{f(x), g(x)\}$ and $H(x) := \max\{f(x), g(x)\}$ for any $x \in [a, b]$. Show that H, $h \in C[a, b]$.

Proof. Since f, g are continuous, for any $x_0 \in [a,b]$ and $\varepsilon > 0$, there is $\delta > 0$ such that for any $x \in [a,b]$ with $|x - x_0| < \delta$, we have

$$f(x_0) - \varepsilon < f(x) < f(x_0) + \varepsilon,$$
$$g(x_0) - \varepsilon < g(x) < g(x_0) + \varepsilon.$$

Hence for any $x \in [a,b]$ with $|x - x_0| < \delta$,

$$\begin{aligned} h(x) &= \min\{f(x), g(x)\} < \min\{f(x_0) + \varepsilon, g(x_0) + \varepsilon\} \\ &= \min\{f(x_0), g(x_0)\} + \varepsilon = h(x_0) + \varepsilon \ , \end{aligned}$$

and

$$\begin{aligned} h(x) &= \min\{f(x), g(x)\} > \min\{f(x_0) - \varepsilon, g(x_0) - \varepsilon\} \\ &= \min\{f(x_0), g(x_0)\} - \varepsilon = h(x_0) - \varepsilon \ . \end{aligned}$$

Therefore, h is continuous at x_0. Since $x_0 \in [a,b]$ is arbitrary, we have $h \in C[a,b]$. Analogously, we also have $H \in C[a,b]$.

6. For any sequence $\{f_n\}_{n\in\mathbb{N}}$ in $C[a,b]$, define

$$h(x) := \inf_{n\in\mathbb{N}}\{f_n(x)\} \text{ and } H(x) := \sup_{n\in\mathbb{N}}\{f_n(x)\} \text{ for any } x \in [a,b] \, .$$

Determine whether H, $h \in C[a,b]$.

Answer. Neither h nor H is necessarily continuous on $[a,b]$.
Example: On $[0,1]$, define, for any $n \in \mathbb{N}$,

$$f_n(x) := \begin{cases} nx & 0 \le x \le \frac{1}{n} \\ 1 & \frac{1}{n} \le x \le 1. \end{cases}$$

Clearly, $f_n \in C[0,1]$ for all $n \in \mathbb{N}$. Since $f_n(0) = 0$ for all n, we have $H(0) = 0$. On the other hand, for any $x_0 \in (0,1]$, there exists $N \in \mathbb{N}$ such that $\frac{1}{N} < x_0$. Hence $f_N(x_0) = 1$ and so $H(x_0) = 1$.

Since this is true for all $x_0 \in (0,1]$, we have $H(x) = 1$ for all $x \in (0,1]$ and so in particular, H is not continuous at 0.
Finally, by considering the sequence of functions $\{-f_n\}_{n\in\mathbb{N}}$ in $C[0,1]$ and by analogous arguments, it is not hard to see that $h \notin C[0,1]$.

7. Let A be a nonempty subset of X. The *characteristic function* of A is the function $\chi_A : X \to \mathbb{R}$ defined by

$$\chi_A(x) = \begin{cases} 1 & \text{if } x \in A, \\ 0 & \text{if } x \in X \setminus A. \end{cases}$$

   Identify the points at which χ_A is discontinuous.

   Answer: χ_A is discontinuous on ∂A.
   Proof. First observe that χ_A is continuous on $X \setminus \partial A$. In fact, since $X = \overline{A} \uplus (X \setminus \overline{A}) = A^\circ \uplus \partial A \uplus (X \setminus \overline{A})$, we have $X \setminus \partial A = A^\circ \uplus (X \setminus \overline{A})$. So for any $x \in X \setminus \partial A$, we have $x \in A^\circ \cup (X \setminus \overline{A})$. If $x \in A^\circ$, then there exists $r > 0$ such that $B(x,r) \subset A^\circ$ and so $\chi_A \equiv 1$ on $B(x,r)$. If $x \in X \setminus \overline{A}$, then there exists $r > 0$ such that $B(x,r) \subset X \setminus \overline{A} \subset X \setminus A$ and so $\chi_A \equiv 0$ on $B(x,r)$. In either case, f is continuous at x and so χ_A is continuous on $X \setminus \partial A$.
   On the other hand, consider $x \in \partial A$. Take $\varepsilon = \frac{1}{2}$. For any $r > 0$, there exist $y \in B(x,r) \cap A$ and $z \in B(x,r) \cap X \setminus A$. If $x \in A$, then $|\chi_A(x) - \chi_A(z)| = |1 - 0| = 1 > \varepsilon$; if $x \in X \setminus A$, then $|\chi_A(x) - \chi_A(y)| = |0 - 1| = 1 > \varepsilon$. Combining, we see that for any $r > 0$, there are points in $B(x,r)$ with images under χ_A far away from that of x. Hence χ_A is not continuous at x.

8. Let $f : (0,1) \to \mathbb{R}$ be given by

$$f(x) := \begin{cases} 0 & x \in (0,1) \setminus \mathbb{Q} \\ \dfrac{1}{q} & x = \dfrac{p}{q} \in (0,1) \cap \mathbb{Q}\,,\ p, q \in \mathbb{N},\ (p,q) = 1. \end{cases}$$

   Determine the set of points $x \in (0,1)$ at which f is continuous.

Answer: f is continuous at all irrational points and discontinuous at all rational points in $(0, 1)$.

Proof. Consider $a \in (0,1) \cap \mathbb{R} \setminus \mathbb{Q}$. For any $\varepsilon > 0$, there are at most finitely many $q \in \mathbb{N}$ with $\frac{1}{q} \geq \varepsilon$, hence there are at most finitely many rational numbers of the form $\frac{p}{q}$ in $(0,1)$ with $p, q \in \mathbb{N}$, $(p,q) = 1$, such that $\frac{1}{q} \geq \varepsilon$. Clearly, each of these finitely many rational numbers $\frac{p}{q}$ has a positive distance from the irrational number a. Let $\delta :=$ the minimum of these finitely many distances. Then $\delta > 0$, and points in $B_{(0,1)}(a, \delta)$ are either irrational, or rational but of the form $\frac{p}{q}$ with p, $q \in \mathbb{N}$, $(p,q) = 1$, and $\frac{1}{q} < \varepsilon$. Hence for any $x \in B_{(0,1)}(a, \delta)$, we have

$$\begin{aligned}|f(x) - f(a)| &= |f(x)| \\ &= \begin{cases} 0 & \text{if } x \in \mathbb{R} \setminus \mathbb{Q} \\ \frac{1}{q} & \text{if } x = \frac{p}{q}, p, q \in \mathbb{N}, (p,q) = 1, \frac{1}{q} < \varepsilon \end{cases} \\ &< \varepsilon\end{aligned}$$

and so f is continuous at a.

Next, consider $a \in (0,1) \cap \mathbb{Q}$. Write $a = \frac{p_a}{q_a}$ with p_a, $q_a \in \mathbb{N}$, $(p_a, q_a) = 1$. Similar to the previous case, there are at most finitely many rational numbers of the form $\frac{p}{q}$ in $(0,1)$ with p, $q \in \mathbb{N}$, $(p,q) = 1$, such that $q \leq q_a$. Let $\delta :=$ the minimum of these finitely many distances. Then $\delta > 0$ and points in $B_{(0,1)}(a, \delta)$ are either irrational, or rational but of the form $\frac{p}{q}$ with p, $q \in \mathbb{N}$, $(p,q) = 1$, and $q > q_a$. Hence for any $x \in B_{(0,1)}(a, \delta)$, we have

$$\begin{aligned}&\left|f(x) - f(a)\right| = \left|f(x) - \frac{1}{q_a}\right| \\ =&\begin{cases} \left|0 - \frac{1}{q_a}\right| & \text{if } x \in \mathbb{R} \setminus \mathbb{Q} \\ \left|\frac{1}{q} - \frac{1}{q_a}\right| & \text{if } x = \frac{p}{q}, p, q \in \mathbb{N}, (p,q) = 1, q > q_a \end{cases} \\ =&\begin{cases} \frac{1}{q_a} & \text{if } x \in \mathbb{R} \setminus \mathbb{Q} \\ \frac{1}{q_a} - \frac{1}{q} & \text{if } x = \frac{p}{q}, p, q \in \mathbb{N}, (p,q) = 1, q > q_a \end{cases} \\ \geq& \frac{1}{q_a} - \frac{1}{q_a + 1} \quad \text{which is a fixed positive number,}\end{aligned}$$

hence f is discontinuous at a.

9. Let f, g be real valued functions on $[0,1]$ defined by

$$f(x) := \begin{cases} 0 & \text{if } x \notin \mathbb{Q} \\ 1 & \text{if } x \in \mathbb{Q}\,; \end{cases} \qquad g(x) := \begin{cases} 0 & \text{if } x \notin \mathbb{Q} \\ x & \text{if } x \in \mathbb{Q}\,. \end{cases}$$

(a) Show that f is not continuous anywhere in $[0,1]$.
(b) Show that g is continuous only at $x = 0$.
(c) If $h : [0,1] \to \mathbb{R}$ is such that $h(x) = c =$ constant for all $x \in \mathbb{Q}$, give a necessary and sufficient condition for h to be continuous on $[0,1]$. Justify your answer.

Proof.

(a) For any $x \in [0,1] \cap \mathbb{Q}$, take $\varepsilon := \frac{1}{2}$. For any $\delta > 0$, there exists $y \in B_{[0,1]}(x,\delta) \cap (\mathbb{R} \setminus \mathbb{Q})$, and we have $d(f(x), f(y)) = |1 - 0| = 1 > \varepsilon$. So f is not continuous at x.
Similarly, for any $x \in [0,1] \cap (\mathbb{R} \setminus \mathbb{Q})$, take $\varepsilon := \frac{1}{2}$. For any $\delta > 0$, there exists $y \in B_{[0,1]}(x,\delta) \cap \mathbb{Q}$, and we have $d(f(x), f(y)) = |0 - 1| = 1 > \varepsilon$. So f is not continuous at x.
Combining, we see that f is not continuous anywhere in $[0,1]$.

(b) For $x = 0$, for any $\varepsilon > 0$, take $\delta := \varepsilon$. Then we have $g(B_{[0,1]}(0,\delta)) = g([0,\delta)) = g([0,\varepsilon)) \subset [0,\varepsilon) \subset B_{\mathbb{R}}(0,\varepsilon) = B_{\mathbb{R}}(g(0),\varepsilon)$. Hence g is continuous at $x = 0$.
For any $x \in (0,1] \cap \mathbb{Q}$, take $\varepsilon := \frac{x}{2}$. For any $\delta > 0$, there exists $y \in B_{[0,1]}(x,\delta) \cap (\mathbb{R} \setminus \mathbb{Q})$, and we have $d(g(x), g(y)) = |x - 0| = x > \varepsilon$. So g is not continuous at x.
Similarly, for any $x \in (0,1] \cap (\mathbb{R} \setminus \mathbb{Q})$, take $\varepsilon := \frac{x}{2}$. For any $\delta \in (0, \frac{x}{2})$, there exists $y \in B_{[0,1]}(x,\delta) \cap \mathbb{Q}$, and we have $d(g(x), g(y)) = |0 - y| = y > x - \delta > \frac{x}{2} = \varepsilon$. So g is not continuous at x.
Combining, we conclude that g is continuous only at $x = 0$.

(c) h is continuous if and only if $h(x) = c$ for all $x \in [0,1]$.
In fact, it is easy to verify that the constant function $h \equiv c$ is continuous. Conversely, let $S = [0,1] \cap \mathbb{Q}$. Then $\overline{S} = [0,1]$ and so for every $x \in [0,1]$, there exists a sequence $\{x_n\}_{n\in\mathbb{N}}$ in S such that $\{x_n\} \to x$ as $n \to \infty$. By the continuity of

h, we have $h(x) = \lim_{n\to\infty} h(x_n) = c$. Hence $h(x) = c$ for all $x \in [0,1]$.

10. Let (X, d) be a metric space. Equip $X \times X$ with metric $D : (X \times X) \times (X \times X) \to \mathbb{R}$ defined by

$$D\big((x,y),(x',y')\big) := \max\{d(x,x'), d(y,y')\} .$$

(a) Show that the metric d, when considered as a real-valued function $d : X \times X \to \mathbb{R}$ on the metric space $X \times X$, is continuous.

(b) Let $f : X \to X$ be a continuous function. Show that $G := \{(x, f(x)) : x \in X\} \subset X \times X$ is closed.

Proof.

(a) It is elementary to verify that D is a well-defined metric on $X \times X$. Let $\{(x_n, y_n)\}_{n\in\mathbb{N}}$ be a sequence in $X \times X$ such that $\{(x_n, y_n)\} \to (x, y)$ as $n \to \infty$. For any $\varepsilon > 0$, there is $N \in \mathbb{N}$ such that $D((x_n, y_n), (x, y)) < \varepsilon$ for all $n \geq N$. Hence by triangle inequality,

$$\begin{aligned}
&|d(x_n, y_n) - d(x, y)| \\
&\leq |d(x_n, y_n) - d(x_n, y)| + |d(x_n, y) - d(x, y)| \\
&\leq d(y_n, y) + d(x_n, x) \\
&\leq 2D\big((x_n, y_n), (x, y)\big) \\
&< 2\epsilon \quad \text{for all } n \geq N .
\end{aligned}$$

Therefore, $\{d(x_n, y_n)\} \to d(x, y)$ as $n \to \infty$ and so d is continuous.

(b) Let $\{(x_n, f(x_n))\}_{n\in\mathbb{N}}$ be a convergent sequence in G, say, $\{(x_n, f(x_n))\} \to (x, y)$ as $n \to \infty$. By the definition of D, we see easily that $\{x_n\} \to x$ and $\{f(x_n)\} \to y$ as $n \to \infty$. As f is continuous, we have $\{f(x_n)\} \to f(x)$ as $n \to \infty$ and so $f(x) = y$. Hence $\{(x_n, f(x_n))\} \to (x, f(x)) \in G$ as $n \to \infty$. Therefore, G is closed in $X \times X$.

11. Suppose $S \subset X$ and $f : S \to Y$. The function f is said to be *Lipschitz continuous* if there is $M > 0$ such that

$$d_Y(f(s_1), f(s_2)) \leq M d_X(s_1, s_2) \quad \text{for all } s_1, s_2 \in S\ .$$

Let $f : S \to Y$ be Lipschitz continuous.

(a) Show that f is continuous.

(b) Prove or disprove that f maps Cauchy sequences in S to Cauchy sequences in Y.

(c) Suppose that Y is complete. Show that there is a continuous function $\tilde{f} : \overline{S} \to Y$ which *extends* f to $\overline{S}$ in the sense that $\tilde{f}\big|_S = f$.

Proof.

(a) Let $s \in S$. Suffices to show that if $\{s_n\}_{n\in\mathbb{N}}$ is a sequence in S convergent to s, then $\lim_{n\to\infty} f(s_n)$ exists and equals $f(s)$. For any $\varepsilon > 0$, as $\{s_n\} \to s$ as $n \to \infty$, there exists $N \in \mathbb{N}$ such that

$$d_X(s_n, s) < \frac{\varepsilon}{M} \qquad \text{for all } n \geq N\ .$$

Using the defining inequality of Lipschitz continuity, we have

$$d_Y(f(s_n), f(s)) \leq M d_X(s_n, s) < M\left(\frac{\varepsilon}{M}\right) = \varepsilon$$

for any $n \geq N$. Hence $\{f(s_n)\} \to f(s)$ as $n \to \infty$.

(b) True. [Compare with Exercise 2.3, Part A, Problem #7.]

Proof. Let $\{s_n\}_{n\in\mathbb{N}}$ be a Cauchy sequence in S. For any $\varepsilon > 0$, there is $N \in \mathbb{N}$ such that

$$d_X(s_n, s_m) < \frac{\varepsilon}{M} \qquad \text{for all } m, n \geq N\ .$$

By the defining inequality of Lipschitz continuity,

$$d_Y(f(s_n), f(s_m)) \leq M d_X(s_n, s_m) < M\left(\frac{\varepsilon}{M}\right) = \varepsilon$$

for any n, $m \geq N$. Hence $\{f(s_n)\}_{n\in\mathbb{N}}$ is Cauchy.

(c) For any $x \in \overline{S}$, there is a sequence $\{s_n\}_{n\in\mathbb{N}}$ in S which converges to x. Define $\tilde{f}(x) := \lim_{n\to\infty} f(s_n)$. Then

(i) $\tilde{f}$ is well-defined:

Observe that since $\{s_n\}_{n\in\mathbb{N}}$ is convergent, it is Cauchy in S. By (b), $\{f(s_n)\}_{n\in\mathbb{N}}$ is Cauchy in Y. Since Y is complete, it is convergent and so $\lim_{n\to\infty} f(s_n)$ exists. On the other hand, let $\{s_n\}_{n\in\mathbb{N}}$ and $\{t_n\}_{n\in\mathbb{N}}$ be sequences in S converging to the same point $x \in \overline{S}$. If $x \in S$, then as f is continuous on S, we have $\lim_{n\to\infty} f(s_n) = f(x) = \lim_{n\to\infty} f(t_n)$. If $x \in \overline{S} \setminus S$, note that all s_n's and t_m's are close to each other when m, n are large and so in particular, the sequence $\{x_n\}_{n\in\mathbb{N}}$ defined by

$$x_n := \begin{cases} s_k & \text{if } n = 2k+1 \\ t_k & \text{if } n = 2k \end{cases}$$

must be Cauchy in S. By (b), $\{f(x_n)\}_{n\in\mathbb{N}}$ is Cauchy in Y and hence is convergent. In particular, being subsequences of $\{f(x_n)\}_{n\in\mathbb{N}}$, both $\{f(s_n)\}_{n\in\mathbb{N}}$ and $\{f(t_n)\}_{n\in\mathbb{N}}$ are convergent and converge to the same limit. Thus $\tilde{f}(x)$ is uniquely defined. Hence $\tilde{f}$ is well-defined.

(ii) $\tilde{f}$ extends f:

Every $s \in S$ can be regarded as the limit of the constant sequence $\{s, s, \dots\}$. Hence $\tilde{f}(s) = \lim_{n\to\infty} f(s) = f(s)$.

(iii) $\tilde{f}$ is continuous:

This follows from the the definition of $\tilde{f}$.

12. Let $f : X \to Y$ be injective and continuous. Determine whether each of the following statements is true or false:

(a) $f^{-1}(\overline{S}) \subset \overline{f^{-1}(S)}$;

(b) $f(S^\circ) \subset f(S)^\circ$.

Proof.

(a) False.

Example: Let $X := [0, \infty)$, $Y := \mathbb{R}$, both with Euclidean metric. Consider $f(x) := x^2$ and $S := (-1, 0) \subset Y$. Clearly, f is continuous and injective on X. However,

$$f^{-1}(\overline{S}) = f^{-1}([-1, 0]) = \{0\} \not\subset \phi = \overline{\phi} = \overline{f^{-1}(S)} .$$

(b) False.

Example: Let $X := \mathbb{R}$, $Y := \mathbb{R}^2$, both with Euclidean metrics. Consider $f(x) := (x, 0)$ and $S := X$. Clearly, f is continuous and injective on X. However,

$$f(S^\circ) = f(\mathbb{R}) = \mathbb{R} \times \{0\} \not\subset \phi = (\mathbb{R} \times \{0\})^\circ = \left(f(\mathbb{R})\right)^\circ .$$

13. Let $f : X \to Y$ be a continuous function and $S \subset X$.
    (a) Show by an example that in general, $f(S') \not\subset f(S)'$.
    (b) Give an additional condition on f which can guarantee that $f(S') \subset f(S)'$.

Solution:

(a) *Example*: Consider the constant function $f : \mathbb{R} \to \mathbb{R}$ defined by $f :\equiv 0$. Then clearly f is continuous. Let $S := \mathbb{R}$. Then $f(S') = f(\mathbb{R}) = \{0\} \not\subset \phi = \{0\}' = f(S)'$.

(b) If $f : X \to Y$ is continuous and is injective, then $f(S') \subset f(S)'$ always holds.

Proof. For any $y \in f(S')$, there exists a unique $x \in S'$ such that $y = f(x)$. Since $x \in S'$, there is an infinite sequence of distinct points $\{s_n\}_{n\in\mathbb{N}}$ in $S\backslash\{x\}$ such that $\{s_n\} \to x$ as $n \to \infty$. Since f is injective and the s_n's are distinct, $\{f(s_n)\}_{n\in\mathbb{N}}$ is a sequence of distinct points in $f(S)$. Since f is continuous, $\{f(s_n)\} \to f(x) = y$ as $n \to \infty$. So $y \in f(S)'$.

Alternatively, for any $y \in f(S')$, there exists a unique $x \in S'$ such that $y = f(x)$. For any $\varepsilon > 0$, since f is continuous at

x, there exists $\delta > 0$ such that $f(B(x,\delta)) \subset B(y,\varepsilon)$. Since $x \in S'$, $B(x,\delta) \cap (S \setminus \{x\}) \neq \phi$. In particular, there exists $x \neq p \in B(x,\delta) \cap S$. Since f is injective, $f(p) \neq f(x) = y$ and so $f(p) \in B(y,\varepsilon) \cap f(S) \setminus \{y\}$. Hence $y \in f(S)'$.

14. Let $f : X \to Y$ be a continuous function and $S \subset X$.
    (a) Show by an example that in general, $f(\partial S) \not\subset \partial f(S)$.
    (b) Give an additional condition on f which can guarantee that $f(\partial S) \subset \partial f(S)$.

Solution:

(a) *Example*: Consider $f : \mathbb{R} \to \mathbb{R}$ defined by $f(x) := x^2$. Then clearly f is continuous. Let $S := [-1, 2]$. Then

$$f(\partial S) = f(\{-1,2\}) = \{1,4\} \not\subset \{0,4\} = \partial([0,4]) = \partial f(S)\,.$$

(b) If $f : X \to Y$ is continuous and is injective, then $f(\partial S) \subset \partial f(S)$ always holds.

Proof. For any $y \in f(\partial S)$, there exists a unique $x \in \partial S = \overline{S} \cap \overline{(X \setminus S)}$ such that $f(x) = y$. Since f is injective, we have

$$y \in f\left(\overline{S} \cap \overline{X \setminus S}\right) = f(\overline{S}) \cap f\left(\overline{X \setminus S}\right)\,.$$

By Exercise 2.3, Part A, Problem #16, we have $f\left(\overline{S}\right) \subset \overline{f(S)}$ and $f\left(\overline{X \setminus S}\right) \subset \overline{f(X \setminus S)}$. Since f is injective,

$$f(X \setminus S) = f(X) \setminus f(S) \subset Y \setminus f(S)\,.$$

Hence

$$\begin{aligned} y \in f(\overline{S}) \cap f\left(\overline{X \setminus S}\right) &\subset \overline{f(S)} \cap \overline{f\,(X \setminus S)} \\ &\subset \overline{f(S)} \cap \overline{Y \setminus f(S)} = \partial f(S)\,. \end{aligned}$$

15. Let $A \subset \mathbb{R}^n$ be any subset and $f : A \to A$ be a continuous function satisfying $d(f(x), f(y)) \geq d(x,y)$ for any $x, y \in A$, where d is usual Euclidean metric.

(a) Prove that f is injective and that the inverse function $f^{-1} : f(A) \to A$ is also continuous.
(b) If A is compact, prove that f is surjective.

Proof.

(a) For any $x, y \in A$, if $f(x) = f(y)$, then $0 = d(f(x), f(y)) \geq d(x, y)$. This forces $x = y$ and so f is injective. Hence the inverse function $f^{-1} : f(A) \to A$ exists.
Next, for any $u, v \in f(A)$, we have

$$d(f^{-1}(u), f^{-1}(v)) \leq d(f(f^{-1}(u)), f(f^{-1}(v))) = d(u, v) ,$$

from which the continuity of f^{-1} follows. [For any $\varepsilon > 0$, let $\delta := \varepsilon$. Then whenever $d(u, v) < \delta$, we have

$$d(f^{-1}(u), f^{-1}(v)) \leq d(u, v) < \delta = \varepsilon .$$

Thus f^{-1} is continuous.]

(b) Assume to the contrary that there exists $x \in A \setminus f(A)$. Since A is compact and f is continuous, $f(A)$ is compact. Hence $d(x, f(A)) > 0$. Write $d := d(x, f(A))$. Construct a sequence $\{x_k\}_{k\in\mathbb{N}}$ in A by

$$x_0 := x; \quad x_k := f(x_{k-1}) \text{ for any } k \in \mathbb{N} .$$

Note that $x_k \in f(A)$ for any $k \in \mathbb{N}$. Since A is compact in $\mathbb{R}^n$, by Exercise 2.1, Part B, Problem #8, it is sequentially compact in $\mathbb{R}^n$. Hence the sequence $\{x_k\}_{k\in\mathbb{N}}$ has a convergent subsequence $\{x_{k_j}\}_{j\in\mathbb{N}}$ which converges in A. This implies that there exists $m \in \mathbb{N}$ such that

$$d(x_{k_{m+1}}, x_{k_m}) < d . \tag{*}$$

On the other hand,

$$\begin{aligned} d(x_{k_{m+1}}, x_{k_m}) &= d(f(x_{k_{m+1}-1}), f(x_{k_m-1})) \\ &\geq d(x_{k_{m+1}-1}, x_{k_m-1}) . \end{aligned}$$

Repeating the process indefinitely, we obtain

$$\begin{aligned} d(x_{k_{m+1}}, x_{k_m}) &\geq d(x_{k_{m+1}-1}, x_{k_m-1}) \\ &\geq \cdots \\ &\geq d(x_{k_{m+1}-k_m}, x_0) \\ &= d(x_{k_{m+1}-k_m}, x) \ . \end{aligned}$$

Since $x_{k_{m+1}-k_m} \in f(A)$, we have

$$\begin{aligned} d(x_{k_{m+1}}, x_{k_m}) &\geq d(x_{k_{m+1}-k_m}, x) \\ &\geq d(x, f(A)) = d \ , \end{aligned}$$

which contradicts to (*).

16. (a) Let A, $B \subset X$ be disjoint nonempty closed sets of X. Define $f : X \to \mathbb{R}$ by

$$f(x) := \frac{d(x, A)}{d(x, A) + d(x, B)} \qquad x \in X \ .$$

Show that f is a continuous function on X, $0 \leq f \leq 1$, $f = 0$ precisely on A, and $f = 1$ precisely on B.

(b) For any $g : X \to \mathbb{R}$, the zero set $Z(g)$ of g is defined as $Z(g) := f^{-1}(\{0\})$. Show that every nonempty closed set $A \subset X$ can be written as the zero set of some continuous real-valued function on X.

(c) Show that for any pair of disjoint closed sets A, $B \subset X$, there is a pair of disjoint open sets V, $W \subset X$ such that $A \subset V$ and $B \subset W$.

Proof.

(a) First of all, by Exercise 2.3, Part B, Problem #2, both $d(x, A)$ and $d(x, B)$ are continuous. Furthermore, as $A \cap B = \phi$, $d(x, A) + d(x, B) > 0$ for all $x \in X$ and so the function f is well-defined and continuous on X. It is also evident that

$0 \le f \le 1$. Finally, by Exercise 1.2, Part B, Problem #7, we have

$$f(x) = 0 \iff d(x, A) = 0 \iff x \in \overline{A} = A$$
$$f(x) = 1 \iff d(x, B) = 0 \iff x \in \overline{B} = B .$$

(b) If $A = X$, then $A = Z(f)$ for $f \equiv 0$. If $A \neq X$, pick $p \in X \setminus A$ and define $f : X \to \mathbb{R}$ by

$$f(x) := \frac{d(x, A)}{d(x, A) + d(x, \{p\})} \qquad x \in X .$$

By (a), $Z(f) = A$.

(c) For any pair of disjoint closed subsets $A, B \subset X$, define $f : X \to [0, 1]$ by the formula in (a). Set $V := f^{-1}\left([0, \frac{1}{2})\right)$ and $W := f^{-1}\left((\frac{1}{2}, 1]\right)$. Then

$$A = f^{-1}(\{0\}) \subset f^{-1}\left([0, \frac{1}{2})\right) = V ,$$
$$B = f^{-1}(\{1\}) \subset f^{-1}\left((\frac{1}{2}, 1]\right) = W .$$

Furthermore, since f is continuous, V, W are open disjoint subsets of X.

17. Let $\pi : \mathbb{R}^2 \to \mathbb{R}$ be the natural projection of $\mathbb{R}^2$ onto the first coordinate, that is, $\pi(x, y) := x$ for any $(x, y) \in \mathbb{R}^2$.
   (a) Determine whether π is open on $\mathbb{R}^2$.
   (b) Determine whether π is closed on $\mathbb{R}^2$.

Solution:

(a) True.

Proof. For any open set $U \subset \mathbb{R}^2$ and any $x \in \pi(U) \subset \mathbb{R}$, there exists $y \in \mathbb{R}$ such that $(x, y) \in U$. Since $U \subset \mathbb{R}^2$ is open, there exists $r > 0$ such that $B_{\mathbb{R}^2}((x, y), r) \subset U$. Hence $x \in (x - r, x + r) = B_{\mathbb{R}}(x, r) \subset \pi\left(B_{\mathbb{R}^2}((x, y), r)\right) \subset \pi(U)$. Therefore, $\pi(U)$ is open and so π is open.

(b) False.
Example: $S := \{(x, \frac{1}{x}) : x > 0\} \subset \mathbb{R}^2$ is closed but $\pi(S) = (0, \infty) \subset \mathbb{R}$ is not closed.

18. If $f : X \to Y$ is open and continuous, must it be closed?

Answer: False.
Example: Let $X := (0,1) \subset \mathbb{R}$ and $Y := \mathbb{R}$ with the usual Euclidean metric. Consider $f(x) := x$, $x \in X$. As open sets in $(0,1)$ are open in $\mathbb{R}$, f is open. It is trivial that f is continuous. However, as the image of the closed set $(0,1) \subset X$ equals $(0,1) \subset Y$ which is not closed in Y, f is not closed.

19. If $f : X \to Y$ is closed and continuous, must it be open?

Answer: False.
Example: Let $X := \{0\} \subset \mathbb{R}$ and $Y := \mathbb{R}$ with the usual Euclidean metric. Consider $f(x) := x$, $x \in X$. Then it is evident that f is closed and continuous. However, as the image of the open set $\{0\} \subset X$ equals $\{0\} \subset Y$ which is not open in Y, f is not open.

20. If $f : X \to Y$ is open and closed, must it be continuous?

Answer: False.
Example: Let $X := \mathbb{R}$ with the usual Euclidean metric and let $Y := \mathbb{R}$ with the discrete metric. Consider $f(x) := x$, $x \in X$. As all subsets of Y are clopen, f is both open and closed. However, as the pre-image of the open set $\{0\} \subset Y$ equals $\{0\} \subset X$ which is not open in X, f is not continuous.

21. Let $f : X \to Y$ be a continuous function. Determine whether each the following statements is true or false:
    (a) If f is injective, then f preserves interior points, i.e., $f(S^\circ) \subset f(S)^\circ$ for all $S \subset X$.
    (b) If f is surjective, then f preserves interior points, i.e., $f(S^\circ) \subset f(S)^\circ$ for all $S \subset X$.

(c) If f is injective, then f preserves accumulation points, i.e., $f(S') \subset f(S)'$ for all $S \subset X$.

(d) If f is surjective, then f preserves accumulation points, i.e., $f(S') \subset f(S)'$ for all $S \subset X$.

(e) If f is injective, then f preserves isolated points, i.e., $f(S \setminus S') \subset f(S) \setminus (f(S))'$ for all $S \subset X$.

(f) If f is surjective, then f preserves isolated points, i.e., $f(S \setminus S') \subset f(S) \setminus (f(S))'$ for all $S \subset X$.

(g) If f is injective, then f preserves boundary points, i.e., $f(\partial S) \subset \partial f(S)$ for all $S \subset X$.

(h) If f is surjective, then f preserves boundary points, i.e., $f(\partial S) \subset \partial f(S)$ for all $S \subset X$.

Solution:

(a) *Answer*: False.

Example: Let $X := \mathbb{R}$ with the discrete metric and $Y := \mathbb{R}$ with the usual Euclidean metric. Consider $f(x) := x$, $x \in X$, and $S := [0, 1]$. It is clear that f is continuous and injective, but $f(S^\circ) = f([0, 1]) = [0, 1] \not\subset (0, 1) = f(S)^\circ$.

(b) *Answer*: False.

Example: The same Example as that in (a) applies (note that the function f is also surjective).

(c) *Answer*: True.

Proof. Let $a \in S'$. Then there is an infinite sequence of distinct points $\{x_n\}_{n\in\mathbb{N}}$ in S such that $\{x_n\} \to a$ as $n \to \infty$. By the injectivity of f, $\{f(x_n)\}_{n\in\mathbb{N}}$ is an infinite sequence of distinct points in $f(S)$. By the continuity of f, we have $\{f(x_n)\} \to f(a)$ as $n \to \infty$. Hence $f(a) \in f(S)'$.

(d) *Answer*: False.

Example: Let $X := \mathbb{R}$ with the usual Euclidean metric and $Y := \{0\}$. Consider $f :\equiv 0$ and $S := [0, 1]$. Then f is continuous and surjective, but $f(S') = f([0, 1]) = \{0\} \not\subset \phi = \{0\}' = f(S)'$.

(e) *Answer*: False.

Example: Let $X := \mathbb{R}$ with the discrete metric and $Y := \mathbb{R}$ with the usual Euclidean metric. Consider $f(x) := x$, $x \in X$, and $S := X$. Then it is clear that f is continuous and surjective, but $f(S \setminus S') = f(S \setminus \phi) = f(S) = \mathbb{R} \not\subset \phi = \mathbb{R} \setminus \mathbb{R} = f(S) \setminus (f(S))'$.

(f) *Answer*: False.

Example: The same Example as that in (e) applies (note that the function f is also surjective).

(g) *Answer*: True.

Proof. Since f is injective, we have

$$f(X \setminus S) = f(X) \setminus f(S) \subset Y \setminus f(S) .$$

As $\partial S = \overline{S} \cap \overline{X \setminus S}$, by Exercise 2.3, Part A, Problem#16, we have

$$\begin{aligned} f(\partial S) = f(\overline{S} \cap \overline{X \setminus S}) &= f(\overline{S}) \cap f(\overline{X \setminus S}) \\ &\subset \overline{f(S)} \cap \overline{f(X \setminus S)} \\ &\subset \overline{f(S)} \cap \overline{Y \setminus f(S)} = \partial f(S) . \end{aligned}$$

(h) *Answer*: False.

Example: Let $X := \mathbb{R}$ with the usual Euclidean metric and $Y := \{0\}$. Consider $f :\equiv 0$ and $S := [0, 1]$. Then f is continuous and surjective, but $f(\partial S) = f(\{0, 1\}) = \{0\} \not\subset \phi = \partial f(S)$.

22. Let $X = S^2 \setminus \{(0, 0, 1)\}$, where as usual, S^2 is the unit sphere in $\mathbb{R}^3$. By considering the function $f : X \to \mathbb{R}^2$ defined by

$$f(x, y, z) := \left(\frac{x}{1-z}, \frac{y}{1-z}\right) \quad \text{for all } (x, y, z) \in X ,$$

or otherwise, show that $X \cong \mathbb{R}^2$.

Proof. Write $f(x, y, z) = (u, v)$. Observe that

$$u^2 + v^2 = \frac{x^2 + y^2}{(1-z)^2} = \frac{1-z^2}{(1-z)^2} = \frac{1+z}{1-z} .$$

Solving for z, we get

$$z = \frac{u^2 + v^2 - 1}{1 + u^2 + v^2}$$

and hence

$$x = u(1-z) = \frac{2u}{1 + u^2 + v^2} ,$$
$$y = v(1-z) = \frac{2v}{1 + u^2 + v^2} .$$

This suggests that we set, for any $(u, v) \in \mathbb{R}^2$,

$$g(u, v) := \left(\frac{2u}{1 + u^2 + v^2} , \frac{2v}{1 + u^2 + v^2} , \frac{u^2 + v^2 - 1}{1 + u^2 + v^2} \right) .$$

It is a routine computation to check that $(f \circ g)(u, v) = (u, v)$ for all $(u, v) \in \mathbb{R}^2$ and $(g \circ f)(x, y, z) = (x, y, z)$ for all $(x, y, z) \in X$. Thus f is a bijection from X onto $\mathbb{R}^2$ with inverse function g. Clearly, both f and g are continuous and so the two spaces are homeomorphic.

23. For each of the following statements, determine whether it is true or false.
    (a) The spaces $(0, 1)$ and $\mathbb{R}$ with Euclidean metric are homeomorphic.
    (b) The unit circle $S^1 \subset \mathbb{R}^2$ with Euclidean metric is homeomorphic to $\mathbb{R}$ with Euclidean metric.
    (c) The spaces $(0, 1) \cup \{2\}$ and $(0, 1]$ with Euclidean metric are homeomorphic.

    Solution:

    (a) *Answer*: True.

    Proof. Consider the function $f : (0, 1) \to \mathbb{R}$ given by

$$f(x) := \tan\left(\pi\left(x - \frac{1}{2}\right)\right), \quad x \in (0, 1) .$$

It is evident that f is $1-1$, onto, continuous, with continuous inverse $f^{-1} : \mathbb{R} \to (0,1)$ given by

$$f^{-1}(y) := \frac{1}{\pi} \tan^{-1} y + \frac{1}{2}\,, \quad y \in \mathbb{R}\,.$$

Thus $(0,1) \cong \mathbb{R}$.

(b) *Answer*: False.

Justification. Being a closed and bounded subset of $\mathbb{R}^2$, S^1 is compact. Since compactness is a topological property, S^1 cannot be homeomorphic to the non-compact space $\mathbb{R}$.

(c) *Answer*: False.

Justification. Write $X = (0,1) \cup \{2\}$. Suppose there were a homeomorphism $f : X \to (0,1\,]$. Then as $(0,1) \subset X$ is closed, $f\big((0,1)\big) \subset (0,1\,]$ should also be closed. However, observe that

$$\begin{aligned} f\big((0,1)\big) &= (0,1\,] \setminus \{f(2)\} \\ &= \begin{cases} (0,1) & \text{if } f(2) = 1 \\ (0,a) \cup (a,1\,] & \text{if } f(2) = a \in (0,1)\,. \end{cases} \end{aligned}$$

In either case, $f\big((0,1)\big)$ is not closed in $(0,1\,]$.

24. Determine which of the following spaces are homeomorphic: (i) $[0,1\,]$, (ii) $(-1,1)$, (iii) $[-1,1\,]$, and (iv) $\mathbb{R}$. All are equipped with the Euclidean metric.

Solution: By Exercise 2.3, Part B, Problem #23, $(0,1) \cong \mathbb{R}$. But then it is clear that $(-1,1) \cong (0,1)$ via the homeomorphism $f(t) := \frac{t+1}{2}$, $t \in (-1,1)$, we have $(-1,1) \cong \mathbb{R}$. On the other hand, $[0,1\,] \cong [-1,1]$ via the homeomorphism $g(t) := 2t-1$, $t \in [0,1\,]$. Finally, $(-1,1) \ncong [-1,1\,]$. In fact, if there was a homeomorphism $h : [-1,1\,] \to (-1,1)$, then $(-1,1) = h[-1,1\,]$ has to be compact, which is absurd. Hence we have

$$[0,1\,] \cong [-1,1\,] \ncong (-1,1) \cong \mathbb{R}\,.$$

25. (a) If $f : [a,b] \to \mathbb{R}$ is one to one and continuous, $f(a) = c \neq d = f(b)$, describe the image set $f([a,b])$.
   (b) Determine whether the intervals $(-1,1)$ and $(-1,1]$ are homeomorphic to each other.

Solution:

(a) Suppose first that $c < d$. As c and d are in the range of f, by Intermediate value Theorem, all points in $[c,d]$ are in the range of f, that is, $f([a,b]) \supset [c,d]$. Suppose there exists $t \in [a,b]$ with $f(t) \notin [c,d]$, say $f(t) = p > d$. By Intermediate-value Theorem, for any $s \in (d,p)$, there exist $u \in (a,t)$ and $v \in (t,b)$ such that $f(u) = s = f(v)$, which contradicts to the injectivity of f. Therefore $f([a,b]) \subset [c,d]$. Hence $f([a,b]) = [c,d]$.

In case $d < c$, by analogous arguments, we get $f([a,b]) = [d,c]$.

(b) Suppose there exists a homeomorphism $f : (-1,1) \to (-1,1]$. Let $c \in (-1,1)$ be the unique point such that $f(c) = 1$. Pick any $x_0 \in (-1,c)$ and write $y_0 := f(x_0)$. Note that $y_0 \neq 1$ and so $y_0 \in (-1,1)$. Then $f : [x_0,c] \to \mathbb{R}$ is one to one, continuous, and with $f(x_0) = y_0 < 1 = f(c)$. By (a), $f([x_0,c]) = [y_0,1]$. On the other hand, if we choose $x_0 \in (c,1)$, we would get $f([c,x_0]) = [y_0,1] = f([x_0,c])$, which contradicts to the injectivity of f.

26. For each of the following statements, determine if it is true or false. If it is true, prove it; if not, give a counterexample.
   (a) Homeomorphisms preserve isolated points.
   (b) Homeomorphisms preserve boundedness.
   (c) Homeomorphisms preserve denseness.
   (d) Homeomorphisms preserve completeness.

Solution:

(a) *Answer*: True.

Proof. Let $f : X \xrightarrow{\cong} Y$ be a homeomorphism and $x \in X$ be an isolated point of X. Then there exists $r > 0$ such that $B_X(x, r) = \{x\}$. Since every homeomorphism is an open mapping, $\{f(x)\} = f(\{x\}) = f(B_X(x, r)) \subset Y$ is an open set in Y. Hence $f(x)$ is an isolated point of Y.

(b) *Answer*: False.

Example: By Exercise 2.3, Part B, Problem #23, $(0, 1) \cong \mathbb{R}$. $(0, 1)$ is bounded but $\mathbb{R}$ is not.

(c) *Answer*: True.

Proof. Let $U \subset X$ be a dense subset of X and $f : X \to Y$ be a homeomorphism. For any nonempty open set $V \subset Y$, since f is continuous, $f^{-1}(V)$ open in X. Since f is onto, $f^{-1}(V) \neq \phi$. Since $U \subset X$ is dense, we have $\overline{U} = X$ and so $U \cap f^{-1}(V) \neq \phi$. Since f is bijective, we have $f(U) \cap V = f(U \cap f^{-1}(V)) \neq \phi$. This shows $f(U)$ intersects every nonempty open subset of Y and so $f(U)$ is dense in Y.

(d) *Answer*: False.

Example: By Exercise 2.3, Part B, Problem #23, $(0, 1) \cong \mathbb{R}$. $\mathbb{R}$ is complete but $(0, 1)$ is not.

27. Let $n \in \mathbb{N}$. Consider the open unit ball

$$B^n := \{x \in \mathbb{R}^n : \|x\| < 1\} \subset \mathbb{R}^n$$

and the function $f : B^n \to \mathbb{R}^n$ defined by $f(x) := \dfrac{1}{1 - \|x\|} x$.

(a) Show that f is a continuous bijection.

(b) Determine whether B^n and $\mathbb{R}^n$ are homeomorphic.

Solution:

(a) By Example 2.3.6, f is continuous. Suppose there exist $x, y \in B^n$ with $f(x) = f(y)$. Observe that

$$\begin{aligned} f(x) = f(y) &\Longrightarrow \frac{x}{1 - \|x\|} = \frac{y}{1 - \|y\|} \\ &\Longrightarrow \frac{\|x\|}{1 - \|x\|} = \frac{\|y\|}{1 - \|y\|} \Longrightarrow \|x\| = \|y\| \,. \end{aligned}$$

Combining $\|x\| = \|y\|$ and $f(x) = f(y)$ yields $x = y$, as desired. Finally, for any $z \in \mathbb{R}^n$, the point

$$x := \frac{z}{1+\|z\|} \in B^n \quad \text{(how to come up with that?)}$$

will satisfy $f(x) = z$.

(b) As f is bijective, the inverse function $f^{-1} : \mathbb{R}^n \to B^n$ exists. By our consideration in (a), $f^{-1}(z) = \dfrac{z}{1+\|z\|}$. By Example 2.3.6 again, f^{-1} is also continuous. Hence $B^n \cong \mathbb{R}^n$.

28. Show that every metric space is homeomorphic to a bounded metric space.

Proof. Let (X, d) be a metric space. It is easily verified that $d_1 := \min\{1, d\}$ is also a well-defined metric on X. We claim that the identity function $f : (X, d) \to (X, d_1)$ is a homeomorphism. In fact, it is obvious that f is a bijection. On the other hand, since $d_1 \leq d$, whenever $\{x_n\}_{n\in\mathbb{N}}$ is a sequence in X which is convergent in d to a point $x \in X$, the image sequence $\{f(x_n)\}_{n\in\mathbb{N}} = \{x_n\}_{n\in\mathbb{N}}$ must be convergent in d_1 to x, which is just $f(x)$. Hence f is continuous. Conversely, if $\{x_n\}_{n\in\mathbb{N}}$ is a sequence in X with $\{x_n\} \to x$ in d_1 as $n \to \infty$, then for any $1 > \varepsilon > 0$, there exists $N \in \mathbb{N}$ such that

$$d_1(x_n, x) < \varepsilon < 1 \quad \text{for all } n \geq N\ .$$

But then as $d_1(x_n, x) = \min\{1, d(x_n, x)\}$, the last inequality implies $d(x_n, x) = d_1(x_n, x) < \varepsilon$ for all $n \geq N$ and so $\{x_n\} \to x$ in d as $n \to \infty$. Therefore, f^{-1} is also continuous.

29. Recall the definitions of metrics d_1 and d_∞ on $C[0, 1]$ as defined in Example 1.1.2. Determine whether the identity function $\iota : (C[0, 1], d_1) \to (C([0, 1], d_\infty))$ defined by $\iota(f) := f$ is a homeomorphism.

Answer: No.

Justification. It is clear that ι is bijective. However, ι is not continuous. In fact, consider, for any $n \in \mathbb{N}$, $f_n : [0,1] \to \mathbb{R}$ defined by $f_n(x) := x^n$ for any $x \in [0,1]$. Clearly, $f_n \in C[0,1]$ for all $n \in \mathbb{N}$. Observe that

$$d_1(f_n, 0) = \int_0^1 x^n\,dx = \frac{1}{n+1} \qquad \text{for all } n \in \mathbb{N}\,,$$
$$d_\infty(f_n, 0) = \sup\left\{|x^n| : x \in [0,1]\right\} = 1 \qquad \text{for all } n \in \mathbb{N}\,.$$

Thus $\{f_n\} \to 0$ in d_1 as $n \to \infty$ but $\{\iota(f_n)\} = \{f_n\} \not\to 0 = \iota(0)$ in d_∞. Therefore, by Theorem 2.3.5, we conclude that ι is not continuous.

[Note, however, that the inverse $\iota^{-1} : (C[0,1], d_\infty) \to (C[0,1], d_1)$ is continuous. In fact, if $\{f_n\} \to f$ in d_∞ as $n \to \infty$, then for any $\varepsilon > 0$, there exists $N \in \mathbb{N}$ such that

$$|f_n(x) - f(x)| \le d_\infty(f_n, f) < \varepsilon \quad \text{for all } x \in [0,1] \text{ and all } n \ge N$$

and so

$$d_1(f_n, f) = \int_0^1 |f_n(x) - f(x)|\,dx < \int_0^1 \varepsilon\,dx = \varepsilon \quad \text{for all } n \ge N\,.$$

Hence $\{\iota^{-1}(f_n)\} = \{f_n\} \to f = \iota^{-1}(f)$ in d_1. By Theorem 2.3.5, ι^{-1} is continuous.]

30. Let M_{22} be the space of all 2×2 matrices over $\mathbb{R}$. It is easily verified that

$$d_m(A, B) := \max_{1 \le i,j \le 2} \left|a_{ij} - b_{ij}\right|\,,$$
$$d_2(A, B) := \left(\sum_{1 \le i,j \le 2} (a_{ij} - b_{ij})^2\right)^{1/2}\,,$$

for any $A = (a_{ij})$ and $B = (b_{ij}) \in M_{22}$, are well-defined metrics on M_{22}. Determine whether (M_{22}, d_m) and (M_{22}, d_2) are homeomorphic.

Answer: Yes.

Proof. Let $A, B \in M_{22}$. On the one hand, since

$$(a_{ij} - b_{ij})^2 \leq d_m(A, B)^2 \qquad \text{for all } 1 \leq i, j \leq 2 ,$$

we have

$$\sum_{1 \leq i,j \leq 2} (a_{ij} - b_{ij})^2 \leq \sum_{1 \leq i,j \leq 2} d_m(A, B)^2 = 4\, d_m(A, B)^2 .$$

It follows that $d_2(A, B) \leq 2d_m(A, B)$. By Exercise 2.3, Part B, Problem #11, the identity function $\iota : (M_{22}, d_m) \to (M_{22}, d_2)$ is continuous.

On the other hand, for any $1 \leq i, j \leq 2$,

$$|a_{ij} - b_{ij}|^2 \leq \sum_{1 \leq p,q \leq 2} (a_{pq} - b_{pq})^2 .$$

It follows that $d_m(A, B) \leq d_2(A, B)$. By Exercise 2.3, Part B, Problem #11 again, $\iota^{-1} : (M_{22}, d_2) \to (M_{22}, d_m)$ is continuous. As ι is clearly $1 - 1$ and onto, it is a homeomorphism and hence the two metric spaces are homeomorphic.

31. Let $\mathcal{B}(X, \mathbb{R})$ be the set of all bounded functions functions of X into $\mathbb{R}$. Define a metric ρ on $\mathcal{B}(X, \mathbb{R})$ by

$$\rho(f, g) := \sup\{|f(x) - g(x)| : x \in X\} .$$

Prove that $(\mathcal{B}(X, \mathbb{R}), \rho)$ is complete.

Proof. Note that the same problem has been proved in Exercise 2.2, Part B, Problem #16, except that over there X was assumed to be a subspace of $\mathbb{R}$. Observe that exactly the same proof works when X is a general metric space. [The only reason why X was assumed to be a subspace of $\mathbb{R}$ over there is that the notion of continuous functions on a general metric space had not been introduced in §2.2.]

32. Show that every surjective isometry is a homeomorphism. That is, if $g : X \to Y$ is an isometry, then $g : X \to g(X)$ is a homeomorphism.
[In this case, X is said to be *isometrically imbedded* into Y by the mapping g, and g is said to be an *isometric imbedding* of *X into Y*.]

Proof. For any p, $q \in X$, if $g(p) = g(q)$, then

$$0 = d_Y\big(g(p), g(q)\big) = d_X(p, q)$$

and so $p = q$. Hence $g : X \to g(X)$ is bijective. In particular, $g^{-1} : g(X) \to X$ exists. As

$$d_Y(r, s) = d_Y\big(g(g^{-1}(r)), g(g^{-1}(s))\big) = d_X(g^{-1}(r), g^{-1}(s))$$

for any r, $s \in g(X)$, g^{-1} is also metric-preserving. Hence for any $x \in X$ and any $\varepsilon > 0$, if $z \in B_X(x, \varepsilon)$, we have

$$d_Y\big(g(z), g(x)\big) = d_X(z, x) < \varepsilon$$

and so $g(z) \in B_{g(X)}\big(g(x), \varepsilon\big)$. Thus

$$g\big(B_X(x, \varepsilon)\big) \subset B_{g(X)}\big(g(x), \varepsilon\big) \ .$$

In particular, this shows $g : X \to g(X)$ is continuous.
Conversely, by analogous arguments we have

$$B_{g(X)}\big(g(x), \varepsilon\big) \subset g\big(B_X(x, \varepsilon)\big)$$

and so $g^{-1} : g(X) \to X$ is also continuous. Hence the assertion.

Remark. In fact, it is not hard to show that actually we have

$$g\big(B_X(x, \varepsilon)\big) = B_{g(X)}\big(g(x), \varepsilon\big) \quad \text{and}$$
$$B_{g(X)}\big(g(x), \varepsilon\big) = g\big(B_X(x, \varepsilon)\big) \ ,$$

although we do not require these in the proof.

33. Let (X, d) be an arbitrary metric space, and $\mathcal{B}(X, \mathbb{R})$ the set of all bounded functions of X into $\mathbb{R}$ as defined in Exercise 2.3, Part B, Problem #31.
    (a) Fix $x_0 \in X$. For any $a \in X$, consider the function $\phi_a : X \to \mathbb{R}$ defined by

$$\phi_a(x) := d(x, a) - d(x, x_0) \qquad \text{for any } x \in X \ .$$

    Show that $\phi_a \in \mathcal{B}(X, \mathbb{R})$.
    (b) Show that X can be isometrically imbedded into $\mathcal{B}(X, \mathbb{R})$ (cf. Exercise 2.3, Part B, Problem #32.)

Proof.

(a) For any $x \in X$, we have

$$d(x, a) \leq d(x, x_0) + d(x_0, a) \ ,$$
$$d(x, x_0) \leq d(x, a) + d(a, x_0) \ ,$$

thus

$$|\phi_a(x)| = \big|d(x, a) - d(x, x_0)\big| \leq d(x_0, a) \ .$$

Hence ϕ_a is bounded on X and so $\phi_a \in \mathcal{B}(X, \mathbb{R})$.

(b) Define $\Phi : X \to \mathcal{B}(X, \mathbb{R})$ by

$$\Phi(a) := \phi_a \qquad \text{for any } a \in X \ .$$

Using the metric ρ on $\mathcal{B}(X, \mathbb{R})$ defined in Exercise 2.3, Part B, Problem #31, namely,

$$\rho(f, g) := \sup\{|f(x) - g(x)| : x \in X\}, \quad f, g \in \mathcal{B}(X, \mathbb{R}) \ ,$$

we have, by triangle inequality,

$$\begin{aligned} \rho(\Phi(a), \Phi(b)) &= \rho(\phi_a, \phi_b) \\ &= \sup\{|\phi_a(x) - \phi_b(x)| : x \in X\} \\ &= \sup\{|d(x, a) - d(x, b)| : x \in X\} \\ &\leq d(a, b) \ . \end{aligned}$$

On the other hand, we have trivially

$$\begin{aligned} d(a,b) &= |d(a,a) - d(a,b)| \\ &\le \sup\{|d(x,a) - d(x,b)| : x \in X\} \\ &= \sup\{|\phi_a(x) - \phi_b(x)| : x \in X\} \\ &= \rho(\phi_a, \phi_b) = \rho(\Phi(a), \Phi(b)) \ . \end{aligned}$$

Combining, we see that $d(a,b) = \rho(\Phi(a), \Phi(b))$ for all $a, b \in X$. That is, Φ is metric-preserving.

34. Prove that every metric space X can be isometrically imbedded into a complete metric space X^* in which X is dense.

Proof. By Exercise 2.3, Part B, Problem #31, $(\mathcal{B}(X,\mathbb{R}), \rho)$ is complete. By Exercise 2.3, Part B, Problem #33, the mapping $\Phi : X \to \mathcal{B}(X,\mathbb{R})$ defined there is metric-preserving. As a closed subset of the complete space $(\mathcal{B}(X,\mathbb{R}), \rho)$, $\overline{\Phi(X)}$ is complete. So

$$\Phi : X \to \Phi(X) \subset \overline{\Phi(X)}$$

is an isometric imbedding of X into the complete space $X^* := \overline{\Phi(X)}$ in which $\Phi(X)$ is dense.

35. (a) Consider two metric subspaces X and Y of $\mathbb{R}$ given by

$$X := \bigcup_{n=0}^{\infty} \Big((3n, 3n+1) \cup \{3n+2\} \Big); \ Y := \big(X \setminus \{2\} \big) \cup \{1\} \ .$$

Define $f : X \to Y$ by $f(x) := \begin{cases} 1 & \text{if } x = 2 \\ x & \text{if } x \neq 2 \ , \end{cases}$

and $g : Y \to X$ by $g(y) := \begin{cases} \frac{y}{2} & \text{if } y \in (0,1] \\ \frac{y-2}{2} & \text{if } y \in (3,4) \\ y - 3 & \text{if } y \in Y \setminus (0,4) \ . \end{cases}$

Show that both f and g are continuous bijections.

(b) If two metric spaces each of which is the image of the other under a bijective continuous function, must they be homeomorphic?

Proof.

(a) It is elementary to check that both f and g are bijections. Note that $f|_{(3n,3n+1)}$ and $f|_{\{3n+2\}}$ are continuous for all $n \geq 0$, so f is continuous on X. Similarly, $g|_{(0,1]}$, $g|_{(3n,3n+1)}$, and $g|_{\{3n+2\}}$ are continuous for all $n \geq 1$, so g is continuous on X.

(b) The statement is false.

Example: Consider the example in (a). So we know that $f : X \to Y$ and $g : Y \to X$ are bijective and continuous. However, X and Y are not homeomorphic. In fact, suppose there were a homeomorphism $h : X \xrightarrow{\cong} Y$. Since $(0,1]$ is a connected component of Y, $h^{-1}\big((0,1]\big)$ is a connected set in X. Furthermore, since h is a bijection, $h^{-1}\big((0,1]\big)$ is not a singleton and so $h^{-1}\big((0,1]\big)$ should lie in an interval of the form $(3k, 3k+1)$, $k \geq 0$. In particular, the point $z := h^{-1}(1) \in (3k, 3k+1)$. Since h is a homeomorphism, it maps intervals to intervals (why?). Hence $h\big((3k, z]\big)$ is some interval in $(0,1]$ containing 1, and $h\big([z, 3k+1)\big)$ is also some interval of $(0,1]$ containing 1. But then as h is $1-1$, this is simply impossible.

Chapter 3

Connectedness

In this chapter, we shall study another important concept of a metric space: connectedness. Naively, we all seem to understand what we mean by saying that two objects are connected or separated. For example, it is clear that the intervals $[0,1]$ and $[2,3]$ are "separated" or, equivalently, "not connected". However, for more subtle situations, the concept of "separatedness" or "connectedness" may not be as clear. For example, the intervals $[0,1]$ and $(1,2]$ are really disjoint. But they are back-to-back close to each other. In fact, the two of them together forms a single interval $[0,2]$ and they are just an artificial splitting of the interval. In that case, are they separated or connected? How about the intervals $[0,1)$ and $(1,2]$? How about the two sets $\{(x, \sin\frac{1}{x}) \in \mathbb{R}^2 : 0 < x \leq 1\}$ and $\{(0,0)\}$? In order to answer such questions, like what we have to do with other seemingly obvious but in reality subtle concepts, we have to rigorously formulate the concept of "connectedness". It turns out that it would lead to some beautiful results which in turn have some important feedback to the equivalence problem.

3.1 Connectedness

Unlike other terminologies, it is rather clumsy to define the concept of connectedness in a direct manner. Instead, we shall first define the concept of disconnectedness or separatedness, and then define connectedness as the state of being not disconnected.

Definition 3.1.1. *A metric space X is said to be disconnected (or separated) if $X = A \cup B$, where $A \neq \phi$, $B \neq \phi$, $A \cap B = \phi$, and both A, B are open in X. In this case, $\{A, B\}$ is called a separation* *of X. X is said to be connected if it is* **not** *disconnected. A subset $S \subset X$ is said to be connected if, when considered as a metric space itself under the induced metric, is connected.*

Examples 3.1.2.

(i) $X := [0,1] \cup [2,3]$ is disconnected.

Proof. As $[0,1]$ and $[2,3]$ are nonempty disjoint open subsets of X whose union is X, they form a separation of X.

(ii) $X := [0,1) \cup (1,2]$ is disconnected.

Proof. As $[0,1)$ and $(1,2]$ are nonempty disjoint open subsets of X whose union is X, they form a separation of X.

(iii) $X := \mathbb{R} \setminus \{0\}$ is disconnected.

Proof. As $(-\infty, 0)$ and $(0, \infty)$ are nonempty disjoint open subsets of X whose union is X, they form a separation of X.

(iv) $\mathbb{Q}$ is disconnected.

Proof. Take any $x \in \mathbb{R} \setminus \mathbb{Q}$. Then $\{(-\infty, x) \cap \mathbb{Q}, (x, \infty) \cap \mathbb{Q}\}$ is a separation of $\mathbb{Q}$.

(v) Every discrete metric space with more than one element is disconnected.

Proof. Let X be a discrete metric space with $\#(X) > 1$ and $a \in X$. Then $X \setminus \{a\} \neq \phi$ and so $\{\{a\}, X \setminus \{a\}\}$ is a separation of X.

(vi) Every metric space contains connected subsets.

Proof. Since a singleton cannot be split into two nonempty subsets, it is clear that every singleton is connected. In particular, every metric space contains connected subsets.

(vii) $S^1 \setminus \{\text{two distinct points on } S^1\} \subset \mathbb{R}^2$ is disconnected.

Proof. Without loss of generality, we consider $S^1 \setminus \{n, s\}$, where $n := (0,1)$ denotes the "north pole" and $s := (0,-1)$ the "south pole" of the unit circle S^1. Then

$$\Big\{S^1 \cap \{(x,y) \in \mathbb{R}^2 : x > 0\},\ S^1 \cap \{(x,y) \in \mathbb{R}^2 : x < 0\}\Big\}$$

is a separation of $S^1 \setminus \{n, s\}$.

(viii) $S^2 \setminus S^1 \subset \mathbb{R}^3$ is disconnected, here by S^1 we mean any great circle (i.e., circle of unit length) on the unit sphere S^2.
Proof. Without loss of generality, we take S^1 as the unit circle $\{(x, y, 0) : x^2 + y^2 = 1\}$ in the xy-plane in $\mathbb{R}^3$. Then

$$\Big\{S^2 \cap \{(x, y, z) \in \mathbb{R}^3 : z > 0\},\ \{S^2 \cap \{(x, y, z) \in \mathbb{R}^3 : z < 0\}\Big\}$$

is a separation of $S^2 \setminus S^1$.

Remark. By taking a careful look into Example 3.1.2 (iv), (v), (vi), (vii), (viii), it is that evident that in general, if a metric space X is disconnected, there could be more than one separation of X.

Corollary 3.1.3. *The following statements are all equivalent:*

(a) *X is disconnected, i.e., $X = A \cup B$, where A, $B \neq \phi$, $A \cap B = \phi$, $A \cup B = X$, and both A, B are open.*
(b) *$X = A \cup B$, where A, $B \neq \phi$, $A \cap B = \phi$, $A \cup B = X$, and both A, B are closed.*
(c) *There exists a proper nonempty clopen subset of X.*

Proof. *(a)* $\Rightarrow$ *(b)* : If $X = A \cup B$, where A, $B \neq \phi$, $A \cap B = \phi$, $A \cup B = X$, and both A, B are open, then as complements of each other, $B = X \setminus A$ and $A = X \setminus B$ are closed. Hence (b).

(b) $\Rightarrow$ *(c)* : If $X = A \cup B$, where A, $B \neq \phi$, $A \cap B = \phi$, $A \cup B = X$, and both A, B are closed, then $A = X \setminus B$ is open. As $B \neq \phi$, $A = X \setminus B \subsetneqq X$ and so A is a proper nonempty clopen subset of X.

(c) $\Rightarrow$ *(a)*: If there exists a proper nonempty clopen subset $A \subset X$, define $B := X \setminus A$. As A is proper, $B \neq \phi$. As A is closed, $B = X \setminus A$ is open. It is also clear that $A \cap B = \phi$ and $X = A \cup B$. Hence (a). □

In general, it is not difficult to prove that a metric space is disconnected, as all that is required is to produce one single separation of it. As expected, it is more difficult to show that a metric space is connected, as we need to show that there cannot be any separation of it, that is, we need to exhaust all possible splittings of the metric space into two portions and show that none could constitute a separation. To this end, the use of a continuous function with only two possible values would simplify the arguments enormously.

Definition 3.1.4. *Any continuous function from X into the space $\{0, 1\}$ with discrete metric is called a 2-valued function on X.*

Theorem 3.1.5. *X is connected if and only if the only possible 2-valued functions on X are constant functions.*

Proof. ($\Rightarrow$): Let $f : X \to \{0, 1\}$ be a 2-valued function. Since f is continuous, both $A := f^{-1}(0)$ and $B := f^{-1}(1)$ are open in X, $A \cup B = X$ and $A \cap B = \phi$. But since X is connected, this can happen only when one of A, B is empty. If $A = \phi$, then $X = B$ and so $f \equiv 1$. If $B = \phi$, then $X = A$ and so $f \equiv 0$.

($\Leftarrow$): We use contra-positive argument. If X is disconnected, then $X = A \cup B$, $A \neq \phi$, $B \neq \phi$, $A \cap B = \phi$, and both A, B are open in X. Define $f : X \to \{0, 1\}$ by

$$f(x) := \begin{cases} 0 & x \in A \\ 1 & x \in B \end{cases}.$$

As the only possible open sets in $\{0, 1\}$ are ϕ, $\{0\}$, $\{1\}$, and $\{0, 1\}$, with inverse images ϕ, A, B, and X, respectively, which are all open in X, the function f is continuous. Since $A \neq \phi$ and $B \neq \phi$, f is a non-constant 2-valued function on X. □

Theorem 3.1.6. Intervals in $\mathbb{R}$ are connected and they are all connected subsets of $\mathbb{R}$, that is, $I \subset \mathbb{R}$ is an interval if and only if it is connected.

Proof. ($\Rightarrow$): Let $I \subset \mathbb{R}$ be an interval. If I is a singleton, there is nothing to prove. So assume that I is a nondegenerate interval, that is, $\#(I) > 1$. Let $f : I \to \{0, 1\}$ be a 2-valued function on I. Suppose there are points $a < b$ in I such that $f(a) \neq f(b)$, say $f(a) = 0$, $f(b) = 1$. Let $t := \inf\{x \in [a, b] : f(x) = 1\}$. Then $t \in (a, b] \subset I$ and $f(x) = 0$ for all $x < t$, hence by the continuity of f, we have $f(t) = \lim_{x\to t^-} f(x) = 0$. On the other hand, by the definition of infimum, for any $n \in \mathbb{N}$, there exists $x_n \in \left[t, t + \frac{1}{n}\right) \cap I$ such that $f(x_n) = 1$. Note that $\{x_n\}_{n\in\mathbb{N}}$ is a sequence in I converging to t as $n \to \infty$. So by the continuity of f, we must have $f(t) = \lim_{x\to t} f(x) = \lim_{n\to\infty} f(x_n) = 1$, which is a contradiction. Hence f = constant and so by Theorem 3.1.5, I is connected.

($\Leftarrow$): If I is not an interval, there exist $a < c < b$ with $a, b \in I$ and $c \notin I$. Then $\{I \cap (-\infty, c),\ I \cap (c, \infty)\}$ is a separation of I and so I is disconnected. $\square$

Example 3.1.7. $S := [0, 1] \cup (1, 2] \subset \mathbb{R}$ is connected.

Proof. Although S is artificially split into two disjoint portions, it is actually the single interval $[0, 2]$ and so in particular, by Theorem 3.1.6, S is connected.

Definition 3.1.8. *Two subsets S, T of a metric space X are said to be separated if $S \cap \overline{T} = \phi$ and $\overline{S} \cap T = \phi$.*

Example 3.1.9.

(a) $[0, 1]$ and $\{2\}$ are separated subsets of $\mathbb{R}$.

(b) $[0, 1)$ and $(1, 2]$ are separated subsets of $\mathbb{R}$.

(c) $[0, 1]$ and $(1, 2]$ are not separated subsets of $\mathbb{R}$.

Theorem 3.1.10. *If S, $T \subset X$ are connected subsets which are not separated, then $S \cup T$ is connected.*

Proof. As S and T are not separated, without loss of generality, we may assume that $\overline{S} \cap T \neq \phi$. Let f be a 2-valued function on $S \cup T$. As S and T are connected, being 2-valued functions on connected sets, $f|_S$ and $f|_T$ are constant functions. Take $a \in \overline{S} \cap T$. There is a sequence $\{x_n\}_{n \in \mathbb{N}}$ in S such that $\{x_n\} \to a$ as $n \to \infty$. As $f|_S$ is constant, say $f \equiv 0$ on S, we have $f(x_n) = 0$ for all n. By the continuity of f, we have $f(a) = \lim_{n\to\infty} f(x_n) = 0$. But then as $a \in T$ and f is constant on T, we also have $f|_T \equiv f(a) = 0$ and so $f \equiv 0$ on $S \cup T$. By Theorem 3.1.5, $S \cup T$ is connected.

□

Example 3.1.11. We re-visit Example 3.1.7. In view of Example 3.1.9 (c), $[0, 1]$ and $(1, 2]$ are connected subsets of $\mathbb{R}$ which are not separated, so by Theorem 3.1.10, $[0, 1] \cup (1, 2]$ is connected.

Theorem 3.1.12. *If $S \subset X$ is connected, then every $T \subset X$ with $S \subset T \subset \overline{S}$ is connected.*

Proof. If $f : T \to \{0, 1\}$ is a 2-valued function on T, $f|_S$ is a 2-valued function on S. Since S is connected, by Theorem 3.1.5, $f|_S =$ constant, say 0. For any $x \in T \subset \overline{S}$, there exists a sequence $\{x_n\}_{n \in \mathbb{N}}$ in S such that $\{x_n\} \to x$ as $n \to \infty$. Since f is continuous, we have $f(x) = \lim_{n\to\infty} f(x_n) = 0$. Thus $f \equiv 0$ on T and so T is connected.

□

Theorem 3.1.13. *Connectedness is preserved by continuous functions, hence in particular, connectedness is a topological property.*

Proof. Let $f : X \to Y$ be continuous and $S \subset X$ a connected set. Let $g : f(S) \to \{0, 1\}$ be any 2-valued function on $f(S)$. Then $g \circ f|_S : S \to \{0, 1\}$ is a 2-valued function on S and hence must be constant on S. Therefore, g is constant on $f(S)$ and thus $f(S)$ is connected.

□

Examples 3.1.14.

(i) S^1 is connected.

Proof. Consider the function $f : [0, 2\pi] \to \mathbb{C}$ given by $f(x) := e^{ix}$, $x \in [0, 2\pi]$. It is clear that f is continuous and $f([0, 2\pi]) = S^1$. By Theorem 3.1.6, $[0, 2\pi]$ is connected. Hence, by Theorem 3.1.13, $S^1 = f\big([0, 2\pi]\big)$ is connected.

(ii) $S^1 \setminus \{\text{a point on } S^1\}$ is connected.

Proof. Without loss of generality, we work on $S^1 \setminus \{(1,0)\} \subset \mathbb{C}$. Consider again the function $f : (0, 2\pi) \to \mathbb{C}$ given by $f(x) := e^{ix}$, $x \in (0, 2\pi)$. As above, f is continuous and $f\big((0, 2\pi)\big) = S^1 \setminus \{(1,0)\}$ [note that it is indeed a homeomorphism of $(0, 2\pi)$ onto $S^1 \setminus \{(1,0)\}$]. By Theorem 3.1.6, $(0, 2\pi)$ is connected. Hence by Theorem 3.1.13, $S^1 \setminus \{(1,0)\} = f\big((0, 2\pi)\big)$ is connected.

(iii) Continuous functions map intervals onto intervals.

Proof. Let $f : \mathbb{R} \to \mathbb{R}$ be continuous and $I \subset \mathbb{R}$ be an interval. By Theorem 3.1.6, I is connected. By Theorem 3.1.13, $f(I) \subset \mathbb{R}$ is also connected. By Theorem 3.1.6 again, $f(I)$ is an interval. Hence continuous functions map intervals onto intervals.

Theorem 3.1.15. *The graph of a continuous function $f : I \to \mathbb{R}$ over an interval I is connected.*

Proof. Equip $I \times \mathbb{R}$ with the Euclidean metric on $\mathbb{R}^2$. It is easy to verify that the function $F : I \to I \times \mathbb{R}$ given by

$$F(x) := \big(x, f(x)\big) , \quad x \in I ,$$

is continuous. Note that the graph of f is precisely the image of I under the continuous function F and hence is connected by Theorem 3.1.13. □

Example 3.1.16. The set

$$T := \Big\{ \Big(x, \sin\frac{1}{x} \Big) : 0 < x \le 1 \Big\} \cup \Big\{ (0,0) \Big\} \subset \mathbb{R}^2$$

is connected.

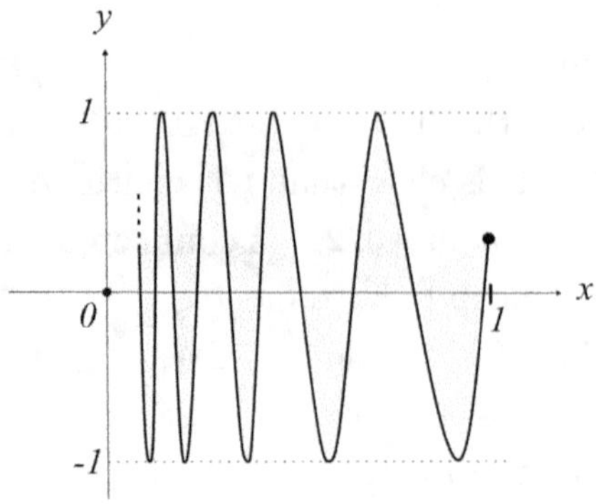

Proof. Observe first that the set

$$S := \left\{\left(x, \sin\frac{1}{x}\right) : 0 < x \leq 1\right\} \subset \mathbb{R}^2$$

is the graph of the continuous function $f : (0, 1] \to \mathbb{R}$ defined by $f(x) := \sin\frac{1}{x}$. By Theorem 3.1.15, S is connected. Note that

$$\overline{S} = \left\{\left(x, \sin\frac{1}{x}\right) : 0 < x \leq 1\right\} \cup \{(0, y) : -1 \leq y \leq 1\}$$

and $S \subset T \subset \overline{S}$. By Theorem 3.1.12, T is connected.

A useful result in elementary Calculus is the Intermediate Value Theorem, which states that if a continuous real-valued function assumes two values a, b in an interval I, then it will assume all values between a and b in I. This result can easily be extended to the context of continuous real-valued functions on a general metric space.

Theorem 3.1.17 (Intermediate Value Theorem). *Let $f : X \to \mathbb{R}$ be a continuous function on a connected metric space X. If f takes on two different values a, b in X, then f takes on every real number between a and b in X.*

Proof. It is obvious in view of Theorem 3.1.13 and Theorem 3.1.6. In fact, as f is continuous and X is connected, $f(X) \subset \mathbb{R}$ is connected and hence it is an interval. So if a, $b \in f(X)$, say $a < b$, then $[a, b] \subset f(X)$. □

Theorem 3.1.18. *The union of any collection of connected subsets of X with nonempty intersection is connected.*

Proof. Let $\{U_\alpha\}_{\alpha\in\Lambda}$ be a collection of connected subsets of X with $\bigcap_{\alpha\in\Lambda} U_\alpha \neq \phi$. Fix $t \in \bigcap_{\alpha\in\Lambda} U_\alpha$. Let $f : \bigcup_{\alpha\in\Lambda} U_\alpha \to \{0,1\}$ be a 2-valued function. Then for any $\beta \in \Lambda$, $f|_{U_\beta} : U_\beta \to \{0,1\}$ is a 2-valued function on U_β. Since U_β is connected, by Theorem 3.1.5, $f|_{U_\beta}$ must be constant. In particular, $f(x) = f(t)$ for all $x \in U_\beta$. Since this is true for all $\beta \in \Lambda$, $f(x) = f(t)$ for all $x \in \bigcup_{\alpha\in\Lambda} U_\alpha$. Hence f is a constant and so by Theorem 3.1.5, $\bigcap_{\alpha\in\Lambda} U_\alpha$ is connected. □

Remark. It is evident that the condition in Theorem 3.1.18 is excessively strong. For example, if there are three connected sets C_1, C_2, $C_3 \subset X$ such that $C_1 \cap C_2 \neq \phi$ and $C_2 \cap C_3 \neq \phi$, then even if $C_1 \cap C_2 \cap C_3 = \phi$, we can still conclude that $C_1 \cup C_2 \cup C_3$ is connected. So the condition for the Theorem could be suitably relaxed (see Exercise 3.1, Part B, #2, 3). However, as any relaxation of the condition would lead to more tedious and technical statements which may blur the vision, it is best to leave it as it is, as the readers could readily apply it to handle various situations.

Examples 3.1.19.

(i) We have seen in Example 3.1.14 (i) that S^1 is connected. An alternative proof goes as follows: By Example 3.1.14 (ii), the subsets $S^1 \setminus \{(0,1)\}$ and $S^1 \setminus \{(0,-1)\}$ are connected. Since $(S^1 \setminus \{(0,1)\}) \cap (S^1 \setminus \{(0,-1)\}) \neq \phi$, by Theorem 3.1.18, $S^1 = (S^1 \setminus \{(0,1)\}) \cup (S^1 \setminus \{(0,-1)\})$ is connected.

(ii) $B(a,r)$ and $\overline{B}(a,r) \subset \mathbb{R}^n$ are connected.

Proof. Every radius of $B(a,r)$ is clearly homeomorphic to the interval $[0,r) \subset \mathbb{R}$, which is connected. Hence all radii of $B(a,r)$ are connected. But then as the intersection of all radii of $B(a,r)$

is clearly nonempty, actually equals the singleton $\{0\} \subset \mathbb{R}^n$ to be precise, by Theorem 3.1.18, the union of all radii of $B(a,r)$, which is the entire $B(a,r)$, is connected.

The connectedness of $\overline{B}(a,r)$ can be derived by similar arguments, or by the observation that since we are working in $\mathbb{R}^n$, we have $\overline{B}(a,r) = \overline{B(a,r)}$ and so by Theorem 3.1.12, $\overline{B}(a,r)$ is connected.

(iii) In $\mathbb{R}^2$, let

$$A := \left\{(x,y) \in \mathbb{R}^2 : 0 < x \leq 1 \;,\; y = \sin\frac{1}{x}\right\},$$
$$B := \{(x,y) \in \mathbb{R}^2 : -1 \leq x \leq 0 \;,\; y = 0\},$$
$$X := A \cup B \;.$$

Then X is connected.

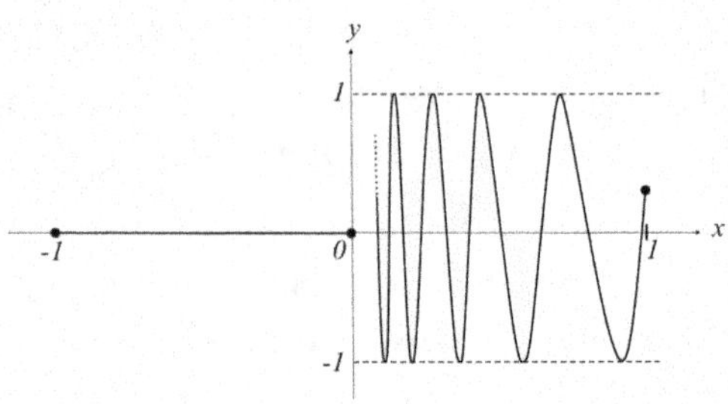

Proof. Let $f : X \to \{0,1\}$ be a 2-valued function. Then $f|_A$ and $f|_B$ are 2-valued functions on A and B, respectively. Since A, B are connected, $f|_A$ and $f|_B$ are constants. Say, $f \equiv 0$ on B. In particular, $f\big((0,0)\big) = 0$. Since f is continuous on X, $f \equiv 0$ in a neighborhood of $(0,0)$ in X. But then as each neighborhood of $(0,0)$ in X contains points in A, so $f = 0$ at these points of A. Since $f|_A =$ constant, we have $f \equiv 0$ in A and so $f \equiv 0$ on X. Thus X is connected.

Alternatively, in view of Example 3.1.16, the set $A \cup \{(0,0)\}$ is connected. On the other hand, as B is connected and

$(A \cup \{(0,0)\}) \cap B \neq \phi$, by Theorem 3.1.18, $X = A \cup B = (A \cup \{(0,0)\}) \cup B$ is connected.

Definition 3.1.20. *A maximal connected set in a metric space is called a (connected) component* of the metric space.

Remark. Fix $x \in X$. Clearly, x is contained in some nonempty connected subset of X. By Theorem 3.1.18, the union $U(x)$ of all connected subsets of X which contain x is also connected. Now as no connected subset of X which contains x could be larger than $U(x)$, we see that $U(x)$ is a connected component of X. Observe that connected components of X are either identical or disjoint. Hence every metric space X is decomposed into a disjoint union of connected components of X.

Corollary 3.1.21. (a) *Every component is closed.*

(b) *Components may not be open.*

(c) *Components of open subsets $S \subset \mathbb{R}^n$ are open in $\mathbb{R}^n$.*

Proof.

(a) If $S \subset X$ is a component, S is connected and so $\overline{S} \supset S$ must also be connected. By the maximality of S, we have $S = \overline{S}$.

(b) *Example*: $\{0\}$ is a component of $X := \left\{\frac{1}{n} : n \in \mathbb{N}\right\} \cup \{0\} \subset \mathbb{R}$ but it is not open in X.

(c) Let $S \subset \mathbb{R}^n$ be open and $T \subset S$ be a component of S. Then for any $x \in T$, there is an open ball $B_{\mathbb{R}^n}(x) \subset S \subset \mathbb{R}^n$. By Example 3.1.13, $B_{\mathbb{R}^n}(x)$ is connected. By the maximality of T, we must have $B_{\mathbb{R}^n}(x) \subset T$. Hence T is open. □

Theorem 3.1.22. *Every open set $S \subset \mathbb{R}^n$ can be expressed in one and only one way as a countable disjoint union of open connected sets, which are precisely the components of S.*

Proof. By the previous Remark and Corollary 3.1.21, S is a disjoint union of its components which are open sets in $\mathbb{R}^n$. By Lindelöf's

Theorem (Theorem 1.2.15), a countable sub-collection of all components of S forms a cover of S. But then as the collection of components are pairwisely disjoint, no proper sub-collection could cover S. Hence the original collection of components must already be countable. Finally, for uniqueness, suppose $S = \bigcup_{n=1}^{\infty} C_n$, where the C_n's are pairwisely disjoint open connected subsets. Observe that being connected, each C_n is lying in some component of S. Since all the C_n's are open, for each $m \in \mathbb{N}$, $C_m = S \setminus \bigcup_{n \neq m} C_n$ is also closed in S. Hence each C_m is clopen in S and so each C_m must be a component of S, for otherwise it would be a proper clopen subset of the component which contains it, which violates the fact that components are connected. Hence the decomposition of S into components is unique up to order. □

Remark. Recall that Theorem 1.2.16 states that every nonempty open subset of $\mathbb{R}$ can be expressed as the union of countably many pairwisely disjoint open intervals in $\mathbb{R}$, and that such a collection of intervals is unique. It has been remarked that since these open intervals are pairwisely disjoint, each of them is indeed maximal. Theorem 3.1.22 is the corresponding result in $\mathbb{R}^n$.

Exercise 3.1

Part A: True or False Questions

For each of the following statements, determine if it is true or false. If it is true, prove it. If it is false, give a counterexample or provide proper justification.

1. If S, $T \subset X$ are disjoint, then $S \cup T$ is disconnected.

   Answer: False.
   Example: Let $X := \mathbb{R}$, $S := [0, 1]$, $T := (1, 2)$. Clearly S and T are disjoint but $S \cup T = [0, 2)$ is connected.

2. If $S \subset X$ is connected, then $S \subset S'$.

   Answer: False.
   Example: Let $X := \mathbb{R}$ and $S := \{0\}$. Then S is connected but $S \not\subset \phi = S'$.

3. If $S \subset X$ is connected, then ∂S is connected.

   Answer: False.
   Example: Let $X := \mathbb{R}$ and $S := (0, 1)$. Then S is connected but $\partial S = \{0, 1\}$ is not.

4. If $S \subset X$ is connected, then $\overline{S} = S'$.

   Answer: False.
   Example: Let $X := \mathbb{R}$ and $S := \{0\}$. Then S is connected but $\overline{S} = \{0\} \neq \phi = S'$.

5. If $S \subset X$ and $\overline{S} = S'$, then S is connected.

   Answer: False.
   Example: Let $X := \mathbb{R}$ and $S :=$ the Cantor Set. Then by Exercise 1.4, Part B, Problem #4, $\overline{S} = S'$. But it is obvious that S is not connected.

6. If $S \subset X$ and S° is connected, then S is connected.

   Answer: False.

   Example: Let $X := \mathbb{R}$ and $S := (0,1) \cup \{2\}$. Then $S^\circ = (0,1)$ is connected but S is not.

7. If $S \subset X$ and $\overline{S}$ is connected, then S is connected.

   Answer: False.

   Example: Let $X := \mathbb{R}$ and $S := \mathbb{Q}$. Then $\overline{S} = \mathbb{R}$ is connected but S is not.

8. If $S \subset X$ and S' is connected, then S is connected.

   Answer: False.

   Example: Let $X := \mathbb{R}$ and $S := \mathbb{Q}$. Then $S' = \mathbb{R}$ is connected but S is not.

9. If $S \subset X$ and the set of isolated points $S \setminus S'$ is connected, then S is connected.

   Answer: False.

   Example: Let $X := \mathbb{R}$ and $S := (0,1) \cup \{2\}$. Then $S \setminus S' = \{2\}$ is connected but S is not.

10. If $S \subset X$ and ∂S is connected, then S is connected.

    Answer: False.

    Example: Let $X := \mathbb{R}$ and $S := \mathbb{Q}$. Then $\partial S = \mathbb{R}$ is connected but S is not.

11. If $S \subset X$ is connected, then S° is connected.

    Answer: False.

    Example: Let $X := \mathbb{R}^2$ and $S := \{(x,y) : xy \geq 0\} \subset X$. Then S is connected but $S^\circ = \{(x,y) : xy > 0\}$ is disconnected, with a separation $\big\{\{(x,y) : x > 0, y > 0\},\ \{(x,y) : x < 0, y < 0\}\big\}$.

12. If S, $T \subset X$ are connected and $S \cap T \neq \phi$, then $S \cap T$ is connected.

    Answer: False.
    Example: Let $X := \mathbb{R}^2$, $S := S^1$, and $T := \{0\} \times \mathbb{R}$. Then S, T are connected and $S \cap T = \{(0,1), (0,-1)\} \neq \phi$, but it is clearly not connected.

13. If S, $T \subset X$ are connected, then $S \cup T$ is connected.

    Answer: False.
    Example: Let $X := \mathbb{R}$, $S := (0,1)$, and $T = (1,2)$. Then S and T are connected but $S \cup T = (0,1) \cup (1,2)$ is not.

14. If $\{A, B\}$ is a separation of X and $S \subset X$ is connected, then either $S \subset A$ or $S \subset B$.

    Answer: True.
    Proof. As $A \cup B = X$, $A \cap B = \phi$ and both A, B are open in X, we have

$$(S \cap A) \cup (S \cap B) = S \cap (A \cup B) = S \cap X = S \ ,$$
$$(S \cap A) \cap (S \cap B) = S \cap (A \cap B) = S \cap \phi = \phi \ ,$$

    and both $S \cap A$ and $S \cap B$ are open in S. Since S is connected, we must have $S \cap A = \phi$ or $S \cap B = \phi$, for otherwise $S \cap A$ and $S \cap B$ will form a separation of S. Hence $S \subset B$ or $S \subset A$.

15. If $A \subset X$ is clopen and $S \subset X$ is connected, then either $S \subset A$ or $S \cap A = \phi$.

    Answer: True.
    Proof. It follows immediately from Exercise 3.1, Part A, Problem #14.

16. For any $S \subset X$, if $S^\circ \neq \phi$ and $\overline{S} \neq X$, then $X \setminus \partial S$ is disconnected.

Answer: True.
Proof. As $X = \overline{S} \cup (X \setminus \overline{S}) = S^\circ \cup \partial S \cup (X \setminus \overline{S})$, we have $(X \setminus \partial S) = S^\circ \cup (X \setminus \overline{S})$. As both S° and $X \setminus \overline{S}$ are open and nonempty, they form a separation of X.

17. Every open ball $B(a, r) \subset X$ is connected.

    Answer: False.
    Example: Let $X := \mathbb{R} \setminus \{0\}$. Then $B(1, 2) = (-1, 0) \cup (0, 3)$ is disconnected.

18. Let $S := \{(x, \sin \frac{1}{x}) : 0 < x \leq 1\} \subset \mathbb{R}^2$. Then $T := S \cup \{(0, 1)\} \subset \mathbb{R}^2$ is connected.

    Answer: True.
    Proof. It is readily seen that $\overline{S} = S \cup \{(0, y) : -1 \leq y \leq 1\}$. Being the graph of a continuous function over an interval, S is connected. Since $S \subset T \subset \overline{S}$, by Theorem 3.1.12, T is connected.

19. If $f : X \to Y$ is continuous and $T \subset Y$ is connected, then $f^{-1}(T)$ is connected.

    Answer: False.
    Example: Consider $f : S^1 \to \mathbb{R}$ defined by $f(x, y) := x$, which is clearly continuous on S^1. The interval $T := \left(-\frac{1}{2}, \frac{1}{2}\right) \subset \mathbb{R}$ is connected but $f^{-1}\left((-\frac{1}{2}, \frac{1}{2})\right)$ is separated into two disjoint open arcs on the unit circle and is hence disconnected.

20. Every finite metric space with at least two elements is disconnected.

    Answer: True.
    Proof. Write $X = \{x_1, \dots, x_n\}$, $n \geq 2$. Let $r := \min_{2 \leq i \leq n} d(x_i, x_1)$. Then $r > 0$ and $x_i \notin B(x_1, r)$ for every $i = 2, \dots, n$. That means $\{x_1\} = B(x_1, r) \subset X$ is open in X. But then being a singleton, $\{x_1\}$ is also closed in X. Hence it is a proper clopen nonempty subset of X and so X must be disconnected.

Alternatively, consider the function: $f : X \to \{0, 1\}$ defined by $f(x_1) = 0$ and $f(x_i) = 1$ for $i = 2, \ldots, n$. Similar to the first proof, it is evident that $\{x_i\}$ is open in X for every $i = 1, \ldots, n$ and so in particular, the inverse image of any open sets in $\{0, 1\}$ under f are open in X. That is, f is continuous. Hence X admits a non-constant 2-valued function and so it is disconnected.

21. Every connected metric space with at least 2 elements is uncountable.

    Answer: True.

    Proof. Let $a, b \in X$ be two distinct elements in X. Let $f : X \to \mathbb{R}$ be defined by

$$f(x) := d(x, a) \qquad \text{for all } x \in X .$$

Then f is continuous. Since X is connected, by Theorem 3.1.13, $f(X) \subset \mathbb{R}$ is also connected, hence an interval. Since $f(a) = 0$ and $f(b) = d(b, a) > 0$, by Theorem 3.1.17, $f(X)$ is a nondegenerate interval containing the interval $[0, d(b, a)]$. In particular, $f(X)$ is uncountable and hence X is uncountable.

Part B: Problems

1. Show that the following sets in $\mathbb{R}^2$ are connected:
   (a) $S := \left\{(x_1, x_2) : \frac{x_1^2}{4} + \frac{x_2^2}{9} = 1\right\} \subset \mathbb{R}^2$;
   (b) $T := \{x^T Ax : x \in \mathbb{R}^2,\ x^T x = 1\}$, where A is a 2×2 real matrix over $\mathbb{R}$. Here, as usual, $x \in \mathbb{R}^2$ is written in Cartesian coordinates as a column vector, and x^T stands for its transpose.

   Proof.
   (a) Let $f : \mathbb{R} \to \mathbb{R}^2$ be defined by $f(x) := (2\cos x, 3\sin x)$, $x \in \mathbb{R}$. It is clear that f is continuous on $\mathbb{R}$ and $f(\mathbb{R}) = S$. Since $\mathbb{R}$ is connected, $f(\mathbb{R})$ is connected. Hence S is connected.

(b) Let $g : \mathbb{R} \to \mathbb{R}^2$ be defined by $g(x) := (\cos x, \sin x)$, $x \in \mathbb{R}$. It is clear that g is continuous on $\mathbb{R}$ and $f(\mathbb{R}) = S^1 = \{x \in \mathbb{R}^2 : x^T x = 1\}$. Since $\mathbb{R}$ is connected, $f(\mathbb{R})$ is connected and so S^1 is connected. On the other hand, the function $h : S^1 \to \mathbb{R}$ given by $h(x) := x^T Ax$, $x \in S^1$, is also readily seen to be continuous on S^1 and $h(S^1) = T$. Since S^1 is connected, $T = h(S^1)$ is connected.

2. (a) Let $A_1, \dots, A_n$ be connected subsets of X such that $A_i \cap A_{i+1} \neq \phi$ for all $i = 1, \dots, n-1$. Show that $\bigcup_{i=1}^n A_i$ is connected.

(b) Does (a) remain true for a countable family of connected sets $\{A_i : i \in \mathbb{N}\}$?

Solution:

(a) Write $B_m := \bigcup_{i=1}^m A_i$ for any $m = 1, \dots, n$. Then $B_1 = A_1$ is connected. Assume that B_m is connected for some $m < n$. Then $B_m \cap A_{m+1} \supset A_m \cap A_{m+1} \neq \phi$. As B_m and A_{m+1} are connected, by Theorem 3.1.18, $B_{m+1} = B_m \cup A_{m+1}$ is connected. Hence by induction, $\bigcup_{i=1}^n A_i = B_n$ is connected.

(b) *Answer*: Yes, (a) remains valid.

Proof. Write $A := \bigcup_{i=1}^\infty A_i$. Suppose to the contrary that A is disconnected. Then A has a proper clopen nonempty subset U. Observe that for any $i \in \mathbb{N}$, $U \cap A_i$ is a clopen subset of A_i. Since A_i is connected, we have $U \cap A_i = \phi$ or $U \cap A_i = A_i$, i.e., $U \cap A_i = \phi$ or $U \supset A_i$. Let $k \in \mathbb{N}$ be the smallest integer such that $A_k \subset U$. Then we have $U \cap A_{k+1} \supset A_k \cap A_{k+1} \neq \phi$ and so $U \supset A_{k+1}$. Inductively, we have $U \supset A_i$ for all $i \geq k$. If $k = 1$, then $U \supset A_i$ for all $i \in \mathbb{N}$. This implies that $A = \bigcup_{i=1}^\infty A_i \subset U \subsetneqq A$, which is absurd. If $k > 1$, then $A_k \subset U$ and $A_{k-1} \cap U = \phi$. But then this implies that $A_k \cap A_{k-1} = \phi$, which is also absurd. Hence A is connected.

3. Let A_α, $\alpha \in \Lambda$, be connected subsets of X, where Λ is an arbitrary index set, such that A_γ and A_δ are not separated for any γ, $\delta \in \Lambda$. Show that $A := \bigcup_{\alpha\in\Lambda} A_\alpha$ is connected.

   Proof. Suppose A is disconnected. Then $A = S \cup T$ for some open nonempty subsets S, $T \subset A$ such that $S \cap T = \phi$. Fix $\alpha_0 \in \Lambda$. Then $A_{\alpha_0} \subset A$ and so by Exercise 3.1, Part A, Problem #14, either $A_{\alpha_0} \subset S$ or $A_{\alpha_0} \subset T$. Without loss of generality, suppose $A_{\alpha_0} \subset S$. For any $\beta \in \Lambda$, by assumption, A_{α_0} and A_β are not separated. By Theorem 3.1.10, $A_{\alpha_0} \cup A_\beta$ is a connected subset of A. By Exercise 3.1, Part A, Problem #14 again, either $A_{\alpha_0} \cup A_\beta \subset S$ or $A_{\alpha_0} \cup A_\beta \subset T$. As we know $A_{\alpha_0} \subset S$, this forces $A_{\alpha_0} \cup A_\beta \subset S$ and so in particular, $A_\beta \subset S$. Since $\beta \in \Lambda$ is arbitrary, we conclude that $A = \bigcup_{\beta\in\Lambda} A_\beta \subset S$ and so $T = \phi$, which is absurd.

4. If $h : [a, b] \to \mathbb{R}$ is continuous and $1 - 1$, show that h is monotone. In particular, $h([a, b]) = [h(a), h(b)]$ if h is monotonic increasing, and $h([a, b]) = [h(b), h(a)]$ if h is monotonic decreasing.

   Proof. Let $D := \{(s, t) : a \leq s < t \leq b\} \subset \mathbb{R}^2$. Geometrically, D is the triangular region in $\mathbb{R}^2$ with vertices (a, a), (a, b), and (b, b), excluding the hypotenuse. Hence in particular, D is connected. Define a function $H : D \to \mathbb{R}$ by $H(s, t) := h(s) - h(t)$. As h is continuous, H is continuous and so $H(D) \subset \mathbb{R}$ is connected, hence an interval. Since h is $1 - 1$, H is never zero and so $H(D) \subset \mathbb{R} \setminus \{0\}$. So either $H(D) \subset (-\infty, 0)$ in which case h is monotonic increasing, or $H(D) \subset (0, \infty)$ in which case h is monotonic decreasing.

5. A set $S \subset X$ is said to be *totally disconnected* if all connected components of S are singletons, that is, every subset $T \subset S$ with $\#(T) > 1$ is disconnected. Show that $\mathbb{Q} \subset \mathbb{R}$ is totally disconnected.

Proof. It is obvious that singletons of $\mathbb{Q}$ are connected. Consider any subset $A \subset \mathbb{Q}$ with at least two elements, say, $a < b$. Pick any $c \in \mathbb{R} \setminus \mathbb{Q}$ such that $a < c < b$. Then $\{(-\infty, c) \cap A, (c, \infty) \cap A\}$ is a separation of A and so A is disconnected. Hence $\mathbb{Q}$ is totally disconnected.

6. If $S \subset X$ is totally disconnected, determine whether $X \setminus S$ is connected.

   Answer: No, $X \setminus S$ need not be connected.

   Example: By Exercise 3.1, Part B, #5, $\mathbb{Q} \subset \mathbb{R}$ is totally disconnected but $\mathbb{R} \setminus \mathbb{Q}$ is clearly disconnected.

7. If $S \subset \mathbb{R}$ is totally disconnected, show that $S^\circ = \phi$.

   Proof. If $S^\circ \neq \phi$, pick $a \in S^\circ$. Then there exists $r > 0$ such that $B_{\mathbb{R}}(a, r) \subset S$. But $B_{\mathbb{R}}(a, r)$ is connected. Thus S contains a connected set which is not a singleton. But that contradicts the assumption that S is totally disconnected.

8. Two points $a, b \in X$ are said to be *connected* if there is a connected subspace of X containing both a and b. Prove that if every pair of points in X are connected, then X is connected.

   Proof. Suppose not, then there exists a proper clopen nonempty subset $U \subset X$. Pick $a \in U$, $b \in X \setminus U$. By assumption, there exists a connected subset $C \subset X$ containing a and b. But then $U \cap C$ is a proper nonempty clopen subset of C, contradicting to the connectedness of C.

   Alternatively, let $f : X \to \{0, 1\}$ be a 2-valued function on X. Fix $a \in X$. For any $x \in X$, there exists a connected set $C \subset X$ containing a and x. The restriction $f\big|_C : C \to \{0, 1\}$ is a 2-valued function on C, and hence $f\big|_C$ is constant. In particular, $f(x) = f(a)$. Since $x \in X$ is arbitrary, we conclude that $f \equiv f(a)$ on X and so X is connected.

9. Let $Y \subset X$.
   (a) For any $A \subset Y$, show that the closure of A in Y is $\overline{A} \cap Y$, where, as usual, $\overline{A}$ stands for the closure of A in X.
   (b) Prove that Y is disconnected if and only if there are nonempty subsets $A, B \subset Y$ such that $A \cup B = Y$, $\overline{A} \cap B = \phi$, and $\overline{B} \cap A = \phi$.

Proof.

(a) Let B be the closure of A in Y. So B is the smallest closed set in Y containing A. Since $\overline{A}$ is closed in X, $\overline{A} \cap Y$ is closed in Y and it is obvious that $A \subset \overline{A} \cap Y$. Hence $B \subset \overline{A} \cap Y$.
On the other hand, since B is closed in Y, we can write $B = C \cap Y$ for some $C \subset X$ that is closed in X. Since $A \subset B \subset C$, we have $\overline{A} \subset \overline{C} = C$. Thus, $B = C \cap Y \supset \overline{A} \cap Y$.
Combining, we conclude that $B = \overline{A} \cap Y$.

(b) ($\Rightarrow$): Suppose Y is disconnected. Then there exist nonempty subsets $A, B \subset Y$, both closed in Y, such that $A \cap B = \phi$ and $A \cup B = Y$. By (a), we have

$$A = \text{ [the closure of } A \text{ in } Y] = \overline{A} \cap Y .$$

Thus

$$\overline{A} \cap B = \overline{A} \cap (Y \cap B) = (\overline{A} \cap Y) \cap B = A \cap B = \phi.$$

Similarly, $\overline{B} \cap A = \phi$.
($\Leftarrow$) Suppose there are nonempty subsets $A, B \subset Y$ such that $A \cup B = Y$, $\overline{A} \cap B = \phi$, and $\overline{B} \cap A = \phi$. Then

$$A = A \cup \phi = A \cup (\overline{A} \cap B) = (A \cup \overline{A}) \cap (A \cup B) = \overline{A} \cap Y$$

and so by (a), $A =$ the closure of A in Y, that is, A is closed in Y. Similar arguments shows that B is closed in Y. Since it is obvious that $A \cap B = \phi$, we conclude that Y is disconnected.

10. Show that X is connected if and only if every nonempty, proper subset S of X has a nonempty boundary.

Proof. Note first that for any $S \subset X$, S is clopen if and only if $S = \overline{S} = S^{\circ}$ if and only if $\partial S = \phi$. Hence X is disconnected if and only if there exists a nonempty proper clopen subset $S \subset X$ if and only if there exists a nonempty proper subset $S \subset X$ with $\partial S = \phi$. So by contrapositive, X is connected if and only if no proper subset S of X can have nonempty boundary if and only if every nonempty proper subset S of X must have nonempty boundary.

11. Show that every connected clopen subset of a metric space is a connected component.

Proof. Let $S \subset X$ be a connected clopen subset. For any $T \subset X$ with $S \subsetneqq T$, as S is clopen in X, $S = T \cap S$ is clopen in T. Having a proper nonempty clopen subset S, the set T must be disconnected. Hence S is a connected component.

12. Identify all the components of the following metric spaces:
    (a) $X = A \cup B \subset \mathbb{R}^2$, where $A := \{(x, 0) : x \in \mathbb{R}\}$ and $B := \{(x, \frac{1}{x}) : x > 0\}$.
    (b) $X = P \cup Q \cup R \subset \mathbb{R}^2$, where $P := \{(x, \sin\frac{1}{x}) : 0 < x \leq 1\}$, $Q := \{(x, 0) : -1 \leq x \leq 0\}$, $R := \{(0, y) : -1 \leq y \leq 1\}$.

Solution:
    (a) Observe first that both A and B are connected. Furthermore, it is elementary to see that both A and B are closed and $A \cap B = \phi$, so $A = X \setminus B$ and $B = X \setminus A$ are also open in X. Being connected clopen subsets of X, by Exercise 3.1, Part B, Problem #11, both A and B are connected components of X.
    (b) By Example 3.1.19 (iii), $P \cup Q$ is connected. Next, being a line segment, R is also connected. Furthermore, since $(P \cup Q) \cap R = \{(0, 0)\} \neq \phi$, by Theorem 3.1.18, $X = P \cup Q \cup R$ is connected. So X has only one connected component, namely, X itself.

13. Let $h : S^1 \to \mathbb{R}$ be continuous. Show that there exists a pair of antipodal points of S^1 which have the same image under h.

That is, there exists $(x, y) \in S^1$ such that $h(x, y) = h(-x, -y)$.

Proof. If $h(1, 0) = h(-1, 0)$, then we are done. So assume that $h(1, 0) \neq h(-1, 0)$. Consider the function $H : S^1 \to S^1$ given by $H(x, y) := h(x, y) - h(-x, -y)$ for any $(x, y) \in S^1$. Clearly we have $H(-1, 0) = -H(1, 0) \neq 0$, so exactly one of $H(1, 0)$ and $H(-1, 0)$ is positive, and the other negative. By Intermediate Value Theorem, there exists $(x_0, y_0) \in S^1$ such that $H(x_0, y_0) = 0$. That is, $h(x_0, y_0) = h(-x_0, -y_0)$.

14. (a) Let $I = [0, 1]$ and $f : I \to I$ be any continuous function. Show that f has a *fixed point* in I, that is, there exists $t \in I$ such that $f(t) = t$.
    (b) Is (a) still valid in case $I = (0, 1]$?
    (c) Is (a) still valid in case $I = [0, \infty)$?

Solution:

(a) First observe that if $f(0) = 0$ or $f(1) = 1$, then there is nothing to prove. So assume $f(0) > 0$ and $f(1) < 1$. Consider the function $g : I \to \mathbb{R}$ given by

$$g(x) := f(x) - x \qquad \text{for any } x \in I\ .$$

Then g is continuous, $g(0) = f(0) - 0 > 0$, and $g(1) = f(1) - 1 < 0$. By Intermediate Value Theorem, there is a point $t \in I$ such that $g(t) = 0$, or $f(t) = t$.

(b) *Answer*: No, (a) will no longer be valid.
Example: The function $f : (0, 1] \to (0, 1]$ given by $f(x) := \frac{x}{2}$, $x \in (0, 1]$, is continuous on $(0, 1]$ but with no fixed point there. In fact, solving the equation $f(x) = x$ we get $x = 0$ which is not in $(0, 1]$. So f has no fixed point in $(0, 1]$.

(c) *Answer*: No, (a) will no longer be valid.
Example: The function $f : [0, \infty) \to [0, \infty)$ given by $f(x) := x + 1$, $x \in [0, \infty)$, is continuous on $[0, \infty)$ but with no fixed point there.

15. If X is a metric space such that every continuous function $f : X \to \mathbb{R}$ has the intermediate value property (i.e., if $y_1, y_2 \in f(X)$ and y is a real number between y_1 and y_2, then there exists an $x \in X$ such that $f(x) = y$), show that X is connected.

Proof. Assume X is disconnected. Then there exist nonempty open subsets $A, B \in X$ such that $A \cup B = X$ and $A \cap B = \phi$. Define $f : X \to \mathbb{R}$ by $f :\equiv 0$ on A and $f \equiv 1$ on B. Since the pre-image of any open set in $\mathbb{R}$ under f must be one of ϕ, A, B, or X, which are all open, f is continuous. However, it is evident that f does not have the intermediate value property. Hence X must be connected.

Alternatively, Let $g : X \to \{0, 1\}$ be a 2-valued function on X. By assumption, g has the intermediate value property. But this is possible only if $g \equiv 0$ or $g \equiv 1$. That is, g must be a constant and so X is connected.

16. Recall that two subsets $A, B \subset X$ are *separated* if each is disjoint from the closure of the other, i.e., if $B \cap \overline{A} = \phi = A \cap \overline{B}$.
    (a) Are disjoint subsets always separated?
    (b) If two sets are separated, must their closures be disjoint?
    (c) If $A, B \subset X$ are separated and $C \subset A \cup B$ is connected, what relation must A, B and C have?

Solution:

(a) *Answer*: No.

Example: $A = (-\infty, 0]$ and $B = (0, \infty) \subset \mathbb{R}$ are disjoint, but they are not separated, as $A \cap \overline{B} = \{0\} \neq \phi$.

(b) *Answer*: No.

Example: $A = (-\infty, 0)$ and $B = (0, \infty) \subset \mathbb{R}$ are separated, but $\overline{A} \cap \overline{B} = \{0\} \neq \phi$.

(c) *Answer*: $C \subset A$ or $C \subset B$.

Proof. Suppose A, $B \subset X$ are separated and $C \subset A \cup B$ is connected. Then

$$\overline{(C \cap A)} \cap (C \cap B) \subset \overline{A} \cap B = \phi\,,$$
$$(C \cap A) \cap \overline{(C \cap B)} \subset A \cap \overline{B} = \phi\,,$$

and so $C \cap A$ and $C \cap B$ are separated. Furthermore, it is obvious that $(C \cap A) \cup (C \cap B) = C$. Hence if both $C \cap A$ and $C \cap B$ are nonempty, then by Exercise 3.1, Part B, Problem #9, C is disconnected, which is absurd. Therefore, we must have either $C \cap A = \phi$ or $C \cap B = \phi$, that is, either $C \subset B$ or $C \subset A$.

17. (a) If $f : X \xrightarrow{\cong} Y$ is a homeomorphism and $x_0 \in X$, then the restriction function $f|_{X \setminus \{x_0\}} : X \setminus \{x_0\} \to Y \setminus \{f(x_0)\}$ remains a homeomorphism.
    (b) Prove (again) that $[0, 1) \ncong (0, 1)$ by using the concept of connectedness.

Proof.

(a) It is obvious that the restriction function

$$f|_{X \setminus \{x_0\}} : X \setminus \{x_0\} \to Y \setminus \{f(x_0)\}$$

is still $1 - 1$, onto, continuous, and with continuous inverse, hence it remains a homeomorphism.

(b) Suppose there were a homeomorphism $f : [0, 1) \xrightarrow{\cong} (0, 1)$. By (a), the restriction

$$f\Big|_{[0,1) \setminus \{0\}} : [0, 1) \setminus \{0\} \to (0, 1) \setminus \{f(0)\}$$

remains a homeomorphism. But then $[0, 1) \setminus \{0\} = (0, 1)$ is connected, while its direct image $(0, 1) \setminus \{f(0)\} = \big(0, f(0)\big) \cup \big(f(0), 1\big)$ is disconnected. This contradicts the fact that connectedness is a topological property. So $[0, 1) \ncong (0, 1)$.

18. Let $f : X \to Y$ be non-constant and continuous. If the direct image $f(X) \subset Y$ of X under f has an isolated point, show that X is disconnected.

Proof. Suppose $y_0 \in f(X)$ is an isolated point of $f(X)$. By definition, there exists $r > 0$ such that $B_Y(y_0, r) \cap f(X) = \{y_0\}$. Since f is continuous, both

$$U := f^{-1}(B_Y(y_0, r)) \quad \text{and} \quad V := f^{-1}(Y \setminus \{y_0\})$$

are open in X. Moreover, it is clear that $U \neq \phi$ and $U \cup V = X$. Since f is non-constant, we have $f(X) \neq \{y_0\}$ and so $V \neq \phi$. Finally, since $B_Y(y_0, r) \cap f(X) = \{y_0\}$, we have

$$(B_Y(y_0, r) \setminus \{y_0\}) \cap f(X) = \phi$$

and so

$$\begin{aligned} U \cap V &= f^{-1}(B_Y(y_0, r)) \cap f^{-1}(Y \setminus \{y_0\}) \\ &= f^{-1}(B_Y(y_0, r) \cap (Y \setminus \{y_0\})) \\ &= f^{-1}(B_Y(y_0, r) \setminus \{y_0\}) = \phi \,. \end{aligned}$$

So X is the disjoint union of two nonempty open subsets and is thus disconnected.

Alternatively, suppose $y_0 \in f(X)$ is an isolated point of $f(X)$. Let $B := \overline{f(X) \setminus \{y_0\}}$. Since y_0 is an isolated point, $\{y_0\}$ and B are disjoint closed subsets of Y. Since f is non-constant, $f(X) \neq \{y_0\}$. Hence $f(X) \setminus \{y_0\} \neq \phi$ and $B \neq \phi$. Define $g : X \to \mathbb{R}$ by

$$g(x) := \frac{d\,(f(x), y_0)}{d\,(f(x), y_0) + d\,(f(x), B)} \quad \text{for any } x \in X \,.$$

Note that

$$\begin{cases} f(x) = y_0 & \iff d(f(x), y_0) = 0 \iff g(x) = 0 \\ f(x) \in f(X) \setminus \{y_0\} & \iff d(f(x), B) = 0 \iff g(x) = 1 \,. \end{cases}$$

Hence $g(X) = \{0, 1\}$. That is, g is a non-constant 2-valued function on X and so X is disconnected.

19. Determine whether the following statements are true or false:
    (a) Every connected subset $S \subset \mathbb{R}^n$ is convex.
    (b) Every convex subset $S \subset \mathbb{R}^n$ is connected.

(a) *Answer*: False.
Example: The unit circle $S^1 \subset \mathbb{R}^2$ is connected but obviously not convex.

(b) *Answer*: True.
Proof. Suppose S is not connected. Then there exist open nonempty subsets U and V of S such that $U \cap V = \phi$ and $S = U \cup V$. Fix $x \in U$ and $y \in V$. Since S is convex, the line segment joining them is contained in S, i.e., the function $f : [0, 1] \to S$ given by $f(t) := (1 - t)x + ty$ is well defined. Moreover, it is evident that f is continuous on $[0, 1]$. In fact, for any $t_0 \in [0, 1]$ and any $\varepsilon > 0$, take $\delta := \varepsilon/\|x - y\|$, then we have

$$d(f(t), f(t_0)) = \|f(t) - f(t_0)\| = |t - t_0| \; \|x - y\| < \varepsilon$$

whenever $|t - t_0| < \delta$. Therefore, f is continuous at t_0.
Hence $f^{-1}(U)$ and $f^{-1}(V)$ are disjoint nonempty open sets in $[0, 1]$ whose union is $[0, 1]$, that is, they form a separation of the connected set $[0, 1]$, which is absurd.
Alternatively, let $f : S \to \{0, 1\}$ be a 2-valued function on S. Fix $a \in S$. Since S is convex, the line segment $\overline{ax}$ joining a and x is lying entirely in S. The restriction of $f\big|_{\overline{ax}} : \overline{ax} \to \{0, 1\}$ is a 2-valued function on the connected set $\overline{ax}$ and so it must be a constant. In particular, $f(x) = f(a)$. Since $x \in S$ is arbitrary, this shows $f \equiv f(a)$ on S and so S is connected.

20. Show that homeomorphisms preserve connected components.

Proof. Let $f : X \xrightarrow{\cong} Y$ be a homeomorphism and $S \subset X$ be a connected component of X. Then $f(S) \subset Y$ is connected and so it is contained in a unique connected component of Y. Call that T. Since f^{-1} is also continuous, $f^{-1}(T) \subset X$ is connected. But since

$f(S) \subset T$, we have $S \subset f^{-1}(T)$. By the maximality of S, this forces $S = f^{-1}(T)$ and so $f(S) = T$ is a connected component.

21. If $X \cong Y$, show that there is a $1-1$ correspondence between connected components of X and those of Y.

Proof. Let $f : X \xrightarrow{\cong} Y$ be a homeomorphism. By Exercise 3.1, Part B, Problem #20, connected components of X are mapped by f to connected components of Y. Furthermore, as f is $1-1$, distinct connected components of X are mapped by f to distinct connected components of Y. Hence there is a $1-1$ correspondence from the collection of connected components of X into the collection of connected components of Y. On the other hand, applying the same arguments to the homeomorphism f^{-1}, we see that it maps distinct connected components of Y to distinct connected components of X. Hence the assertion.

22. Determine whether the two subsets of $\mathbb{R}^2$:
$A := \{(x,y) : xy = 0\}$, $B := \{(x,y) : xy(x^2 - y^2) = 0\}$
are homeomorphic.

Answer: No.

Justification. Observe that A is the union of 2 intersecting straight lines $y = 0$ and $x = 0$, and B is the union of 4 concurrent straight lines $y = 0$, $x = 0$, and $y = \pm x$. If there were a homeomorphism $f : B \xrightarrow{\cong} A$, then by Exercise 3.1, Part B, Problem #17,

$$f\Big|_{B\setminus\{(0,0)\}} : B \setminus \{(0,0)\} \longrightarrow A \setminus \{f(0,0)\}$$

remains a homeomorphism. However, observe that $B \setminus \{(0,0)\}$ has 8 components, while $A \setminus \{f(0,0)\}$ has either 4 or 2 components, depending on whether $f(0,0) = (0,0)$ or not. That violates Exercise 3.1, Part B, Problem #21.

23. Determine whether each pair of the following subsets of $\mathbb{R}^2$ are homeomorphic.

(a) The digit 8 and the letter O.
(b) The symbols $\times$ and $\perp$.
(c) The symbols $\exists$ and $\forall$.
(d) The objects ∇ and $\square$.
(e) The symbols $\oslash$ and $\otimes$.
(f) The symbols $\longrightarrow$ and $\Longrightarrow$.

Answer:

(a) No.

Justification. Removing any point from the letter O we have one connected component, while removing the middle point from figure 8 we have two connected components. In view of Exercise 3.1, Part B, Problem #21, the two objects are not homeomorphic.

(b) No.

Justification. Removing the middle point from symbol $\times$ we have four connected components, while removing a point from symbol $\perp$ we have either two of three connected components. Hence the two objects are not homeomorphic.

(c) No.

Justification. Removing a point from the symbol $\forall$ we have one or two connected components, while removing the middle point on the vertical edge of the symbol $\exists$ we have three connected components. Hence the two objects cannot not be homeomorphic.

(d) Yes.

Proof. Both objects are homeomorphic to the circle. Details are left with the readers.

(e) No.

Justification. Removing the two antipodal points at the end of the diameter from the symbol $\oslash$ we have three connected components, while removing any two points from the symbol $\otimes$ we have no more than two connected components. By repeated arguments of Exercise 3.1, Part B, Problems #17, and by Exercise 3.1, Part B, Problems #21, we conclude that the two objects are not homeomorphic.

(f) No.

Justification. Removing the two points in the symbol $\Longrightarrow$ which are the intersecting points of the horizontal lines with the slanted lines, we have 5 connected components, while removing any two points from the symbol $\longrightarrow$ we are left with no more than four connected components. Hence the two objects are not homeomorphic.

24. Let $\Lambda =$ be some index set. If V and V_α, $\alpha \in \Lambda$, are connected subsets of X such that $V \cap V_\alpha \neq \phi$ for all $\alpha \in \Lambda$, then $V \cup \bigcup_{\alpha\in\Lambda} V_\alpha$ is connected.

Answer: True.

Proof. For every $\alpha \in \Lambda$, since V, V_α are connected and $V \cap V_\alpha \neq \phi$, by Theorem 3.1.18, $V \cup V_\alpha$ is connected. Since

$$\bigcap_{\alpha\in\Lambda} \left(V \cup V_\alpha\right) = V \cap \left(\bigcup_{\alpha\in\Lambda} V_\alpha\right) \neq \phi\,,$$

$\{V \cup V_\alpha\}_{\alpha\in\Lambda}$ is a collection of connected sets with nonempty intersection, by Theorem 3.1.12 again,

$$V \cup \left(\bigcup_{\alpha\in\Lambda} V_\alpha\right) = \bigcup_{\alpha\in\Lambda} \left(V \cup V_\alpha\right)$$

is connected.

25. Let (X, d_X) and (Y, d_Y) be metric spaces. It is easy to see that $d : X \times Y \to \mathbb{R}$ given by

$$d\big((x_1, y_1), (x_2, y_2)\big) := \max\big\{d_X(x_1, x_2), d_Y(y_1, y_2)\big\}$$

for any $(x_i, y_i) \in X \times Y$, $i = 1, 2$, is a well-defined metric, making $(X \times Y, d)$ a metric space called the *product space* of X and Y.

(a) For any $(x_0, y_0) \in X \times Y$, prove that $\{x_0\} \times Y$ and $X \times \{y_0\}$ are homeomorphic to Y and X, respectively.

(b) For any $\phi \neq A \subset X$ and $\phi \neq B \subset Y$, if $A \times B$ is closed in $X \times Y$, show that A is closed in X and B is closed in Y.

(c) Prove that the product space $X \times Y$ is connected if and only if both X and Y are connected.

Proof.

(a) Observe that $\{x_0\} \times Y$ and $X \times \{y_0\}$ are metric subspaces of $X \times Y$. The projection maps $\pi_1 : X \times \{y_0\} \to X$ and $\pi_2 : \{x_0\} \times Y \to Y$ given by $\pi_1(x, y_0) := x$ and $\pi_2(x_0, y) := y$ are isometries, hence homeomorphisms.

(b) Pick $x_0 \in A$ and $y_0 \in B$. Since $A \times B$ is closed in $X \times Y$, $A \times \{y_0\} = (A \times B) \cap (X \times \{y_0\})$ is closed in $X \times \{y_0\}$. By (a), the projection map $\pi_1 : X \times \{y_0\} \to X$ is a homeomorphism. Since homeomorphisms preserve closedness, it follows that $A = \pi_1(A \times \{y_0\})$ is closed in X. Similarly, $B = \pi_2(\{x_0\} \times B)$ is closed in Y.

(c) ($\Rightarrow$) Consider the projection maps $\pi_1 : X \times Y \to X$ and $\pi_2 : X \times Y \to Y$ given by $\pi_1(x, y) := x$ and $\pi_2(x, y) := y$ for all $(x, y) \in X \times Y$. It is elementary to check that both π_1 and π_2 are continuous on $X \times Y$. So if $X \times Y$ is connected, then $X = \pi_1(X \times Y)$ and $Y = \pi_2(X \times Y)$ are connected.

($\Leftarrow$) Suppose X and Y are connected. Fix $x_0 \in X$. Write $V := \{x_0\} \times Y$. For any $y \in Y$, write $U_y := X \times \{y\}$. By (a), $V \cong Y$ and for all $y \in Y$, $U_y \cong X$. In particular, V and all U_y's are connected. Moreover, $V \cap U_y = \{(x_0, y)\} \neq \phi$ for every $y \in Y$ and $X \times Y = V \cup \bigcup_{y \in Y} U_y$. Hence by Exercise 3.1, Part B, Problem #24, $X \times Y$ is connected.

26. If a function preserves connectedness, determine whether it is continuous.

Answer: No.

Example: Consider the function $f : [0, \infty) \to \mathbb{R}$ defined by

$$f(x) = \begin{cases} 0 & x = 0 \\ \sin \dfrac{1}{x} & x \neq 0 \,. \end{cases}$$

Then f preserves connectedness. In fact, suppose $S \subset [0, \infty)$ is connected. If S is a singleton, $f(S)$ is also a singleton and hence is connected. If S is not a singleton, then it is a nondegenerate interval.
If $0 \notin S$, then f is continuous on S and so $f(S)$ is connected.
If $0 \in S$, then S is a nondegenerate interval in $[0, \infty)$ containing the point 0. No matter how small it is, by the definition of f, we always have $f(S) = [-1, 1]$ which is connected.
Combining, we see that f preserves connectedness. However, it is elementary to see that f is discontinuous at 0.

27. Let S be a subset of $\mathbb{R}^3$ consisting of the four points $(0, 0, 0)$, $(1, 0, 0)$, $(0, 1, 0)$, $(0, 0, 1)$ together with the straight line segments joining each pair of these points. If $f : S \to S$ is a self-homeomorphism, how many possibilities are there for $f(0, 0, 0)$?

Answer: 4.

Prove. By the continuity of f, there exists $\delta_1 > 0$ such that

$$f(B_S((0, 0, 0), \delta_1)) \subset B(f(0, 0, 0), 0.1) \,.$$

Take $\delta := \min\{\delta_1, 0.1\}$. Observe that we have $B_S((0, 0, 0), \delta) \cong f(B_S((0, 0, 0), \delta))$ and hence in particular,

$$B_S\big((0, 0, 0), \delta\big) \setminus \{(0, 0, 0)\} \cong f\big(B_S((0, 0, 0), \delta)\big) \setminus \{f(0, 0, 0)\} \,.$$

Note that $B_S\big((0, 0, 0), \delta\big) \setminus \{(0, 0, 0)\}$ has 3 connected components. By Exercise 3.1, Part B, Problem #21,

$$f\big(B_S((0, 0, 0), \delta)\big) \setminus \{f(0, 0, 0)\}$$

should also have 3 connected components. Observe that the subset $f\big(B_S((0,0,0),\delta)\big) \subset S$ is connected and the only possible scenario of removing a point from a connected subset of S would result in 3 connected components is when the point removed is one of the four given points. Thus any self-homeomorphism of S must permute the four given points and so there are exactly 4 possibilities for $f(0,0,0)$.

3.2 Path-connectedness

In the last section, we defined connectedness of a metric space as the state of not separated. This approach is rather indirect and causes uneasiness. At least it owes us an intuitive geometrical feeling. In this section we would discuss yet another concept of "connectedness" which is in certain sense stronger and provides a simple intuitive geometrical meaning. That is called *path-connectedness*. It turns out that path-connectedness is on the one hand stronger than connectedness, and on the other hand easier to establish, hence facilitating us an efficient way of proving the connectedness of a geometric object.

Definition 3.2.1. *A path (or a curve) in a metric space X is a continuous function $\gamma : [0,1] \to X$. If we write $p := \alpha(0)$ and $q := \alpha(1)$, we would refer to α as a path in X joining p to q, and the independent variable $t \in [0,1]$ is called the* parameter *of the path α. Sometimes it would be convenient to identify the direct image of a path with the function itself and regard the parameter t as time. Under such an identification, α (the direct image of the function α in the space X) is the trajectory in X when we travel from the point p at time $t = 0$ to the point q at time $t = 1$. Observe also that under this identification, every path in S is a (1-dimensional, although we would not make it clear here) connected subset of X.*

Let $\alpha : [0,1] \to X$ and $\beta : [0,1] \to X$ be paths in X with $\alpha(0) = p$, $\alpha(1) = q = \beta(0)$, $\beta(1) = r$. Then $\gamma : [0,1] \to X$ defined by

$$\gamma(t) = \begin{cases} \alpha(2t) & 0 \le t \le \frac{1}{2} \\ \beta(2t-1) & \frac{1}{2} \le t \le 1 \end{cases}$$

is a path in X joining p to r via q through α and β, called the product *of the paths α and β, and is denoted as $\gamma = \beta \cdot \alpha$.*

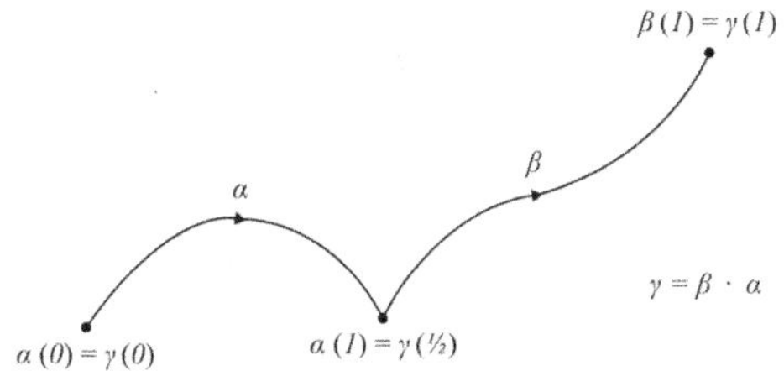

Geometrically, what γ does is that we travel along the trajectory of α, followed by that of β, but at a speed twice as fast as that of α and β.

Examples 3.2.2

(i) Let $X := \mathbb{R}^n$ and $a,\ b \in X$. The image of the function $\gamma : [0,1] \to \mathbb{R}$ defined by $\gamma(t) := (1-t)a + tb$, $t \in [0,1]$, is a straight line segment in $\mathbb{R}^n$ joining a to b. It is clear that γ is continuous and so it is a path in $\mathbb{R}^n$ joining a and b.

(ii) Let $X := \mathbb{R}^n$, and $\alpha,\ \beta,\ \gamma,\ \delta : [0,1] \to X$ be functions defined by

$$\begin{aligned} \alpha(t) &:= (-1+2t, -1) \\ \beta(t) &:= (1, -1+2t) \\ \gamma(t) &:= (1-2t, 1) \\ \delta(t) &:= (-1, 1-2t) \end{aligned}$$

for all $t \in [0,1]$. Then α, β, γ, and δ are paths in $\mathbb{R}^2$ which form the boundary of the rectangular region $D := \{(x,y) \in \mathbb{R}^2 : |x| \leq 1, |y| \leq 1\}$. In particular, the boundary ∂D is the image of the single path $\delta \cdot \gamma \cdot \beta \cdot \alpha : [0,1] \to \mathbb{R}^2$.

Remark. The more matured readers would notice from the definition that the requirement that the domain of a path has to be $[0,1]$ is a bit artificial and not strictly necessary. In fact, for any $a < b \in \mathbb{R}$, and any continuous function $\gamma : [a,b] \to X$, if we define

$\alpha : [0,1] \to [a,b]$ by $\alpha(t) := a+(b-a)t$ for any $t \in [0,1]$, then clearly α is continuous and so the composition function $\beta := \gamma \circ \alpha : [0,1] \to X$ is a continuous function of $[0,1]$ into X, hence it is a path in X. Furthermore, it is easy to see that the direct image of β in X is exactly the same as that of the given continuous function γ. This shows that any continuous function $\gamma : [a,b] \to X$ can be converted into a path $\beta : [0,1] \to X$ with identical direct image. In this respect, any continuous function $\gamma : [a,b] \to X$ can be regarded as a path in X in a broad sense, and the process of changing variables of the "path" in the broad sense is called a *reparametrization*. So in the broad sense, a path in X is simply a continuous function from a closed and bounded interval into X. Having said that, however, in order to standardize things, in the sequel we still require a path in X to have domain $[0,1]$.

Definition 3.2.3. *A metric space X is said to be path-connected (or arcwise-connected) if any two points p, $q \in X$ can be joined by a path in X, that is, there is a path $\gamma : [0,1] \to X$ with $\gamma(0) = p$, $\gamma(1) = q$.*

Examples 3.2.4.

(i) Let $X := \mathbb{R}^n$ and p, $q \in X$. By Example 3.2.2 (i), the straight line segment $\gamma(t) := (1-t)p + tq$, $t \in [0,1]$, is a path in $\mathbb{R}^n$ joining p to q. Hence $\mathbb{R}^n$ is path-connected.

(ii) Every convex set in $\mathbb{R}^n$ is path-connected.
Proof. Let $S \subset \mathbb{R}^n$ be convex and a, $b \in S$. Similar to (i) above, the straight line segment $\gamma(t) := (1-t)a + tb$, $t \in [0,1]$, is a path in $\mathbb{R}^n$ joining a to b. Since S is convex, the whole line segment joining a, b is lying entirely in S. Hence γ is a path in S joining a to b and so S is path-connected.

(iii) All open balls and closed balls in $\mathbb{R}^n$ are path-connected.
Proof. As open balls and closed balls in $\mathbb{R}^n$ are convex, the assertion follows immediately from (ii).

(iv) The unit circle $S^1 \subset \mathbb{R}^2$ is path-connected.
Proof. The function $\gamma : [0, 2\pi] \to S^1$ given by

$$\gamma(t) := (\cos t, \sin t) \ , \quad t \in [0, 2\pi] \ ,$$

is clearly continuous and it passes through all points in S^1. Hence γ is, in the broad sense, a path in S^1 which goes through every point of S^1, and so S^1 should be path-connected.
To put everything into absolute rigor, we argue as follows: For any p, $q \in S^1$, write $p = (\cos a, \sin a)$ and $q = (\cos b, \sin b)$ for some unique a, $b \in [0, 2\pi)$. Without loss of generality, we assume $a < b$. Define $\alpha : [0,1] \to [a,b]$ by

$$\alpha(t) := a + (b-a)t \ , \quad t \in [0,1] \ .$$

Then clearly α is continuous and so the composition function $\beta := \gamma \circ \alpha : [0,1] \to S^1$ is a path in S^1 with

$$\begin{cases} \beta(0) &= \gamma\big(\alpha(0)\big) = \gamma(a) = p \\ \beta(1) &= \gamma\big(\alpha(1)\big) = \gamma(b) = q \ . \end{cases}$$

Thus S^1 is path-connected.

(v) A torus with two generating circles removed is path-connected.
Proof. Let T be a torus with two generating circles removed and a, $b \in T$. We have seen that T is homeomorphic to an open rectangular region $R \subset \mathbb{R}^2$. Let $f : T \xrightarrow{\cong} R$ be a homeomorphism. Since R is convex, by (ii), it is path-connected. Hence there is a path $\gamma : [0,1] \to R$ joining $f(a)$ and $f(b) \in R$. Then $f^{-1} \circ \gamma : [0,1] \to T$ is a path in T joining a and b.

(vi) If S and $T \subset X$ are path-connected subsets of X such that $S \cap T \neq \phi$, then $S \cup T$ is also path-connected.
Proof. Fix $a \in S \cap T$. For any $p \in S$ and any $q \in T$, since S and T are path-connected, there exist a path $\alpha : [0,1] \to S$ joining p to a, and a path $\beta : [0,1] \to T$ joining a to q. Then the product $\beta \cdot \alpha : [0,1] \to S \cup T$ is a path in $S \cup T$ joining p to q via a. Hence $S \cup T$ is path-connected.

(vii) A torus in $\mathbb{R}^3$ is path-connected.

Proof. Let T be a torus in $\mathbb{R}^3$. Let $\{C_1, C_2\}$, $\{\tilde{C}_1, \tilde{C}_2\}$ be two different pairs of generating circles of T, and $S := T \setminus (C_1 \cup C_2)$, $\tilde{S} := T \setminus (\tilde{C}_1 \cup \tilde{C}_2)$. By (v) above, both S and $\tilde{S}$ are path-connected. Furthermore, it is obvious that $S \cap \tilde{S} \neq \phi$ and so by (vi), $S \cup \tilde{S}$ is path-connected. On the other hand, by (iv) above, C_1 and C_2 are path-connected. Since $(S \cup \tilde{S}) \cap C_1 \neq \phi$, by (vi), $S \cup \tilde{S} \cup C_1$ is path-connected. Finally, since $(S \cup \tilde{S} \cup C_1) \cap C_2 \neq \phi$, by (vi) again, $S \cup \tilde{S} \cup C_1 \cup C_2$ is path-connected. Note that $S \cup \tilde{S} \cup C_1 \cup C_2 = T$.

Theorem 3.2.5. *Every path-connected set is connected.*

Proof. Let $S \subset X$ be path-connected, and $f : S \to \{0, 1\}$ a 2-valued function on S. Fix $a \in S$. For any $x \in S$, there is a path $\gamma : [0, 1] \to S$ in S joining a to x. Since γ is continuous and $[0, 1]$ is connected, $\gamma([0, 1])$ is connected. The 2-valued function

$$f\big|_{\gamma([0,1])} : \gamma\big([0, 1]\big) \to \{0, 1\}$$

must therefore be a constant and so $f(x) = f(\gamma(1)) = f(\gamma(0)) = f(a)$. Since $x \in S$ is arbitrary, $f \equiv$ constant on S. That is, the only possible 2-valued function on S must be constant. So by Theorem 3.1.5, S is connected. □

Remark. In general, the converse of Theorem 3.2.5 is not valid. So path-connectedness is strictly stronger than connectedness.

Example 3.2.6 (Connectedness $\not\Rightarrow$ path-connectedness).
In $\mathbb{R}^2$, consider

$$A := \{(x, y) \in \mathbb{R}^2 : 0 < x \leq 1 \ , \ y = \sin \frac{1}{x}\} \ ,$$
$$B := \{(x, y) \in \mathbb{R}^2 : -1 \leq x \leq 0 \ , \ y = 0\} \ ,$$
$$X := A \cup B \ .$$

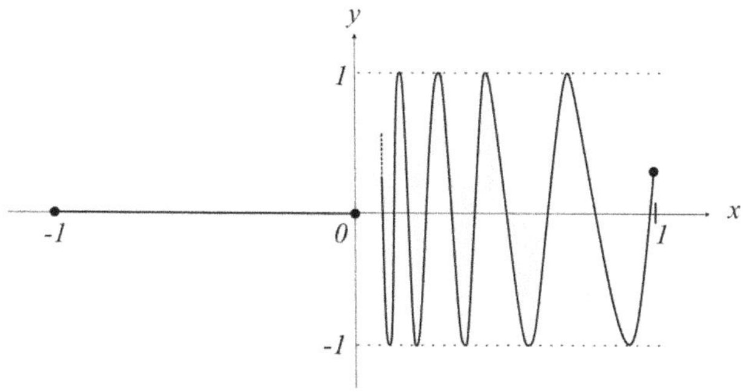

It turns out that X is connected but not path-connected.

Proof. In Example 3.1.16, we saw already that X is connected. To show that X is not path-connected, we first observe that both A and B are path-connected. Every point in A can be joined to another point in A by a path in $A \subset X$, and every point in B can be joined to another point in B by a path in $B \subset X$. So we only need to show that points in A cannot be joined to points in B by any path in X. In other words, it suffices to show that any path $\alpha : [0,1] \to X$ which passes through a point in B must stay forever in B. Equivalently, we need to show $\alpha([0,1]) \subset B$, or $\alpha^{-1}(B) = [0,1]$. Now as α passes through a point in B, we have $\phi \neq \alpha^{-1}(B) \subset [0,1]$. Since $[0,1]$ is connected, it remains to show $\alpha^{-1}(B) \subset [0,1]$ is clopen.

Since B is closed in $\mathbb{R}^2$, it is closed in X. As α is continuous, $\alpha^{-1}(B)$ is closed in $[0,1]$. On the other hand, for any $t_0 \in \alpha^{-1}(B) \subset [0,1]$, since α is continuous, there exists $\delta > 0$ s.t.

$$\alpha\big(B_{[0,1]}(t_0,\delta)\big) \subset B_X\left(\alpha(t_0), \frac{1}{2}\right) .$$

Since $B_{[0,1]}(t_0,\delta) = (t_0-\delta, t_0+\delta) \cap [0,1]$ is connected, $\alpha\big(B_{[0,1]}(t_0,\delta)\big)$ is also connected and contains the point $\alpha(t_0)$, so it is contained in the unique connected component of $B_X\big(\alpha(t_0), \frac{1}{2}\big)$ which contains $\alpha(t_0)$. Now observe that the set $B_X\big(\alpha(t_0), \frac{1}{2}\big)$ splits into a number of parts, including the horizontal piece $B_X\big(\alpha(t_0), \frac{1}{2}\big) \cap B$ and the "almost vertical" pieces of the curve $y = \sin \frac{1}{x}$ in $B_X\big(\alpha(t_0), \frac{1}{2}\big)$.

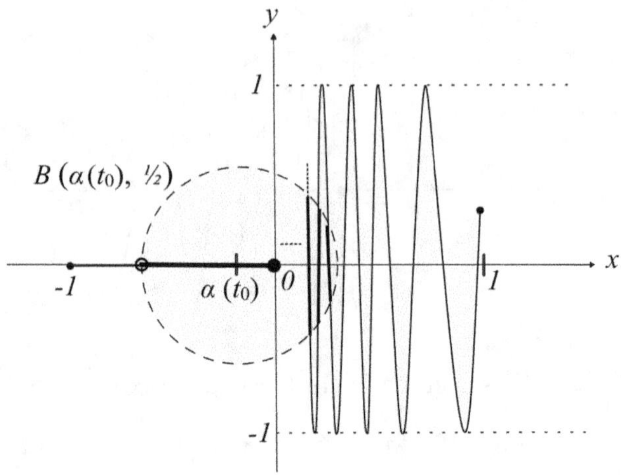

It is evident that each of the "almost vertical" pieces of the curve $y = \sin \frac{1}{x}$ in $B_X(\alpha(t_0), \frac{1}{2})$ is a connected component of $B_X(\alpha(t_0), \frac{1}{2})$. On the other hand, as $B_X(\alpha(t_0), \frac{1}{2}) \cap B$ is an interval, it is connected and we claim that it is indeed another connected component of $B_X(\alpha(t_0), \frac{1}{2})$. In fact, for any $S \subset B_X(\alpha(t_0), \frac{1}{2})$ which is a strict superset of $B_X(\alpha(t_0), \frac{1}{2}) \cap B$, S contains at least one point in one of the "almost vertical" pieces of $y = \sin \frac{1}{x}$. Name one of such almost vertical pieces V. It is clear that $\{S \cap V, S \setminus (S \cap V)\}$ forms a separation of S and so S cannot be connected. Thus $B_X(\alpha(t_0), \frac{1}{2}) \cap B$ is a connected component of $B_X(\alpha(t_0), \frac{1}{2})$ and hence it is indeed the unique connected component of $B_X(\alpha(t_0), \frac{1}{2})$ that contains $\alpha(t_0)$. Therefore, we have

$$\alpha(B_{[0,1]}(t_0, \delta)) \subset B_X\left(\alpha(t_0), \frac{1}{2}\right) \cap B .$$

Hence

$$B_{[0,1]}(t_0, \delta) \subset \alpha^{-1}\left(B_X\left(\alpha(t_0), \frac{1}{2}\right) \cap B\right) \subset \alpha^{-1}(B)$$

and so $\alpha^{-1}(B)$ is open in $[0, 1]$. Since $[0, 1]$ is connected, this forces $\alpha^{-1}(B) = [0, 1]$, or $\alpha([0, 1]) \subset B$.

Remark. The curious readers should wonder why we chose radius $\frac{1}{2}$ in the proof. In fact, $\frac{1}{2}$ is not such a magical number. Any number smaller than 1 would do. The point is, if we happen to choose a radius $r > 1$, then $B_X\big(\alpha(t_0), r\big)$ will be so large that the sine curve will no longer be cut into those "almost vertical" pieces. Instead, when we are looking at a vicinity near the y-axis, the whole piece of the sine curve is still there and so the problem remains the same.

Although in general, connectedness is weaker than path-connectedness, under some special circumstances, they turn out to be the same.

Theorem 3.2.7. *Every open connected set in* $\mathbb{R}^n$ *is path-connected.*

Proof. Let $S \subset \mathbb{R}^n$ be open and connected. Fix $a \in S$. Let

$$\begin{aligned} A &:= \big\{x \in S: \ x \text{ and } a \text{ are joined by a path in } S\big\}, \\ B &:= S \setminus A\,. \end{aligned}$$

As $a \in A$, we have $A \neq \phi$. Note that by construction, A is path-connected. Furthermore, it is obvious that $A \cup B = S$ and $A \cap B = \phi$. So if we can show that both A, B are open in S, then since S is connected and $A \neq \phi$, we must have $B = \phi$ and we will be done, because in this case $S = A$ is path-connected. So the proof boils down to showing the openness of A and B in S. Interestingly, the proofs are based on the same trick.

For any $x \in A$, there is a path $\alpha : [0,1] \to S$ joining a to x. Since $x \in S$ and S is open in $\mathbb{R}^n$, there is an open ball $B_{\mathbb{R}^n}(x)$ centered at x with $B_{\mathbb{R}^n}(x) \subset S$. Since $B_{\mathbb{R}^n}(x)$ is convex, by Example 3.2.4 (ii), it is path-connected. So for any $y \in B_{\mathbb{R}^n}(x)$, there is a path (for example, a line segment) $\beta : [0,1] \to B_{\mathbb{R}^n}(x) \subset S$ joining x to y.

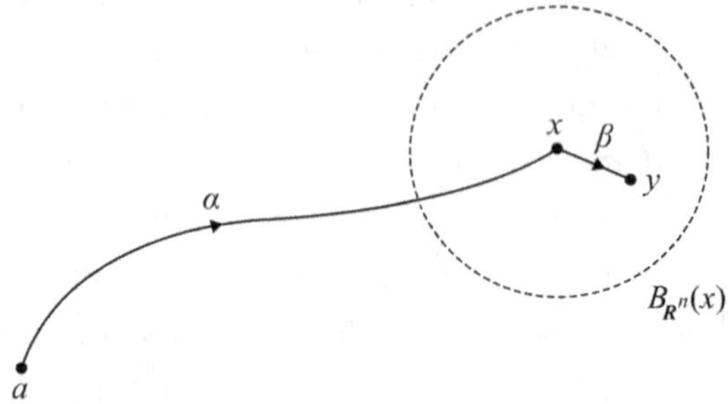

Therefore, $\beta \cdot \alpha : [0, 1] \to S$ is a path in S joining a to y via x. Hence $y \in A$. Since $y \in B_{\mathbb{R}^n}(x)$ is arbitrary, we conclude that $B_{\mathbb{R}^n}(x) \subset A$ and so A is open in $\mathbb{R}^n$, hence open in S.

Similarly, for any $z \in B \subset S$, there is an open ball $B_{\mathbb{R}^n}(z)$ centered at z with $B_{\mathbb{R}^n}(z) \subset S$. For any $w \in B_{\mathbb{R}^n}(z)$, since $B_{\mathbb{R}^n}(z)$ is convex and hence path-connected, there is a path γ (for example, a line segment) in $B_{\mathbb{R}^n}(z) \subset S$ joining w to z. Observe that $w \in B$. For if $w \in X \setminus B = A$, then there is a path δ in S joining a to w and so the composition $\gamma \cdot \delta$ would be a path in S joining a to z via w, which means $z \in A$, contradicting to the choice of z. Thus $w \in B$. Since w in $B_{\mathbb{R}^n}(y)$ is arbitrary, we have $B_{\mathbb{R}^n}(y) \subset B$ and so B is open in $\mathbb{R}^n$, hence open in S. □

Remark. Observe that in the proof of Theorem 3.2.7, the only place where $\mathbb{R}^n$ comes into play is that at every point $x \in S$, we can find a small open ball $B_{\mathbb{R}^n}(x) \subset S$ and the only property of such an open ball that we need is that it is path-connected. Hence Theorem 3.2.7 can be relaxed to a more general situation that at each point of S we can find arbitrarily small path-connected open neighborhood. For the details, the readers are referred to Exercise 3.2, Part B, Problem #12.

Remark. A path in $\mathbb{R}^n$ is said to be *polygonal* if its image in $\mathbb{R}^n$ is the union of a finite number of straight line segments.

Observe that precisely the same arguments used in the proof of Theorem 3.2.7 shows that every open connected set in $\mathbb{R}^n$ is *polygonally connected*, that is, any two points can be joined by a polygonal path lying entirely in the set.

Remark. As connectedness and path-connectedness have rather intimate connections, it would lead to speculations that results on connectedness can be carried forward to analogous results on path-connectedness. Certainly some of them can, but not all.

Example 3.2.8 (A is path-connected $\not\Leftrightarrow$ $\overline{A}$ is path-connected).
In $\mathbb{R}^2$, consider

$$\begin{aligned} A &:= \left\{(x,y) \in \mathbb{R}^2 : 0 < x \le 1 \ , \ y = \sin\frac{1}{x}\right\} , \\ B &:= \left\{(x,y) \in \mathbb{R}^2 : x = 0 \ , \ -1 \le y \le 1\right\}, \\ X &:= A \cup B \ . \end{aligned}$$

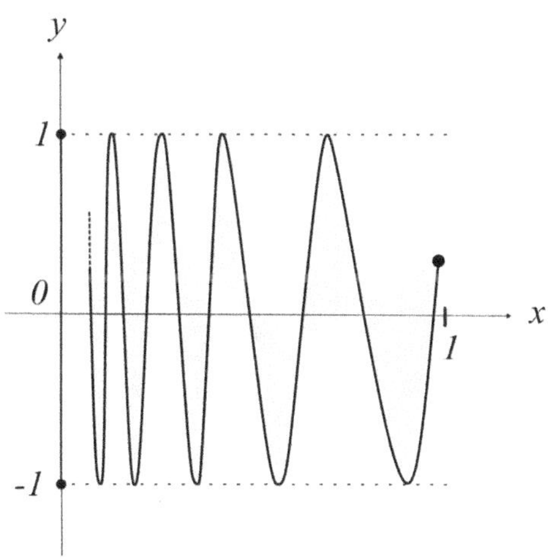

It turns out that A is path-connected but $\overline{A}$ is not.

Proof. That A is path-connected is clear. On the other hand, it is readily seen that $\overline{A} = A \cup B = X$. To show that $\overline{A} = X$ is not path-connected, we use arguments similar to those in Example 3.2.6: It suffices to show that points in A cannot be joined to points in B by a path in X. In other words, we need to show that any path $\gamma : [0,1] \to X$ which passes through a point in B must stay forever in B, that is, γ should satisfy $\gamma([0,1]) \subset B$, or $\gamma^{-1}(B) = [0,1]$. Now as γ passes through a point in B, we have $\phi \neq \gamma^{-1}(B) \subset [0,1]$. Since $[0,1]$ is connected, it remains to show $\gamma^{-1}(B) \subset [0,1]$ is clopen.

Since B is closed in $\mathbb{R}^2$, it is closed in X. As γ is continuous, $\gamma^{-1}(B)$ is closed in $[0,1]$. On the other hand, for any $t_0 \in \gamma^{-1}(B) \subset [0,1]$, since γ is continuous, there exists $\delta > 0$ s.t.

$$\gamma\big(B_{[0,1]}(t_0,\delta)\big) \subset B_X\Big(\gamma(t_0), \frac{1}{2}\Big) \,.$$

Since $B_{[0,1]}(t_0,\delta) = (t_0-\delta, t_0+\delta) \cap [0,1]$ is connected, $\gamma\big(B_{[0,1]}(t_0,\delta)\big)$ is also connected and contains the point $\gamma(t_0)$, so it is contained in the unique connected component of $B_X\big(\gamma(t_0), \frac{1}{2}\big)$ which contains $\gamma(t_0)$.

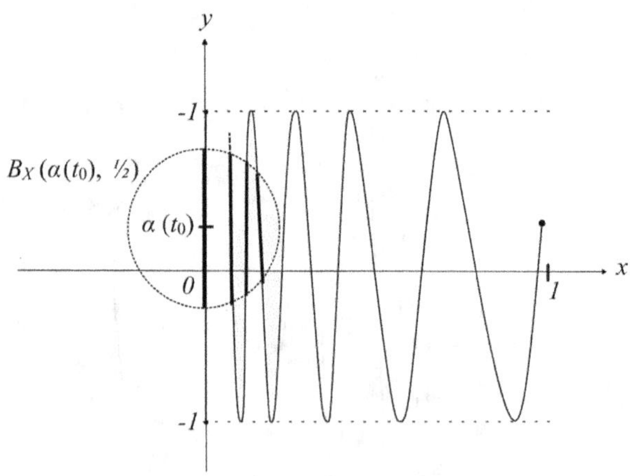

Observe that the set $B_X\big(\gamma(t_0), \frac{1}{2}\big)$ splits into a number of parts, including the vertical piece $B_X\big(\gamma(t_0), \frac{1}{2}\big) \cap B$ and the "almost vertical" pieces of the curve $y = \sin\frac{1}{x}$ in $B_X\big(\gamma(t_0), \frac{1}{2}\big)$. It is evident that each of the "almost vertical" pieces of the curve $y = \sin\frac{1}{x}$ in $B_X\big(\gamma(t_0), \frac{1}{2}\big)$ is a connected component of $B_X\big(\gamma(t_0), \frac{1}{2}\big)$.

On the other hand, as $B_X\left(\gamma(t_0), \frac{1}{2}\right) \cap B$ is an interval, it is connected and we claim that it is indeed another connected component of $B_X\left(\gamma(t_0), \frac{1}{2}\right)$. In fact, for any $S \subset B_X\left(\gamma(t_0), \frac{1}{2}\right)$ which is a strict superset of $B_X\left(\gamma(t_0), \frac{1}{2}\right) \cap B$, S contains at least one point in one of the "almost vertical" pieces of $y = \sin \frac{1}{x}$. Name one of such almost vertical pieces V. Then it is clear that $\{S \cap V, S \setminus (S \cap V)\}$ is a separation of S and so S cannot be connected. Thus $B_X\left(\gamma(t_0), \frac{1}{2}\right) \cap B$ is a connected component of $B_X\left(\gamma(t_0), \frac{1}{2}\right)$ and so it is the unique connected component of $B_X\left(\gamma(t_0), \frac{1}{2}\right)$ that contains $\gamma(t_0)$. Hence we have

$$\gamma\big(B_{[0,1]}(t_0, \delta)\big) \subset B_X\Big(\gamma(t_0), \frac{1}{2}\Big) \cap B\ .$$

Therefore,

$$B_{[0,1]}(t_0, \delta) \subset \gamma^{-1}\left(B_X\Big(\gamma(t_0), \frac{1}{2}\Big) \cap B\right) \subset \gamma^{-1}(B)$$

and so $\gamma^{-1}(B)$ is open in $[0,1]$. Since $[0,1]$ is connected, this forces $\gamma^{-1}(B) = [0,1]$, or $\gamma\big([0,1]\big) \subset B$.

Alternatively, by Example 3.2.6, it is evident that $\{(0,0)\}$ cannot be joined to any point in A by a path in $A \cup \{(0,0)\}$. That is, $A \cup \{(0,0)\}$ is not path-connected. Similar argument shows that $A \cup \{p\}$ for any $p \in B$ is not path-connected. Now if $A \cup B$ were path-connected, then there exists a path γ joining a point in A to a point in B. In particular, there exists a point in B which is a point at which γ "enters" B. In other words, there is a point $p \in B$ such that $A \cup \{p\}$ is path-connected. That is a contradiction. Hence $A \cup B$ is not path-connected.

Exercise 3.2

Part A: True or False Questions

For each of the following statements, determine if it is true or false. If it is true, prove it. If it is false, give a counterexample or provide proper justification.

1. If S, $T \subset X$ are disjoint, then $S \cup T$ is not path-connected.

   Answer: False.

   Example: Let $X := \mathbb{R}$, $S := [0,1]$, $T := (1,2)$. Clearly S and T are disjoint but $S \cup T = [0,2)$ is path-connected.

2. If $S \subset X$ is path-connected, then $S \subset S'$.

   Answer: False.

   Example: Let $X := \mathbb{R}$ and $S := \{0\}$. Then S is path-connected but $S \not\subset \phi = S'$.

3. If $S \subset X$ is path-connected, then ∂S is connected.

   Answer: False.

   Example: Let $X := \mathbb{R}$ and $S := (0,1)$. Then S is path-connected but $\partial S = \{0,1\}$ is not.

4. If $S \subset X$ is path-connected, then $\overline{S} = S'$.

   Answer: False.

   Example: Let $X := \mathbb{R}$ and $S := \{0\}$. Then S is path-connected but $\overline{S} = \{0\} \neq \phi = S'$.

5. If $S \subset X$ and $\overline{S} = S'$, then S is path-connected.

   Answer: False.

   Example: Let $X := \mathbb{R}$ and $S :=$ the Cantor Set. Then by Exercise 1.4, Part B, Problem #4, $\overline{S} = S'$. But it is obvious that S is not connected, hence by Theorem 3.2.5, not path-connected.

6. If $S \subset X$ and S° is path-connected, then S is path-connected.

   Answer: False.

   Example: Let $X := \mathbb{R}$ and $S := (0,1) \cup \{2\}$. Then $S^\circ = (0,1)$ is path-connected but S is not.

7. If $S \subset X$ and $\overline{S}$ is path-connected, then S is path-connected.

   Answer: False.

   Example: Let $X := \mathbb{R}$ and $S := \mathbb{Q}$. Then $\overline{S} = \mathbb{R}$ is path-connected but S is not.

8. If $S \subset X$ and S' is path-connected, then S is path-connected.

   Answer: False.

   Example: Let $X := \mathbb{R}$ and $S := \mathbb{Q}$. Then $S' = \mathbb{R}$ is path-connected but S is not.

9. If $S \subset X$ and the set of isolated points $S \setminus S'$ is path-connected, then S is path-connected.

   Answer: False.

   Example: Let $X := \mathbb{R}$ and $S := (0,1) \cup \{2\}$. Then $S \setminus S' = \{2\}$ is path-connected but S is not.

10. If $S \subset X$ and ∂S is path-connected, then S is path-connected.

    Answer: False.

    Example: Let $X := \mathbb{R}$ and $S := \mathbb{Q}$. Then $\partial S = \mathbb{R}$ is path-connected but S is not.

11. If $S \subset X$ is path-connected, then S° is path-connected.

    Answer: False.

    Example: Let $X := \mathbb{R}^2$ and $S := \{(x,y) : xy \geq 0\} \subset X$. Then S is path-connected but $S^\circ = \{(x,y) : xy > 0\}$ is not path-connected.

12. If S, $T \subset X$ are path-connected and $S \cap T \neq \phi$, then $S \cap T$ is path-connected.

Answer: False.

Example: Let $X := \mathbb{R}^2$, $S := S^1$, and $T := \{0\} \times \mathbb{R}$. Then S, T are path-connected and $S \cap T = \{(0,1), (0,-1)\} \neq \phi$, but it is clearly not path-connected.

13. If S, $T \subset X$ are path-connected, then $S \cup T$ is path-connected.

    Answer: False.

    Example: Let $X := \mathbb{R}$, $S := (0,1)$, and $T = (1,2)$. Then S and T are path-connected but $S \cup T = (0,1) \cup (1,2)$ is not.

14. The union of any collection of path-connected subsets of X with nonempty intersection is path-connected.

    Answer: True.

    Proof. Let $\{S_\lambda\}_{\lambda \in \Lambda}$ be any collection of path-connected subsets of X with $\bigcap_{\lambda \in \Lambda} S_\lambda \neq \phi$, and $S := \bigcup_{\lambda \in \Lambda} S_\lambda$. Let $a \in \bigcap_{\lambda \in \Lambda} S_\lambda$. For any $x_1, x_2 \in S$, there exists $\lambda_1, \lambda_2 \in \Lambda$ such that $x_i \in S_{\lambda_i}$, $i = 1, 2$. For each $i = 1, 2$, since S_{λ_i} is path-connected, there exists a path $\alpha_i : [0,1] \to S_{\lambda_i} \subset S$ joining x_i to a. The composition $\alpha_2^{-1} \cdot \alpha_1$ is then a path in S joining x_1 to x_2. Thus S is path-connected.

15. Every open ball $B(a,r) \subset X$ is path-connected.

    Answer: False.

    Example: Let $X := \mathbb{R} \setminus \{0\}$. Then $B(1,2) = (-1,0) \cup (0,3)$ is not path-connected.

16. For any $n \geq 2$, $\mathbb{R}^n \setminus \{0\}$ is path-connected hence also connected.

    Answer: True.

    Proof. For any $x \neq y \in S := \mathbb{R}^n \setminus \{0\}$, if $x \neq -y$, then the function $\gamma : [0,1] \to S$ given by $\gamma(t) := (1-t)x + ty$, $t \in [0,1]$ is a path in S joining x to y. If $x = -y$, as $n \geq 2$, there is a point $z \in S \setminus \overline{xy}$, where, as usual, $\overline{xy}$ stands for the line segments joining x and y. Then $\alpha : [0,1] \to S$ given by $\alpha(t) := (1-t)x + tz$, and $\beta : [0,1] \to S$

given by $\beta(t) := (1-t)z + ty$, $t \in [0,1]$, are paths in S joining x to z and z to y, respectively. Hence the composition $\beta \cdot \alpha$ is a path in S joining x to y via z.

17. For each $n = 0, 1, 2, \ldots$, the unit sphere $S^n := \{x \in \mathbb{R}^{n+1} : \|x\| = 1\} \subset \mathbb{R}^{n+1}$ is path-connected.

    Answer: All S^n's are path-connected except S^0.

    Proof. It is clear that $S^0 = \{-1, 1\}$ is not path-connected. For any $n \geq 1$, for any x, $y \in S^n$, there is a "great circle" (a circle of radius 1) on S^n passing through x and y. It is evident that either arc on such a great circle determined by x and y is the image of a path on S^n. Hence S^n is path-connected.

18. Every finite metric space with at least two elements is not path-connected.

    Answer: True.

    Proof. Let X be a finite metric space. As the image of any path in X must be connected, and the only connected subsets of X are singletons, every path in X must be a constant path. Hence if X contains at least two elements, it cannot be path-connected.

19. Every path-connected metric space with at least 2 elements is uncountable.

    Answer: True.

    Proof. This follows from Exercise 3.1, Part A, Problem #21, and Theorem 3.2.5.

Part B: Problems

1. Let a, $b \in \mathbb{R}^n$ be points such that $\|a\| < 1$ and $\|b\| > 1$. If $\alpha : [0,1] \to \mathbb{R}^n$ is a path joining a to b, show that there exists $t \in (0,1)$ with $\|\alpha(t)\| = 1$.

Proof. Observe first that the norm function $\|\cdot\| : \mathbb{R}^n \to \mathbb{R}$ is continuous. Hence the composition $g = \|\cdot\| \circ \alpha : [0,1] \to \mathbb{R}$ is continuous, with $g(0) = \|\alpha(0)\| < 1$, $g(1) = \|\alpha(1)\| > 1$. By the ordinary Intermediate Value Theorem of real-valued functions, there exists $t \in (0,1)$ such that $\|\alpha(t)\| = g(t) = 1$.

2. Show that continuous functions preserve path-connectedness. That is, let X, Y be metric spaces and $f : X \to Y$ be continuous. If $S \subset X$ is path-connected, then $f(S)$ is also path-connected.

   Proof. Fix $y_1, y_2 \in f(S)$. Let $s_1, s_2 \in S$ be such that $f(s_i) = y_i$, $i = 1, 2$. Since S is path-connected, there exists a path $\alpha : [0,1] \to S$ such that $\alpha(0) = s_1$, $\alpha(1) = s_2$. As f and α are continuous, the composition $f \circ \alpha : [0,1] \to f(S)$ is also continuous. Moreover, we have $f \circ \alpha(0) = f(s_1) = y_1$, $f \circ \alpha(1) = f(s_2) = y_2$ and so $f(S)$ is path-connected.

3. Determine whether the pre-image of a path-connected set under a continuous function must be path-connected or not. That is, if $f : X \to Y$ is continuous and $T \subset Y$ is path-connected, determine whether $f^{-1}(T)$ is path-connected.

   Answer: No.
   Example: Consider $f : S^1 \to \mathbb{R}$ defined by $f(x,y) := x$, which is clearly continuous on S^1. The interval $T := \left(-\frac{1}{2}, \frac{1}{2}\right) \subset \mathbb{R}$ is connected but $f^{-1}\left(\left(-\frac{1}{2}, \frac{1}{2}\right)\right)$ is separated into two disjoint open arcs on the unit circle and is hence not-path-connected.

4. Show that the graph $\Gamma := \{(x, f(x)) : x \in I\}$ of a continuous function $f : I \to \mathbb{R}$ on an interval I is path-connected.

   Proof. For any $(a, f(a))$, $(b, f(b)) \in \Gamma$, say, $a < b$. Define $\gamma : [0,1] \to \Gamma$ by

$$\gamma(t) := \Big((1-t)a + bt, f\big((1-t)a + bt\big)\Big).$$

Obviously $\gamma([0,1]) = \Gamma$, $\gamma(0) = (a, f(a))$, $\gamma(1) = (b, f(b))$, and γ is continuous. Thus Γ is path-connected.

5. If $S, T \subset X$ are path-connected, with $\overline{S} \cap T \neq \phi$ or $S \cap \overline{T} \neq \phi$, determine whether $S \cup T$ is path-connected.
[Compare with Theorem 3.1.10.]

Answer: False.

Example: Consider $X := S \cup T \subset \mathbb{R}^2$, where
$$S := \left\{(x,y) \in \mathbb{R}^2 : 0 < x \leq 1 \ , \ y = \sin\frac{1}{x}\right\} ,$$
$$T := \left\{(x,y) \in \mathbb{R}^2 : -1 \leq x \leq 0 \ , \ y = 0\right\} .$$
By Exercise 3.2, Part B, Problem #4, S and T are path-connected. Furthermore, it is easy to see that in the metric space X, $\overline{S} = S \cup \left\{(0,0)\right\} \subset X$. Hence $\overline{S} \cap T = \{(0,0)\} \neq \phi$. However, by Example 3.2.6, $X = S \cup T$ is not path-connected.

6. For any $S \subset X$, if $S^\circ \neq \phi$ and $\overline{S} \neq X$, determine whether $X \setminus \partial S$ is path-connected.
[Compare with Exercise 3.1, Part A, Problem #16.]

Answer: No.

Example. Consider $X := S \cup T \subset \mathbb{R}^2$, where
$$S := \left\{(x,y) \in \mathbb{R}^2 : 0 < x \leq 1 \ , \ y = \sin\frac{1}{x}\right\} ,$$
$$T := \left\{(x,y) \in \mathbb{R}^2 : -1 \leq x \leq 0 \ , \ y = 0\right\} .$$
It is clear that in the metric space X, $S^\circ = S \neq \phi$ and $\overline{S} = S \cup \left\{(0,0)\right\} \neq X$. However, it is elementary to see that $\partial S = \{(0,0)\}$ and so $X \setminus \partial S = S \cup \left\{(x,y) \in \mathbb{R}^2 : -1 \leq x < 0 \ , \ y = 0\right\}$ which is clearly not path-connected.

7. Prove that the set
$$A := \left\{f \in C[0,1] : \int_0^1 f(x)dx = 0\right\} \subset \left(C[0,1], d_\infty\right) ,$$
where $d_\infty(f,g) := \sup_{x\in[0,1]} \left|f(x) - g(x)\right|$, is path-connected.

Proof. For any fixed $f, g \in A$, define a function $\gamma : [0,1] \to C[0,1]$ by $\gamma(t) := tg + (1-t)f$ for all $t \in [0,1]$. Clearly γ is well-defined. Furthermore, as

$$\begin{aligned} d_\infty(\gamma(t), \gamma(s)) &= \sup_{x\in[0,1]} \Big|\gamma(t)(x) - \gamma(s)(x)\Big| \\ &= \sup_{x\in[0,1]} \Big|(t-s)\big(g(x) - f(x)\big)\Big| \\ &= |t-s| \sup_{x\in[0,1]} \Big|g(x) - f(x)\Big| = |t-s|\ d_\infty(g, f)\ , \end{aligned}$$

$\gamma : [0,1] \to C[0,1]$ is continuous and so it is a path in $C[0,1]$. Clearly, $\gamma(0) = f$ and $\gamma(1) = g$. Furthermore, for every $t \in [0,1]$, we have

$$\int_0^1 (\gamma(t))(x)dx = t\int_0^1 g(x)dx + (1-t)\int_0^1 f(x)dx = 0\ .$$

So $\gamma(t) \in A$ for all $t \in [0,1]$. Hence γ is a path in A joining f to g and so A is path-connected.

8. In $\mathbb{R}^2$, consider the "Deleted Comb Space" $S := A \cup B \cup C$, where

$$\begin{aligned} A &:= \left\{\left(\frac{1}{n},\ y\right) : 0 \le y \le 1, n \in \mathbb{N}\right\}\ , \\ B &:= \{(x, 0) : 0 \le x \le 1\}\ , \\ C &:= \{(0, 1)\}\ . \end{aligned}$$

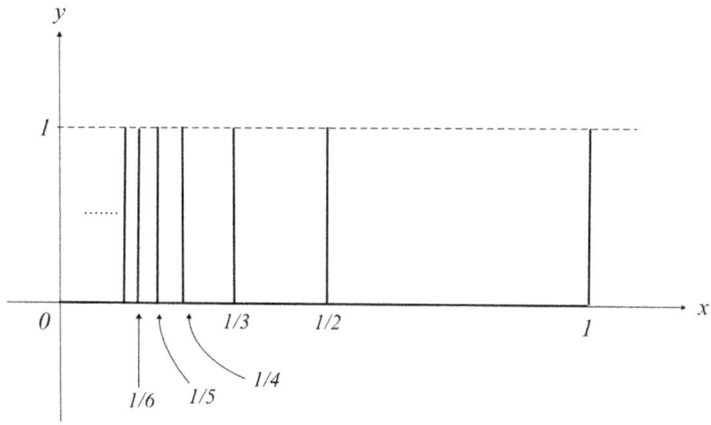

(a) Determine whether S is connected.
(b) Determine whether S is path-connected.

Solution:

(a) *Answer*: S is connected.

Proof. It is clear that $A \cup B$ is path-connected and hence connected. Furthermore, as

$$A \cup B \subset S \subset A \cup B \cup \{(0, y) : 0 \leq y \leq 1\} = \overline{A \cup B}\ ,$$

S is connected.

(b) *Answer*: S is not path-connected.

Justification. It suffices to show that the point $P := (0, 1)$ cannot be joined to any point in $A \cup B$ by a path in S. Let $\gamma : [0, 1] \to S$ be a path through P. Then $\phi \neq \gamma^{-1}(C) \subset [0, 1]$. So the problem boils down to showing $\gamma^{-1}(C) = [0, 1]$, for then γ will always stay in C and cannot go to any point in $A \cup B$.

Observe that as C is closed in $\mathbb{R}^2$, it is closed in S and so $\gamma^{-1}(C) \subset [0, 1]$ is closed. It thus remains to show that $\gamma^{-1}(C) \subset [0, 1]$ is also open, for then $\gamma^{-1}(C)$ is a nonempty clopen subset of the connected space $[0, 1]$ and so it forces $\gamma^{-1}(C) = [0, 1]$.

Pick any $t_0 \in \gamma^{-1}(C)$. As γ is continuous at t_0, there exists $\delta > 0$ such that $\gamma(B_{[0,1]}(t_0, \delta)) \subset B_S(P, \frac{1}{2})$. Since $B_{[0,1]}(t_0, \delta) = (t_0 - \delta, t_0 + \delta) \cap [0, 1]$ is connected, $\gamma(B_{[0,1]}(t_0, \delta))$ is a connected subset of $B_S(P, \frac{1}{2})$ which contains P, hence it is lying in the connected component of P in $B_S(P, \frac{1}{2})$. But then it is evident that the connected component of P in $B_S(P, \frac{1}{2})$ is $C = \{P\}$, so we have $\gamma(B_{[0,1]}(t_0, \delta)) \subset C$ and hence $B_{[0,1]}(t_0, \delta) \subset \gamma^{-1}(C)$. So $\gamma^{-1}(C)$ is open.

9. In $\mathbb{R}^2$, consider the "Deleted Infinite Broom"

$$S := \{(1,0)\} \cup \left\{ \left(s, \frac{s}{n}\right) : n \in \mathbb{N},\ 0 \leq s \leq 1 \right\}.$$

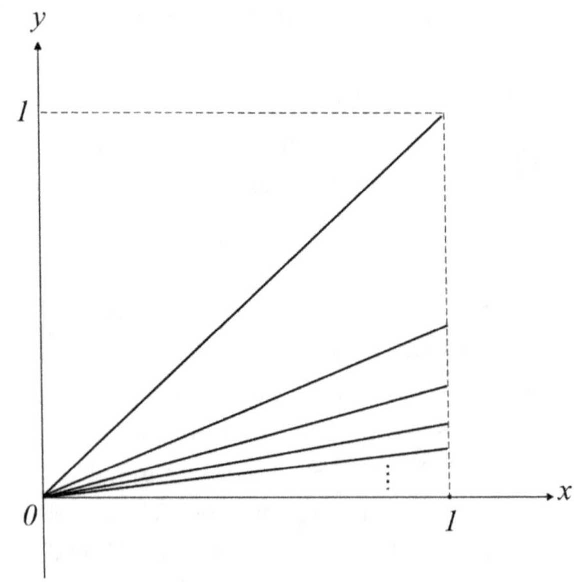

(a) Determine whether S is connected.
(b) Determine whether S is path-connected.

Solution:

(a) *Answer*: S is connected.

Proof. For each $n \in \mathbb{N}$, write $L_n := \{(s, \frac{s}{n}) : 0 \le s \le 1\}$. Then each L_n is a line segment and so each L_n is connected. As $\bigcap_{n\in\mathbb{N}} L_n = \{(0,0)\} \neq \phi$, $\bigcup_{n\in\mathbb{N}} L_n$ is connected. On the other hand, as it is evident that

$$(1,0) \in \overline{\bigcup_{n\in\mathbb{N}} L_n} \, ,$$

we have

$$\bigcup_{n\in\mathbb{N}} L_n \subset S = \{(1,0)\} \cup \bigcup_{n\in\mathbb{N}} L_n \subset \overline{\bigcup_{n\in\mathbb{N}} L_n}$$

and so S is connected.

(b) *Answer*: S is not path-connected.

Justification. As the L_n's defined in (a) are line segments, each of them is path-connected and they all pass through the point $O := (0,0)$. Hence $\bigcup_{n\in\mathbb{N}} L_n$ is path-connected. It thus remains to show that any path $\gamma : [0,1] \to S$ passing through $P := (1,0)$ cannot join any point in any L_n.

Let $\gamma : [0,1] \to S$ be a path passing through P. Without loss of generality, we assume that $\gamma(0) = P$. Let $C := \gamma^{-1}(\{P\})$. It suffices to show that $C = [0,1]$. Now since $0 \in C$, $C \neq \phi$. So the problem boils down to showing that C is clopen in $[0,1]$, for then C is a nonempty clopen subset of the connected space $[0,1]$ and so if forces $C = [0,1]$. As γ is continuous and $\{P\} \subset \mathbb{R}^2$ is closed, $C = \gamma^{-1}(\{P\})$ is closed in $[0,1]$. On the other hand, pick any $t_0 \in C$. By the continuity of γ, there exists $\delta > 0$ such that $\gamma(B_{[0,1]}(t_0, \delta)) \subset B_S(P, \frac{1}{2})$.

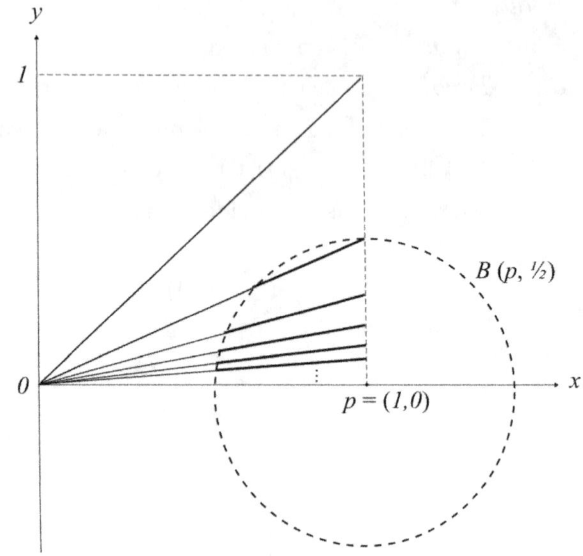

Since $B_{[0,1]}(t_0, \delta) = (t_0 - \delta, t_0 + \delta) \cap [0, 1]$ is connected, $\gamma(B_{[0,1]}(t_0, \delta))$ is a connected subset of $B_S(P, \frac{1}{2})$ which passes through P, hence it is lying in the connected component of P in $B_S(P, \frac{1}{2})$. But then it is clear that the connected component of $B_S(P, \frac{1}{2})$ is $\{P\}$, so we have $\gamma(B_{[0,1]}(t_0, \delta)) \subset \{P\}$ and hence $B_{[0,1]}(t_0, \delta) \subset \gamma^{-1}(\{P\}) = C$. So C is open in $[0, 1]$.

10. If a function preserves path-connectedness, determine whether it is continuous.

Answer: False.

Example: Consider the function $f : [0, \infty) \to \mathbb{R}$ defined by

$$f(x) = \begin{cases} 0 & x = 0 \\ \sin \dfrac{1}{x} & x > 0\,. \end{cases}$$

Then f preserves path-connectedness. In fact, suppose $S \subset [0, \infty)$ is path-connected. If S is a singleton, $f(S)$ is also a singleton and hence is path-connected. If S is not a singleton, then since S must be connected, it is a nondegenerate interval. Let x, $y \in f(S)$ be two distinct points. Then there exists u, $v \in S$ such that $f(u) = x$ and $f(v) = y$. Assume without loss of generality that $u < v$.

Case 1: $u > 0$.

Since S is an interval, $0 < (1-t)u + tv \in S$ for any $t \in [0,1]$. The function $\alpha : [0,1] \to f(S) \subset \mathbb{R}$ given by $\alpha(t) := f\big((1-t)u + tv\big)$, $t \in [0,1]$, is thus a well-defined continuous function with $\alpha(0) = f(u) = x$, and $\alpha(1) = f(v) = y$. That is, it is a path in $f(S)$ joining x to y.

Case 2: $u = 0$.

Note that as S is a nondegenerate interval in $[0,\infty)$ containing the point $u = 0$, by the definition of f, we have $f(S) = [-1,1]$ which is obviously path-connected.

Combining, we see that f preserves path-connectedness. However, it is elementary to see that f is discontinuous at 0.

11. For any $n \geq 2$, show that $\mathbb{R}^n \setminus S$ is path-connected for any countable subset $S \subset \mathbb{R}^n$.

Proof. Since $\mathbb{R}^n$ is uncountable, $\mathbb{R}^n \setminus S$ is uncountable. Let x, $y \in \mathbb{R}^n \setminus S$ be any two distinct points. We need to construct a path in $\mathbb{R}^n \setminus S$ from x to y. Note first that there are uncountably many lines passing through x. Now for any $z \in S$, there is exactly one line passing through x and z. It follows (by Pigeon-hole principle) that there must exists at least one line L passing through x, but not any point of S, that is, $L \subset \mathbb{R}^n \setminus S$. If L passes through y, then we are done. So now suppose L does not pass through y. Similar to considerations above, there are uncountably many lines passing through y, and for any $z \in S$, there is exactly one line passing through y and z. As there are uncountably many points on L, it follows (by Pigeon-hole principle) that there must exist at least one line L' passing through y and some point, say u, of L but not any point of S. Thus $L' \subset \mathbb{R}^n \setminus S$. This gives rise to a polygonal path from x to y via u in $\mathbb{R}^n \setminus S$. The result follows.

12. A metric space X is said to be *locally path-connected* if for each point $x \in X$ and for each neighborhood N of x, there is a path-connected open set $U \subset X$ such that $x \in U \subset N$. If X is connected and locally path-connected, show that X is path-connected.

Proof. Fix $a \in X$. It suffices to show that every point in X can be joined by a path in X to a. Imitating the proof of Theorem 3.2.7, let

$$A := \{x \in X : x \text{ and } a \text{ are joined by a path in } X\} .$$

As $a \in A$, we have $A \neq \phi$. So if we can show that A is clopen in X, then since X is connected and $A \neq \phi$, we must have $X = A$ is path-connected.

For any $x \in A$, there is a path α in X joining a to x. Choose a path-connected open set $U \subset X$ containing x. Then for any $y \in U$, there is a path β in U joining x to y and so the composition $\beta \cdot \alpha$ is a path in X joining a to y via x. Hence $y \in A$ and so $U \subset A$. That is, A is open.

On the other hand, for any $z \notin A$, choose a path-connected open set $V \subset X$ containing z. Then for any $w \in V$, there is a path γ in V joining w to z. Note that $w \notin A$. In fact, if $w \in A$, there would be a path δ in X joining a to w and so the composition $\gamma \cdot \delta$ would be a path in X joining a to z via w, contradicting to the choice of z. Hence we have $w \notin A$ and so $V \subset X \setminus A$, that is, A is closed.

13. Prove or disprove the following statements:
   (a) Local path-connectedness implies path-connectedness.
   (b) Path-connectedness implies local path-connectedness.

Solution:

(a) *Answer*: False.

Example: $S := [0,1] \cup [2,3] \subset \mathbb{R}$ is locally path-connected but not path-connected.

(b) *Answer*: False.

Example: The "Comb Space"

$$C = \left(\left(\{0\} \cup \{\frac{1}{n} : n \in \mathbb{N}\}\right) \times [0,1]\right) \cup ([0,1] \times \{0\})$$

in $\mathbb{R}^2$ is clearly path-connected. But at the point $p := (0,1)$, the neighborhood $B_C(p, \frac{1}{2})$ does not contain any path-connected open subset and so C is not locally path-connected.

14. Let X be a metric space and $S \subset X$. If X and ∂S are path-connected, determine whether each of the following is true:
    (a) S is path-connected.
    (b) $\overline{S}$ is path-connected.

Solution:

(a) *Answer*: False.
Example: Let $X := \mathbb{R}^2$ and $S := \{(x,y) \in \mathbb{R}^2 : x \neq 0\}$. Then X and $\partial S = \{(x,y) \in \mathbb{R}^2 : x = 0\}$ are path-connected, but S is not.

(b) *Answer*: True.
Proof. Let $x,\ y \in \overline{S}$. Since X is path-connected, there exists a path $\gamma : [0,1] \to X$ such that $\gamma(0) = x$ and $\gamma(1) = y$. If $\gamma([0,1]) \subset \overline{S}$, then we are done. Otherwise, there exists some $t_0 \in (0,1)$ such that $\gamma(t_0) \notin \overline{S}$. Let

$$T := \{t \in [0,1] : \gamma(t) \in X \setminus \overline{S}\}\ .$$

Write $a := \inf T \in [0,1]$. Observe that $a \notin T$. In fact, suppose $a \in T$. Then $\gamma(a) \in X \setminus \overline{S}$. Since $\gamma(0) \in \overline{S}$, we have $a \neq 0$. Meanwhile, since $X \setminus \overline{S}$ is open in X, by the continuity of γ, there exists $\varepsilon > 0$ such that $\gamma(B_{[0,1]}(a,\varepsilon)) \subset X \setminus \overline{S}$ and so $B_{[0,1]}(a,\varepsilon) \subset T$. This contradicts that $a = \inf T$. Hence $a \notin T$ and so $\gamma(a) \in \overline{S}$.

On the other hand, by the definition of infimum, there is a sequence $\{t_n\}_{n \in \mathbb{N}}$ in T which converges to a as $n \to \infty$. Hence

$$\gamma(a) = \lim_{n \to \infty} \gamma(t_n) \in \overline{(X \setminus \overline{S})} = X \setminus (\overline{S})^\circ \subset X \setminus S^\circ\ .$$

Combining, we have $\gamma(a) \in \overline{S} \cap (X \setminus S^\circ) = \overline{S} \setminus S^\circ = \partial S$.
Similarly, setting $b := \sup T$, we have $\gamma(b) \in \partial S$.
Finally, since ∂S is path-connected, there exists a path $\sigma : [0,1] \to \partial S$ such that $\sigma(0) = \gamma(a)$ and $\sigma(1) = \gamma(b)$. So the path $\tau : [0,1] \to \overline{S}$ defined by

$$\tau(t) := \begin{cases} \gamma(t) & \text{if } t \in [0,a] \cup [b,1] \\ \sigma\left(\frac{t-a}{b-a}\right) & \text{if } t \in (a,b) \end{cases}$$

joins x to y. Hence $\overline{S}$ is path-connected.

15. Let M_n be the set of all $n \times n$ matrices with real entries. For any A, $B \in M_n$, define

$$d(A,B) := \left[\sum_{i,j=1}^{n} |a_{ij} - b_{ij}|^2\right]^{1/2},$$

where as usual, a_{ij} and b_{ij} denote the ij-th entry of A and B, respectively.

(a) Show that d is a metric on M_n.

(b) Show that the determinant function $\det : M_n \to \mathbb{R}$ is continuous.

(c) The *orthogonal group* $O(n) \subset M_n$ is defined as

$$O(n) := \{A \in M_n : A^T A = I\},$$

where A^T denotes the transpose of A and $I \in M_n$ denotes the $n \times n$ identity matrix. Determine whether each of the following is true or false:

(i) $O(n)$ is compact.

(ii) $O(n)$ is complete.

(iii) $O(n)$ is connected.

(iv) $O(n)$ is path-connected.

(v) When n is odd, there is a matrix $A \in M_n$ such that $\det\left(\sum_{i=0}^{n}(i+1)A^i\right) = 0$.

Solution:

(a) Let $\Phi : M_n \to \mathbb{R}^{n^2}$ be defined by

$$\Phi(A) := (a_{11}, a_{12}, \dots, a_{1n}, a_{21}, a_{22}, \dots, a_{2n}, \dots, a_{nn}) .$$

Clearly, Φ is a bijection. Observe that

$$d(A, B) = \|\Phi(A) - \Phi(B)\| \quad \text{for all } A, B \in M_n .$$

So d is a well-defined metric on M_n induced by Φ from the Euclidean metric $\| \cdot \|$ on $\mathbb{R}^{n^2}$.

(b) For any $x \in \mathbb{R}^{n^2}$, the function $f(x) := \det\left(\Phi^{-1}(x)\right)$ is a polynomial of n^2 variables, which is continuous. It follows that $\det = f \circ \Phi$ is also continuous.

(c) (i) *Answer*: True.

Proof. Note that

$$O(n) = \bigcap_{i,j=1}^{n} F_{ij}^{-1}\left(\{\delta_{ij}\}\right) ,$$

where for any i, $j = 1, \dots, n$ and $F_{ij} : M_n \to \mathbb{R}$ is defined by

$$F_{ij}(A) := \sum_{k=1}^{m} a_{ik} a_{jk} \quad \text{for all } A \in M_n ,$$

and δ_{ij} is the Kronecker delta function. Note that for all i, $j = 1, \dots, n$, F_{ij} is continuous and being a singleton, $\{\delta_{ij}\} \subset \mathbb{R}$ is closed. Hence $O(n)$ is closed. On the other hand, by definition, for all $A \in O(n)$, we have

$$\sum_{j=1}^{n} a_{ij}^2 = 1 \quad \text{for all } i = 1, \dots, n .$$

Thus if we denote by $O \in M_n$ the zero $n \times n$ matrix, then

$$d(A, O) = \sum_{i,j=1}^{n} a_{ij}^2 = n \quad \text{for all } A \in O(n)$$

and so $O(n)$ is bounded in M_n. As Φ defined in (a) is by construction an isometry isomorphism, $\Phi(O(n))$ is closed and bounded in $\mathbb{R}^{n^2}$. Hence $\Phi(O(n))$ is compact. Since Φ is a homeomorphism, $O(n)$ is also compact.

(ii) *Answer*: True.

Proof. As $O(n)$ is compact, by Theorem 2.2.7, it is complete.

(iii) *Answer*: False.

Justification. For any $A \in O(n)$, since $A^T A = I$, we have $(\det(A))^2 = 1$ and hence $\det A = \pm 1$. So $\det\big(O(n)\big) \subset \{\pm 1\}$. On the other hand, it is obvious that $I \in O(n)$ with $\det(I) = 1$ and the diagonal matrix J with diagonal entries $(-1, 1, 1, \dots, 1)$ is also in $O(n)$ but with $\det(J) = -1$. Hence $\det\big(O(n)\big) = \{\pm 1\}$. That is, the continuous function det maps $O(n)$ onto a disconnected set $\{\pm 1\}$. Since continuous functions must preserve connectedness, $O(n)$ must be disconnected.

(iv) *Answer*: False.

Justification. By (iii), $O(n)$ is not connected, hence it is not path-connected.

(v) *Answer*: True.

Proof. As the function $\Phi : M_n \to \mathbb{R}^{n^2}$ defined in (a) is a homeomorphism and $\mathbb{R}^{n^2}$ is connected, M_n is also connected. Define $F : M_n \to \mathbb{R}$ by

$$F(A) := \det\left(\sum_{i=0}^{n} (i+1)A^i\right), \quad A \in M_n\,.$$

Since det is continuous, F is continuous. Observe that $F(O) = \det(I) = 1 > 0$ and

$$F(-I) = \det\left[\sum_{i=0}^{n} (i+1)(-I)^i\right] = \left[-\frac{n+1}{2}\right]^n < 0\,.$$

So by the Intermediate Value Theorem, there exists an $A \in M_n$ such that $F(A) = 0$.

Chapter 4

Uniform Continuity

In Chapter 2, the notion of continuity of a function $f : X \to Y$ at a point $a \in X$ has been introduced. Conceptually, f is continuous at a if points $x \in X$ which are close to a are being mapped by f to points $f(x) \in Y$ which are close to $f(a)$. Using the ε-δ language, f is continuous at $a \in X$ if for every given $\varepsilon > 0$, we can find a $\delta > 0$ such that

$$d_Y\big(f(x), f(a)\big) < \varepsilon \quad \text{whenever} \quad d_X(x, a) < \delta\ ,$$

or equivalently,

$$f\big(B_X(a, \delta)\big) \subset B_Y\big(f(a), \varepsilon)\big)\ .$$

Obviously, the δ we choose depends on the given ε. In general, if we are given a smaller $\varepsilon > 0$, we would need to find correspondingly a smaller δ. So to be absolutely precise, δ should be written as $\delta(\varepsilon)$ to indicate its dependence on ε. To avoid unnecessarily complicated symbols, we would simply say that "we can find a $\delta = \delta(\varepsilon) > 0$ such that ..." to indicate the dependence and then use the symbol δ instead of $\delta(\varepsilon)$ to simplify the notations afterwards.

Globally speaking, $f : X \to Y$ is continuous on X if it is continuous at every point $a \in X$. That is, for every given $\varepsilon > 0$, for every $a \in X$, we can find a $\delta > 0$ such that

$$d_Y\big(f(x), f(a)\big) < \varepsilon \quad \text{whenever} \quad d_X(x, a) < \delta\ . \tag{*}$$

An important point to note here is that in general, for a specific $\varepsilon > 0$ and a specific $a \in X$, there is a corresponding δ. That is, the δ we choose depends not only on the given ε but also on the point $a \in X$ in question. So to be absolutely precise, δ should be written as $\delta(\varepsilon, a)$ to indicate its dependence on both ε and a.

In this Chapter, we shall investigate what types of continuous functions $f : X \to Y$ would have the property that the same δ is good for all points of X, that is, given any $\varepsilon > 0$, there is a "uniform" $\delta = \delta(\varepsilon) > 0$ such that $(*)$ holds for every $a \in X$. We shall also investigate what properties such continuous functions would possess.

4.1 Uniform Continuity

Example 4.1.1. Let us verify that the function $f : (0, \infty) \to \mathbb{R}$ given by $f(x) := \frac{1}{x}$ is continuous on $(0, \infty)$.

The thinking process goes as follows: Fix $a \in (0, \infty)$. For any given $\varepsilon > 0$, we need to find a $\delta > 0$ such that whenever $|x - a| < \delta$, we have

$$|f(x) - f(a)| = \left|\frac{1}{x} - \frac{1}{a}\right| = \frac{|x-a|}{ax} < \frac{\delta}{ax} < \varepsilon\ .$$

The last inequality is true whenever $x > \frac{\delta}{a\varepsilon}$. Note that we are restricting x to $|x - a| < \delta$ and so in particular, we have $x > a - \delta$. Hence in case $a - \delta > \frac{\delta}{a\varepsilon}$, or equivalently, $\delta < \frac{a^2\varepsilon}{1+a\varepsilon}$, we would have $x > a - \delta > \frac{\delta}{a\varepsilon}$ and we are done. Hence the formal proof goes as follows:

Fix $a \in (0, \infty)$. For any given $\varepsilon > 0$, choose a $\delta > 0$ satisfying $0 < \delta < \frac{a^2\varepsilon}{1+a\varepsilon}$. Then $\delta < a\varepsilon(a - \delta)$ and so

$$|f(x) - f(a)| = \left|\frac{1}{x} - \frac{1}{a}\right| = \frac{|x-a|}{ax} < \frac{\delta}{ax} < \frac{a\varepsilon(a-\delta)}{a(a-\delta)} = \varepsilon$$

whenever $|x - a| < \delta$. Hence f is continuous at a. Since $a \in (0, \infty)$ is arbitrary, we conclude that f is continuous on $(0, \infty)$.

Note, however, that for this particular function f, for any given $\varepsilon > 0$, the $\delta > 0$ we need to choose depends on the position of the point a. In particular, when a is large, we can choose a relatively large δ. But then when a is getting closer and closer to 0, we need to pick a smaller and smaller δ. In the limiting case when $a \to 0^+$, it is easy to see that $\delta \to 0^+$ and so there does not exist a fixed $\delta > 0$ that is good for every point in $(0, \infty)$.

Definition 4.1.2. *A function $f : X \to Y$ is said to be uniformly continuous on a subset $S \subset X$ if for any $\varepsilon > 0$, there exists $\delta = \delta(\varepsilon) > 0$ such that*

$$d\big(f(x_1), f(x_2)\big) < \varepsilon \quad \textit{whenever } x_1, x_2 \in S \textit{ with } d(x_1, x_2) < \delta \ ,$$

or, equivalently, for any $\varepsilon > 0$, there exists $\delta = \delta(\varepsilon) > 0$ such that

$$f\big(B_X(x, \delta) \cap S\big) \subset B_Y\big(f(x), \varepsilon\big) \quad \textit{for every } x \in S \ .$$

That is to say, f is uniformly continuous on S if f is pointwisely continuous on S and associated to a given $\varepsilon > 0$, there is a "uniform δ" which is good for (i.e., satisfies the definition of continuity of f at) every point in S.

Example 4.1.3. If X is discrete and Y is any metric space, then every function $f : X \to Y$ is uniformly continuous on X. [Compare with Exercise 2.3, Part A, Problem #10.]

Proof. For any $\varepsilon > 0$, let $\delta := \frac{1}{2}$. Then for any x_1, $x_2 \in X$ with $d(x_1, x_2) < \delta$, since X is discrete, we have $x_1 = x_2$ and so $d\big(f(x_1), f(x_2)\big) = 0 < \varepsilon$. Hence f is uniformly continuous on X.

Example 4.1.4. $f : \mathbb{R} \to \mathbb{R}$ defined by $f(x) := x^2$ is uniformly continuous on $(0, 1]$.

Proof. For any $\varepsilon > 0$ and any x_1, $x_2 \in (0, 1]$, we have

$$|f(x_1) - f(x_2)| = |x_1^2 - x_2^2| = |x_1 + x_2| \ |x_1 - x_2| \le 2|x_1 - x_2| \ .$$

Hence whenever $2|x_1 - x_2| < \varepsilon$, we have $|f(x_1) - f(x_2)| < \varepsilon$. This suggests that $\delta := \varepsilon/2$ is a uniform δ that works everywhere on $(0, 1]$.

So the formal proof goes as follows:
For any $\varepsilon > 0$, pick $\delta := \varepsilon/2$. Then we have

$$|f(x_1) - f(x_2)| = |x_1^2 - x_2^2| = |x_1 + x_2|\,|x_1 - x_2| \leq 2|x_1 - x_2| < 2\delta = \varepsilon$$

for all x_1, $x_2 \in S$ with $|x_1 - x_2| < \delta$. Hence f is uniformly continuous on $(0, 1]$.

Remark. The careful readers should notice that in Example 4.1.4, the only reason why we restrict to the subset $(0, 1] \subset \mathbb{R}$ is to make sure that all points in there are bounded. Hence by similar arguments, it is easy to see that the function f is actually uniformly continuous on every bounded subset of $\mathbb{R}$.

Remark. It is obvious that uniformly continuous functions are automatically continuous, but the converse need not be true. In order to prove from (first principle) that a function $f : X \to Y$ is *not* uniformly continuous on $S \subset X$, we need to establish the negation of Definition 4.1.2, namely, there exists an $\varepsilon > 0$ such that for any $\delta > 0$, the condition

$$\text{“}d\big(f(x_1), f(x_2)\big) < \varepsilon \quad \text{whenever } x_1, x_2 \in S \text{ with } d(x_1, x_2) < \delta\text{”}$$

fails to hold, that is, there are points x_1, $x_2 \in S$ with $d(x_1, x_2) < \delta$ but $d\big(f(x_1), f(x_2)\big) \geq \varepsilon$.

Example 4.1.5. $f : \mathbb{R} \to \mathbb{R}$ defined by $f(x) := x^2$ is not uniformly continuous on $(0, \infty)$.

Proof. It is clear that f is continuous everywhere on $\mathbb{R}$. Also, by the Remark following Example 4.1.4, f is uniformly continuous on every bounded subset of $\mathbb{R}$. However, it is not uniformly continuous on $(0, \infty)$. In fact, in view of the preceding Remark, let us take $\varepsilon := 1$. For any $\delta > 0$, take $x_1 := \frac{1}{\delta}$ and $x_2 := \frac{1}{\delta} + \frac{\delta}{2}$. Then x_1, $x_2 \in (0, \infty)$

and $d(x_1, x_2) = \frac{\delta}{2} < \delta$, but $d(f(x_1), f(x_2)) = 1 + \frac{\delta^2}{4} > 1 = \varepsilon$. Hence the assertion.

Example 4.1.6. $f : (0, \infty) \to \mathbb{R}$ defined by $f(x) := \frac{1}{x}$ is not uniformly continuous on $(0, \infty)$.

Proof. In Example 4.1.1, we saw that f is continuous on $(0, \infty)$ and we argued that there is no uniform δ for all points in $(0, \infty)$. For the more matured readers, that is accepted to be a rigorous proof that f is not uniformly continuous on $(0, \infty)$. A more concrete proof using the preceding Remark is as follows:

Take $\varepsilon := 1$. For any $\delta > 0$, without loss of generality we can assume that $1 > \delta > 0$, we need to pick points x_1, $x_2 \in (0, \infty)$ which are close to each other (within a distance $< \delta$) but with images under f far apart from each other ($\geq \varepsilon$). In view of the considerations in Example 4.1.1, we see that the graph of f blows up very quickly as $x \to 0^+$ and so this is exactly the place where x_1 and x_2 should be in. This suggests that we pick $x_1 := \delta$ and $x_2 := \frac{\delta}{2}$. Then both x_1, $x_2 \in (0, 1) \subset (0, \infty)$, and we have $|x_1 - x_2| = \frac{\delta}{2} < \delta$, but $\big|f(x_1) - f(x_2)\big| = |\frac{1}{\delta} - \frac{2}{\delta}| = \frac{1}{\delta} > \varepsilon$. Hence by the preceding Remark, f is not uniformly continuous on $(0, \infty)$.

Theorem 4.1.7 (Heine). *If $f : X \to Y$ is continuous on a compact subset $C \subset X$, then f is uniformly continuous on C.*

Proof. Let $\varepsilon > 0$ be given. Since f is continuous on C, for any $c \in C$, there exists $\delta_c > 0$ such that

$$d\big(f(x), f(c)\big) < \frac{\varepsilon}{2} \quad \text{whenever} \quad x \in B(c, \delta_c) \cap C .$$

Collectively, $\Big\{B\big(c, \frac{\delta_c}{2}\big) : c \in C\Big\}$ is an open cover of the compact set C and so it has a finite subcover, say, $\Big\{B\big(c_i, \frac{\delta_i}{2}\big) : i = 1, \dots , n\Big\}$. Let $\delta := \min\big\{\frac{\delta_1}{2}, \dots , \frac{\delta_n}{2}\big\}$. Then $\delta > 0$. For any p, $q \in C$ with $d(p, q) < \delta$, there is an $i \in \{1, \dots, n\}$ such that $p \in B\big(c_i, \frac{\delta_i}{2}\big)$. Observe that by triangle inequality,

$$d(q, c_i) \le d(q, p) + d(p, c_i) < \delta + \frac{\delta_i}{2} \le \frac{\delta_i}{2} + \frac{\delta_i}{2} = \delta_i \ .$$

Hence $q \in B(c_i, \delta_i)$ and so both $p, q \in B(c_i, \delta_i)$. Therefore, by triangle inequality again, we have

$$d\big(f(p), f(q)\big) \le d\big(f(p), f(c_i)\big) + d\big(f(q), f(c_i)\big) < \frac{\varepsilon}{2} + \frac{\varepsilon}{2} = \varepsilon \ .$$

This completes the proof of the Theorem. □

Note that in view of Heine's Theorem, Example 4.1.4 and its immediate Remark become trivial.

Theorem 4.1.8. *Composition of uniformly continuous functions is uniformly continuous.*

Proof. The proof is similar to that for Theorem 2.3.8. Suppose $f : X \to Y$ and $g : Y \to Z$ are uniformly continuous. For any $\varepsilon > 0$, by the uniform continuity of g on Y, there exists $\delta_1 = \delta_1(\varepsilon) > 0$ such that

$$g\big(B_Y(y, \delta_1))\big) \subset B_Z\big(g(y), \varepsilon\big) \quad \text{for all} \quad y \in Y \ .$$

By the uniform continuity of f, there exists $\delta = \delta(\delta_1(\varepsilon)) > 0$ such that

$$f\big(B_X(x, \delta)\big) \subset B_Y\big(f(x), \delta_1\big) \quad \text{for all} \quad x \in X$$

and so

$$\begin{aligned} g \circ f\big(B_X(x, \delta)\big) &\subset g\left(B_Y\big(f(x), \delta_1\big)\right) \\ &\subset B_Z\big(g(f(x)), \varepsilon\big) = B_Z\big(g \circ f(x), \varepsilon\big) \end{aligned}$$

for all $x \in X$. Hence $g \circ f$ is uniformly continuous on X. □

Exercise 4.1

Part A: True or False Questions

For each of the following statements, determine if it is true or false. If it is true, prove it. If it is false, give a counterexample or provide proper justification.

1. If X is discrete and Y is any metric space, then every function $f : X \to Y$ is uniformly continuous.
   [Compare with Exercise 2.3, Part A, Problem #10.]

   Answer: True.

   Proof. For any $x \in X$ and any $\varepsilon > 0$,

   $$f(B_X(x,1)) = f(\{x\}) = \{f(x)\} \subset B_Y(f(x),\varepsilon) \ .$$

   Hence f is uniformly continuous on X.

2. $f : \mathbb{R} \to \mathbb{R}$ given by $f(x) := e^x$, $x \in \mathbb{R}$, is uniformly continuous on $[0,1]$.

   Answer: True.

   Proof. First, by elementary Calculus, f is differentiable everywhere in $\mathbb{R}$ with derivative $f'(x) = e^x$, $x \in \mathbb{R}$. For any x, $y \in [0,1]$, say, $x < y$, by Mean Value Theorem, there is $z \in (x,y)$ such that $\dfrac{e^y - e^x}{y - x} = e^z < e^y \leq e$. Hence for any $\varepsilon > 0$, take $\delta := \frac{\varepsilon}{e}$, then $e^y - e^x < (y-x)e < \delta e = \varepsilon$ whenever $x < y \in [0,1]$ with $y - x < \delta$. Hence f is uniformly continuous on $[0,1]$.

3. $f : \mathbb{R} \to \mathbb{R}$ given by $f(x) := e^x$, $x \in \mathbb{R}$, is uniformly continuous on every bounded subset of $\mathbb{R}$.

   Answer: True.

   Proof. Let $S \subset \mathbb{R}$ be a bounded set. Then there exists $M > 0$ such that $S \subset [-M,M]$. Similar to the proof of Exercise 4.1, Part A, Problem #2, it is evident that f is uniformly continuous on $[-M,M]$, hence uniformly continuous on S.

4. $f : \mathbb{R} \to \mathbb{R}$ given by $f(x) := e^x$, $x \in \mathbb{R}$, is uniformly continuous on $\mathbb{R}$.

   Answer: False.

   Justification. By elementary Calculus, f is differentiable everywhere in $\mathbb{R}$. Let $\varepsilon := 1$. For any x, $y \in \mathbb{R}$, say, $x < y$, by Mean Value Theorem, there is $z \in (x, y)$ such that $\dfrac{e^y - e^x}{y - x} = e^z > e^x$. Hence for any $1 > \delta > 0$, the points $x := -\ln \frac{\delta}{2}$ and $y := \frac{\delta}{2} - \ln \frac{\delta}{2}$ will satisfy $0 < y - x = \frac{\delta}{2} < \delta$ but $e^y - e^x > (y - x)e^x = \frac{\delta}{2} e^{-\ln \frac{\delta}{2}} = \frac{\delta}{2} \cdot \frac{2}{\delta} = 1 = \varepsilon$. Hence f is not uniformly continuous on $\mathbb{R}$.

5. The function $f(x) := \ln x$ is uniformly continuous on $(0, \infty)$.

   Answer: False.

   Justification. Take $\varepsilon := \ln 2$. For all $\delta > 0$, let n be large enough such that $\frac{1}{2n} < \delta$ and $p := \frac{1}{n}$, $q := \frac{1}{2n}$. Then $|p - q| = \frac{1}{2n} < \delta$ but $|f(p) - f(q)| = \ln 2 \not< \varepsilon$.

6. The function $f(x) := \ln x$ is uniformly continuous on $(1, 2)$.

   Answer: True.

   Proof. By Mean Value Theorem, for any $x \neq y \in (1, 2)$, we have $|f(x) - f(y)| = |\ln x - \ln y| = \frac{1}{c}\,|x - y|$ for some c between x and y. As $c \in (1, 2)$, we have $\frac{1}{c} < 1$ and so $|f(x) - f(y)| < |x - y|$. It is then evident that f is uniformly continuous on $(1, 2)$.

7. The function $f(x) := \tan^{-1} x$ is uniformly continuous on $\mathbb{R}$.

   Answer: True.

   Proof. By mean value theorem, for any x, $y \in \mathbb{R}$,

   $$\big|f(x) - f(y)\big| = \big|\tan^{-1} x - \tan^{-1} y\big| = \frac{|x - y|}{(1 + c^2)} \leq |x - y|$$

   for some c between x and y. It is then evident that f is uniformly continuous on $\mathbb{R}$.

8. The function $f(x) := \sqrt{x}$ is uniformly continuous on $[0, \infty)$.

Answer: True.

Proof. For any $\varepsilon > 0$, take $\delta := \varepsilon^2$. For any $x,\ y \in [0,\infty)$ with $|x-y| < \delta$, we have

$$\begin{aligned}|f(x) - f(y)| &= |\sqrt{x} - \sqrt{y}| = \sqrt{\left|\sqrt{x} - \sqrt{y}\right|^2} \\ &\le \sqrt{\left(\left|\sqrt{x} - \sqrt{y}\right|\right)\left(\left|\sqrt{x}\right| + \left|\sqrt{y}\right|\right)} \\ &= \sqrt{\left|\sqrt{x} - \sqrt{y}\right|\,\left|\sqrt{x} + \sqrt{y}\right|} \\ &= \sqrt{|x-y|} < \sqrt{\delta} = \varepsilon\,.\end{aligned}$$

Hence f is uniformly continuous on $[0,\infty)$.

9. For any $n \in \mathbb{N}$, the function $f : (0,\infty) \to \mathbb{R}$ defined by $f(x) := \frac{1}{x^n}$, $x \in (0,\infty)$, is uniformly continuous on $[1,\infty)$.

   Answer: True.

   Proof. By elementary Calculus, f is differentiable everywhere in $(0,\infty)$. For any $x,\ y \in [1,\infty)$, say, $x < y$, by Mean Value Theorem, there is $z \in (x,y)$ such that $\dfrac{\frac{1}{y^n} - \frac{1}{x^n}}{y-x} = -\dfrac{n}{z^{n+1}}$. For any $\varepsilon > 0$, take $\delta := \frac{\varepsilon}{n}$. Noting that $z > 1$, we have

$$\left|\frac{1}{y^n} - \frac{1}{x^n}\right| = |y-x| \cdot \left|-\frac{n}{z^{n+1}}\right| < |y-x|\,n < \delta\,n = \varepsilon$$

   whenever $|y-x| < \delta$. Hence f is uniformly continuous on $[1,\infty)$.

10. For some $n \in \mathbb{N}$, the function $f : (0,\infty) \to \mathbb{R}$ defined by $f(x) := x^n$, $x \in (0,\infty)$, is uniformly continuous on $[1,\infty)$.

    Answer: False. f is not uniformly continuous for any $n \in \mathbb{N}$.

    Justification. Fix $n \in \mathbb{N}$. Take $\varepsilon = 1$. For any $\delta > 0$, take $x \in [1,\infty)$ be any point satisfying $x^{n-1} > \frac{2}{n\delta}$ and $y := x + \frac{\delta}{2}$. Then $|y - x| = \frac{\delta}{2} < \delta$. By elementary Calculus, f is differentiable everywhere in $(0,\infty)$ and so by Mean Value Theorem, there is $z \in (x,y)$ such that $\dfrac{f(y) - f(x)}{y-x} = nz^{n-1} > nx^{n-1} > \frac{2}{\delta}$ and hence we have

$|f(y) - f(x)| > |y - x| \cdot \frac{2}{\delta} = 1 = \varepsilon$. Thus f is not uniformly continuous on $[1, \infty)$.

11. For some $n \in \mathbb{N}$, the function $f : (0, \infty) \to \mathbb{R}$ defined by $f(x) := \frac{1}{x^n}$, $x \in (0, \infty)$, is uniformly continuous on $(0, 1]$.

Answer: False. f is not uniformly continuous for any $n \in \mathbb{N}$.
Justification. Fix $n \in \mathbb{N}$. Take $\varepsilon := 1$. For any $1 > \delta > 0$, pick $x := \delta$ and $y := \frac{\delta}{2}$. Then both $x, y \in (0, 1) \subset (0, \infty)$, and we have $|x - y| = \frac{\delta}{2} < \delta$ but $|f(x) - f(y)| = |\frac{1}{\delta^n} - \frac{2^n}{\delta^n}| = \frac{2^n - 1}{\delta^n} > 1 = \varepsilon$. Hence f is not uniformly continuous on $(0, 1]$.

12. The function $f : \mathbb{R} \to \mathbb{R}$ given by $f(x) := \dfrac{1}{1 + x^4}$ is uniformly continuous.

Answer: True.
Proof. For any $\varepsilon > 0$, let $\delta := \dfrac{2\varepsilon}{3}$, then for any $x, y \in \mathbb{R}$ with $|x - y| < \delta$, since geometric mean $\leq$ algebraic mean, we have

$$|x^2 y| \leq \frac{1}{27}(8|x|^3 + 12|x|^2|y| + 6|x|\,|y|^2 + |y|^3)$$
$$|xy^2| \leq \frac{1}{27}(|x|^3 + 6|x|^2|y| + 12|x|\,|y|^2 + 8|y|^3)$$

and so

$$|x^2 y| + |xy^2| \leq |x|^3 + |y|^3 .$$

Hence

$$\begin{aligned} |f(x) - f(y)| &= \left| \frac{y^4 - x^4}{(1 + x^4)(1 + y^4)} \right| \\ &= \frac{|y^3 + y^2 x + y x^2 + x^3|\,|y - x|}{(1 + x^4)(1 + y^4)} \\ &\leq \frac{(|y^3| + |y^2 x| + |y x^2| + |x^3|)\,|y - x|}{(1 + x^4)(1 + y^4)} \\ &\leq \frac{2|y|^3 + 2|x|^3}{(1 + x^4)(1 + y^4)}\,|y - x| . \end{aligned}$$

Also because geometric mean $\leq$ algebraic mean, we have

$$|y|^3 \leq \frac{|y|^4 + |y|^4 + |y|^4 + 1}{4}, \ |x|^3 \leq \frac{|x|^4 + |x|^4 + |x|^4 + 1}{4},$$

and so

$$\left|f(x) - f(y)\right| \leq \frac{2 + 3y^4 + 3x^4}{2(1 + x^4 + y^4 + x^4y^4)} \ |y - x| < \frac{3}{2}\delta = \varepsilon .$$

Alternative Proof. It is elementary to check that

$$\begin{aligned} \left|y^3 + y^2x + yx^2 + x^3\right| &\leq \begin{cases} 4|x|^3 & \text{if } |y| \leq |x| \\ 4|y|^3 & \text{if } |x| \leq |y| \end{cases} \\ &\leq 4(|x|^3 + |y|^3) , \end{aligned}$$

while

$$|x|^3 \leq \begin{cases} 1 & \text{if } |x| \leq 1 \\ |x|^4 & \text{if } |x| \geq 1 \end{cases} \leq (1 + |x|^4)(1 + |y|^4) ,$$

and similarly,

$$|y|^3 \leq (1 + |x|^4)(1 + |y|^4) .$$

Hence

$$\begin{aligned} |f(x) - f(y)| &= \left|\frac{y^4 - x^4}{(1 + x^4)(1 + y^4)}\right| \\ &= \frac{\left|y^3 + y^2x + yx^2 + x^3\right| \ |y - x|}{(1 + x^4)(1 + y^4)} \\ &\leq 8|x - y| . \end{aligned}$$

So for any $\varepsilon > 0$, by taking $\delta := \varepsilon/8$, the uniform continuity of f is readily seen.

13. $f : [a, \infty) \to \mathbb{R}$ defined by $f(x) := 1/(1 + x)$ is uniformly continuous for every $a > -1$.

Answer: True.

Proof. For any $\varepsilon > 0$, take $\delta := \varepsilon(1+a)^2 > 0$, then

$$|f(x) - f(y)| = \frac{1}{(1+x)(1+y)} |x - y| \leq \frac{|x-y|}{(1+a)^2} < \varepsilon$$

whenever $|x - y| < \delta$.

14. The function $f : (1, \infty) \to \mathbb{R}$ given by $f(x) := \dfrac{x}{(1-x)^2}$ is uniformly continuous.

Answer: False.

Justification. Consider the sequences $\{x_n\}_{n\in\mathbb{N}}$, $\{y_n\}_{n\in\mathbb{N}}$ in $(1, \infty)$ defined by $x_n := 1 + \frac{1}{n}$, $y_n := 1 + \frac{2}{n}$, $n \in \mathbb{N}$. We have

$$\left|f(x_n) - f(y_n)\right| = \left|\left(1 + \frac{1}{n}\right)n^2 - \left(1 + \frac{2}{n}\right)\frac{n^2}{4}\right| = \frac{3n^2 + 2n}{4}.$$

Take $\varepsilon := 1$. For any $\delta > 0$, and any $N \in \mathbb{N}$ with $N > \frac{1}{\delta}$, we have $|x_N - y_N| = \frac{1}{N} < \delta$ but

$$\left|f(x_N) - f(y_N)\right| = \frac{3N^2 + 2N}{4} \geq \frac{5}{4} > 1 = \varepsilon$$

and so f is not uniformly continuous.

Alternative Justification. Take $\varepsilon := 1$. For any $1 > \delta > 0$, take $x_1 := 1 + \delta$ and $x_2 := 1 + \frac{\delta}{2}$. Then

$$\left|f(x_1) - f(x_2)\right| = \frac{3+\delta}{\delta^2} = \frac{3}{\delta^2} + \frac{1}{\delta} > 4 > \varepsilon$$

and so f is not uniformly continuous.

15. If $f : \mathbb{R} \to \mathbb{R}$ satisfies the following conditions
    (i) f is continuous at $x = 0$,
    (ii) $f(0) = 0$,
    (iii) f is "sub-additive", that is, $f(x_1 + x_2) \leq f(x_1) + f(x_2)$ for all $x_1, x_2 \in \mathbb{R}$,

    then f is uniformly continuous on $\mathbb{R}$.

Answer: True.

Proof. For any $\varepsilon > 0$, since f is continuous at 0 and $f(0) = 0$, there exists $\delta > 0$ such that $|f(t)| = |f(t) - f(0)| < \varepsilon$ whenever $|t| = |t - 0| < \delta$. So for any x, $t \in \mathbb{R}$ with $|t| < \delta$, we have, by (iii),

$$f(x+t) - f(x) \leq f(t) < \varepsilon$$

and

$$f(x) - f(x+t) \leq f(-t) < \varepsilon\ .$$

Combining, we have $|f(x+t) - f(x)| < \varepsilon$ and so f is uniformly continuous on $\mathbb{R}$.

16. Every differentiable function $f : \mathbb{R} \to \mathbb{R}$ is uniformly continuous.

    Answer: False.

    Example: The function $f : \mathbb{R} \to \mathbb{R}$ given by $f(x) := x^2$ is differentiable everywhere but by Example 4.1.5, it is not uniformly continuous on $(0, \infty)$ and hence neither is it uniformly continuous on $\mathbb{R}$.

17. Every uniformly continuous function $f : \mathbb{R} \to \mathbb{R}$ is differentiable.

    Answer: False.

    Example: The function $f : \mathbb{R} \to \mathbb{R}$ given by $f(x) := |x|$ is clearly uniformly continuous but it is not differentiable at 0.

18. If $f : S \subset \mathbb{R} \to \mathbb{R}$ is uniformly continuous, then so is $|f|$.

    Answer: True.

    Proof. For any $\varepsilon > 0$, there exists $\delta > 0$ such that $|f(y) - f(x)| < \varepsilon$ for all x, $y \in S$ with $|y - x| < \delta$. Hence by triangle inequality,

$$\Big| |f|(y) - |f|(x) \Big| = \Big| |f(y)| - |f(x)| \Big| \leq \Big| f(y) - f(x) \Big| < \varepsilon$$

for all x, $y \in S$ with $|y - x| < \delta$. Hence the assertion.

19. If $f : X \to Y$ is continuous on a closed and bounded subset $S \subset X$, then f is uniformly continuous on S.

Answer: False.

Example: Let $X := (0, 1]$ and $f : X \to \mathbb{R}$ be given by $f(x) := 1/x$. Then $S := X$ is closed (in X) and bounded, f is continuous but not uniformly continuous.

20. The sum of two uniformly continuous functions $f, g : S \subset \mathbb{R} \to \mathbb{R}$ is uniformly continuous.

Answer: True.

Proof. For any $\varepsilon > 0$, there exists δ_f, $\delta_g > 0$ such that

$$\begin{aligned} \big|f(y) - f(x)\big| &< \frac{\varepsilon}{2} \quad \text{for all } x, y \in S \text{ with } |y - x| < \delta_f \ , \\ \big|g(y) - g(x)\big| &< \frac{\varepsilon}{2} \quad \text{for all } x, y \in S \text{ with } |y - x| < \delta_g \ . \end{aligned}$$

Take $\delta := \min\{\delta_f, \delta_g\} > 0$. For all x, $y \in S$ with $|y - x| < \delta$, we have

$$\begin{aligned} \big|(f + g)(y) - (f + g)(x)\big| &= \big|f(y) + g(y) - f(x) - g(x)\big| \\ &\le \big|f(y) - f(x)\big| + \big|g(y) - g(x)\big| \\ &< \varepsilon/2 + \varepsilon/2 = \varepsilon \end{aligned}$$

and so $f + g$ is uniformly continuous on S.

21. The product of two uniformly continuous functions $f, g : S \subset \mathbb{R} \to \mathbb{R}$ is uniformly continuous.

Answer: False.

Example: The functions $f, g : \mathbb{R} \to \mathbb{R}$ given by $f(x) := g(x) := x$, $x \in \mathbb{R}$, are clearly uniformly continuous on $\mathbb{R}$. However, as has been shown in Example 4.1.5, the function $(fg)(x) = x^2$ is not.

Part B: Problems

1. If $f : \mathbb{R} \to \mathbb{R}$ is differentiable with f' bounded on an interval $I \subset \mathbb{R}$, show that f is uniformly continuous on I.

   Proof. Suppose $|f'| \leq M$ on I. For any $x \neq y \in I$, say, $x < y$, by Mean Value Theorem, there exists $z \in (x, y)$ such that

   $$|f(x) - f(y)| = |f'(z)(y - x)| = |f'(z)|\,|y - x| \leq M\,|y - x|\,.$$

   For any $\varepsilon > 0$, let $\delta := \varepsilon/M$. Then

   $$|f(x) - f(y)| < \varepsilon \quad \text{whenever } x, y \in I \text{ with } |x - y| < \delta$$

   and so f is uniformly continuous on I.

2. Show that every polynomial function is uniformly continuous on every bounded subset of $\mathbb{R}$.

   Proof. Let p be a polynomial and $S \subset \mathbb{R}$ be a bounded subset. Then certainly there is a bounded interval $I \subset \mathbb{R}$ such that $S \subset I$. Observe that every polynomial is bounded on every bounded subset of $\mathbb{R}$, so p' which is again a polynomial, is bounded on I. Hence by Exercise 4.1, Part B, Problem #1, p is uniformly continuous on I and hence it is uniformly continuous on S.

3. Determine whether it is true that every non-constant polynomial function is uniformly continuous on $\mathbb{R}$.

   Answer: False.
   Justification. Let

   $$p(x) := a_n x^n + a_{n-1} x^{n-1} + \cdots + a_1 x + a_0$$

   be a non-constant polynomial. Without loss of generality, we assume that $a_n > 0$. We have

   $$p'(x) = n a_n x^{n-1} + (n-1) a_{n-1} x^{n-2} + \cdots + a_1\,.$$

As the leading term na_nx^{n-1} is the dominating term of p', there exists $R \gg$ such that $p'(x) > \frac{na_n}{2}x^{n-1}$ for all $x \geq R$.
Let $\varepsilon := 1$. For any $\delta > 0$, by taking a larger R if necessary, we may assume that $R^{n-1} \geq \frac{4}{na_n\delta}$. Let $x := R$ and $y := R + \frac{\delta}{2}$. Then $|y - x| = \frac{\delta}{2} < \delta$. Since p is differentiable everywhere, by Mean Value Theorem, there exists $z \in (x, y)$ such that

$$p(y) - p(x) = (y - x)p'(z)$$

and so

$$\begin{aligned}|p(y) - p(x)| &= |y - x|\,|p'(z)| = \frac{\delta}{2}\,|p'(z)| \\ &> \frac{\delta}{2}\,\frac{na_n}{2}\,z^{n-1} \geq \frac{\delta}{2}\,\frac{na_n}{2}\,R^{n-1} \geq 1 = \varepsilon\,.\end{aligned}$$

Hence p is not uniformly continuous on $\mathbb{R}$.

4. Prove that the sine function is uniformly continuous on $\mathbb{R}$.

   Proof. Since the sine function is differentiable in $\mathbb{R}$ and with derivative (the cosine function) bounded on $\mathbb{R}$, by Exercise 4.1, Part B, Problem #1, the sine function is uniformly continuous on $\mathbb{R}$.

5. Prove that the cosine function is uniformly continuous on $\mathbb{R}$.

   Proof. Since the cosine function is differentiable in $\mathbb{R}$ and with derivative (negative of the sine function) bounded on $\mathbb{R}$, by Exercise 4.1, Part A, Problem #1, the cosine function is uniformly continuous on $\mathbb{R}$.

6. Prove or disprove: $f : (0, 1) \to \mathbb{R}$ given by $f(x) := \cos\frac{1}{x}$, $x \in (0, 1)$, is uniformly continuous.

   Answer: False.

   Justification. Let $\varepsilon = 1$. For any $\delta > 0$, let $n \in \mathbb{N}$ be such that $\frac{1}{2n(2n+1)\pi} < \delta$. Then $d\left(\frac{1}{2n\pi}, \frac{1}{(2n+1)\pi}\right) < \delta$ but

$$d\left(f\left(\frac{1}{2n\pi}\right), f\left(\frac{1}{(2n+1)\pi}\right)\right) = 2 > 1 = \varepsilon\,.$$

   Thus f is not uniformly continuous.

7. Prove or disprove: $f : (0,1] \to \mathbb{R}$ given by $f(x) := x \cos \frac{1}{x}$, $x \in (0,1]$, is uniformly continuous.

   Answer: True.

   Proof. Let $\tilde{f} : [0,1] \to \mathbb{R}$ be the trivial extension of f to $[0,1]$, that is,

$$\tilde{f}(x) := \begin{cases} f(x) = x \cos \frac{1}{x} & \text{for } x \in (0,1] \\ 0 & \text{for } x = 0. \end{cases}$$

   It is elementary to check that $\tilde{f}$ is continuous on the compact set $[0,1]$ and so it is uniformly continuous there. In particular, its restriction to $(0,1]$ is also uniformly continuous. That means f is uniformly continuous on $(0,1]$.

8. The product of two uniformly continuous functions $f,\ g : S \subset \mathbb{R} \to \mathbb{R}$, where g is bounded, is uniformly continuous. [Compare with Exercise 4.1, Part A, Problem #21.]

   Answer: False.

   Example: The function $f : \mathbb{R} \to \mathbb{R}$ given by $f(x) := x$, $x \in \mathbb{R}$, is clearly uniformly continuous on $\mathbb{R}$. On the other hand, the function $g : \mathbb{R} \to \mathbb{R}$ given by $f(x) := \sin x$, $x \in \mathbb{R}$, is clearly bounded on $\mathbb{R}$ and by Exercise 4.1, Part B, Problem #4, uniformly continuous on $\mathbb{R}$. However, $(fg)(x) = x \sin x$ is not uniformly continuous on $\mathbb{R}$. In fact, take $\varepsilon := 1$. Then for any $1 > \delta > 0$, whenever $n \in \mathbb{N}$ is large, we have $2n\pi\ \sin(\delta/2) > 1$ and so

$$\left| (fg)\Big(2n\pi + \frac{\delta}{2}\Big) - (fg)(2n\pi) \right| = \Big(2n\pi + \frac{\delta}{2}\Big) \sin \frac{\delta}{2} > 1 = \varepsilon\,.$$

9. Prove or disprove: Every uniformly continuous function is bounded.

Answer: False.

Example: The identity function $f : \mathbb{R} \to \mathbb{R}$ given by $f(x) := x$ for all $x \in \mathbb{R}$ is clearly uniformly continuous on $\mathbb{R}$. However, it is obvious that f is unbounded.

10. Prove or disprove: If $S \subset \mathbb{R}$ is bounded and $f : S \subset \mathbb{R} \to \mathbb{R}$ is uniformly continuous, then f is bounded on S.

Answer: True.

Proof. By the uniform continuity of f on S, for any $\varepsilon > 0$, there exists $\delta > 0$ such that

$$f(z) \in B_{\mathbb{R}}(f(x), \varepsilon) \quad \text{whenever } z \in B_S(x, \delta) \; .$$

By Exercise 1.3, Part B, Problem #10(c)(ii), $S \subset \mathbb{R}$ is totally bounded and so it is covered by finitely many open balls of radius δ, say $B_S(x_i, \delta)$, $i = 1, \ldots, n$. Hence for any $x \in S$, $x \in B_S(x_i, \delta)$ for some $i = 1, \ldots, n$ and so $f(x) \in B_{\mathbb{R}}\big(f(x_i), \varepsilon\big)$. Therefore, $f(S) \subset \bigcup_{i=1}^{n} B\big(f(x_i), \varepsilon\big)$ and so f is bounded on S.

11. Prove or disprove: The tangent function is uniformly continuous on $\left(-\frac{\pi}{2}, \frac{\pi}{2}\right)$.

Answer: False.

Justification. Intuitively, it is evident that the tangent function blows up as x approaches $\frac{\pi}{2}$, so there is no hope that there could be a uniform δ for all $x \in \left(-\frac{\pi}{2}, \frac{\pi}{2}\right)$.

To prove this in more concrete terms, we let $f(x) := \tan x$ for $x \in \left(-\frac{\pi}{2}, \frac{\pi}{2}\right)$. Let $\varepsilon := 1$. For any $0 < \delta < \frac{\pi}{2}$, let $x_1 := \frac{\pi}{2} - \delta > 0$. Then $f(x_1)$ is a finite positive number. Since $\lim_{x \to \frac{\pi}{2}} f(x) = \infty$, there exists $x_2 \in (x_1, \frac{\pi}{2})$ such that $f(x_2) > f(x_1) + 1$. Thus x_1, $x_2 \in (0, \frac{\pi}{2})$ are such that $|x_1 - x_2| < \delta$ but $|f(x_1) - f(x_2)| > 1 = \varepsilon$. Thus f is not uniformly continuous.

Alternatively, here is a simple indirect proof by Exercise 4.1, Part B, Problem #10: In fact, since the tangent function is unbounded on the bounded set $\left(-\frac{\pi}{2}, \frac{\pi}{2}\right)$, it cannot be uniformly continuous there.

12. Prove or disprove: Every linear map $T : \mathbb{R}^n \to \mathbb{R}^m$ is uniformly continuous.

Answer: True.

Proof. By elementary Linear Algebra, the linear map $T : \mathbb{R}^n \to \mathbb{R}^m$ is represented by an $m \times n$ matrix $A = (a_{ij})_{m\times n}$ of real entries a_{ij}, $1 \le i \le m$, $1 \le j \le n$. Then for all x, $y \in \mathbb{R}^n$, it is easily verified that

$$\|Tx - Ty\| = \|T(x-y)\| = \|A(x-y)\| \le M\|x-y\| ,$$

where $M := \sup\{|a_{ij}| : 1 \le i \le m, 1 \le j \le n\}$, from which it is now obvious that T is uniformly continuous.

13. Let $A \subset X$ be any subset. Show that the function $f : X \to \mathbb{R}$ given by $f(x) := d(x, A)$, $x \in X$, is uniformly continuous. [Compare with Exercise 2.3, Part B, Problem #2.]

Proof. The proof is exactly the same as that for Exercise 2.3, Part B, Problem #2, noting that the δ chosen is independent of the choice of $x \in X$.

14. Let $\{x_n\}_{n\in\mathbb{N}}$ be a fixed sequence in X. Show that the function $f : X \to \mathbb{R}$ given by $f(x) := \inf\{d(x, x_n) : n \in \mathbb{N}\}$ is uniformly continuous on X.

Proof. For any x, $y \in X$ and any $n \in \mathbb{N}$, by triangle inequality, we have

$$|d(x,x_n) - d(y,x_n)| \le d(x,y)$$

or

$$d(y,x_n) - d(x,y) \le d(x,x_n) \le d(y,x_n) + d(x,y) .$$

Thus

$$\begin{aligned} f(y) - d(x,y) &= \inf_{n\in\mathbb{N}} d(y,x_n) - d(x,y) \le \inf_{n\in\mathbb{N}} d(x,x_n) \\ &= f(x) \\ &\le \inf_{n\in\mathbb{N}} d(y,x_n) + d(x,y) \\ &= f(y) + d(x,y) \end{aligned}$$

and so

$$|f(x) - f(y)| \leq d(x, y) ,$$

from which it is readily seen that f is uniformly continuous on X.

15. A function $f : X \to Y$ is said to be *Lipschitz* if there exists $M > 0$ called a *Lipschitz constant* of f such that

$$d\big(f(x_1), f(x_2)\big) \leq M\, d(x_1, x_2) \quad \text{for all } x_1, x_2 \in X .$$

If $f : \mathbb{R} \to \mathbb{R}$ is differentiable and f' is bounded on $\mathbb{R}$, show that f is Lipschitz.

Proof. Suppose $f : \mathbb{R} \to \mathbb{R}$ is differentiable with $|f'| \leq M$ for some $M > 0$. Then for any x_1, $x_2 \in \mathbb{R}$, by Mean Value Theorem, there exists t between x_1 and x_2 such that

$$|f(x_2) - f(x_1)| = |f'(t)(x_2 - x_1)| = |f'(t)|\, |x_2 - x_1| \leq M\, |x_2 - x_1| .$$

Hence f is Lipschitz.

16. Show that every Lipschitz function is uniformly continuous.

Proof. Suppose $f : X \to Y$ is Lipschitz with Lipschitz constant M. For any $\varepsilon > 0$, let $\delta := \varepsilon/M$. Then for any x_1, $x_2 \in X$ with $d(x_1, x_2) < \delta$, we have $d(f(x_1), f(x_2)) \leq M\, d(x_1, x_2) < M\delta = \varepsilon$ and so f is uniformly continuous.

17. Let (X, d) be a metric space. Equip the product space $X \times X$ with the metric

$$\rho\big((x_1, y_1), (x_2, y_2)\big) := d(x_1, x_2) + d(y_1, y_2)$$

for any x_i, $y_i \in X$, $i = 1, 2$. The the metric d on X can be considered as a function $d : X \times X \to \mathbb{R}$. Show that d is uniformly continuous.

Proof. By Example 2.3.7, ρ is a well-defined metric on $X \times X$. Furthermore, from the first proof of the continuity of d on $X \times X$ in the same Example, it is easy to see that associated to any $\varepsilon > 0$, a uniform δ ($= \frac{\varepsilon}{2}$) can be chosen. Hence d is indeed uniformly continuous on $X \times X$.

18. Show that Cauchyness is preserved by uniformly continuous functions, that is, uniform continuous functions map Cauchy sequences to Cauchy sequences.
[Compare with Exercise 2.3, Part A, Problem #7.]

Proof. Suppose $f : X \to Y$ is uniformly continuous and $\{x_n\}_{n\in\mathbb{N}}$ is a Cauchy sequence in X. Then for any $\varepsilon > 0$, there exists $\delta = \delta(\varepsilon) > 0$ such that

$$d_Y\big(f(x_1), f(x_2)\big) < \varepsilon \quad \text{whenever } d_X(x_1, x_2) < \delta\ .$$

Since $\{x_n\}_{n\in\mathbb{N}}$ is Cauchy, there exists $N \in \mathbb{N}$ such that $d_X(x_n, x_m) < \delta$ for all n, $m \geq N$. Hence we have

$$d_Y\big(f(x_n), f(x_m)\big) < \varepsilon \quad \text{for all } n, m \geq N$$

and so $\{f(x_n)\}_{n\in\mathbb{N}}$ is Cauchy.

19. Determine whether the converse of Exercise 4.1, Part B, Problem #18 is valid. That is, whether a function is uniformly continuous if it preserves Cauchyness.
[Compare with Exercise 2.3, Part A, Problem #8.]

Answer: No.

Example: Consider the exponential function $f : \mathbb{R} \to \mathbb{R}$, $f(x) := e^x$, $x \in \mathbb{R}$. If $\{x_n\}_{n\in\mathbb{N}}$ is a Cauchy sequence in $\mathbb{R}$, then since $\mathbb{R}$ is complete, $\{x_n\} \to$ some $x \in \mathbb{R}$ as $n \to \infty$. Since f is continuous, $\{f(x_n)\} \to f(x)$ as $n \to \infty$ and so in particular, $\{f(x_n)\}_{n\in\mathbb{N}}$ is Cauchy. Thus f preserves Cauchyness but we have seen in Exercise 4.1, Part A, #4 that it is not uniformly continuous on $\mathbb{R}$.

20. Determine whether boundedness is preserved by uniformly continuous functions.
[Compare with Example 2.3.22.]

Answer: No.

Example: Consider the identity mapping $f : (\mathbb{N}, d_1) \to (\mathbb{N}, d_2)$, where d_1 is the discrete metric, and d_2 is the Euclidean metric. By Example 4.1.3, f is uniformly continuous on $(\mathbb{N}, d_1)$. However, $\mathbb{N}$ is bounded in d_1 but $f(\mathbb{N}) = \mathbb{N}$ is unbounded in d_2. So boundedness is not preserved by the uniformly continuous function f.

21. Prove that total boundedness (cf. Exercise 1.3, Part B, Problem #10) is preserved by uniformly continuous functions.

Proof. Let $f : X \to Y$ be uniformly continuous and $S \subset X$ be totally bounded. Fix $r > 0$. Suppose $f(S) \subset Y$ is not totally bounded. Then $f(S)$ cannot be covered by any finite collection of open balls in Y of radius r. Pick $y_1 \in f(S)$. Then $B(y_1, r) \not\supset f(S)$ and so there exists $y_2 \in f(S) \setminus B(y_1, r)$. In particular, we have $d(y_2, y_1) \geq r$. Since $B(y_1, r) \cup B(y_2, r) \not\supset f(S)$, there exists $y_3 \in f(S) \setminus \bigcup_{i=1}^{2} B(y_i, r)$. In particular, we have $d(y_i, y_j) \geq r$ for any $i \neq j \in \{1, 2, 3\}$. Inductively, we have a sequence $\{y_n\}_{n \in \mathbb{N}}$ in $f(S)$ such that $d(y_i, y_j) \geq r$ for any $i \neq j$. Let $\{x_n\}_{n \in \mathbb{N}}$ be a sequence in S with $f(x_n) = y_n$ for all $n \in \mathbb{N}$. As the y_i's are all distinct, the x_i's must also be all distinct. Since S is totally bounded, by Exercise 2.2, Part B, Problem #3, $\{x_n\}_{n \in \mathbb{N}}$ contains a Cauchy subsequence, say, $\{x_{n_k}\}_{k \in \mathbb{N}}$. By Exercise 4.1, Part B, Problem #18, $\{y_{n_k}\}_{k \in \mathbb{N}} = \{f(x_{n_k})\}_{k \in \mathbb{N}}$ is a Cauchy sequence, which is impossible. Hence $f(S)$ must be totally bounded.

22. Determine whether the converse of Exercise 4.1, Part B, Problem #21 is valid. That is, whether a function is uniformly continuous if it preserves total boundedness.

Answer: No.

Example: First of all, it is elementary to verify that every bounded subset of $\mathbb{R}$ is totally bounded.

Consider the exponential function $f : \mathbb{R} \to \mathbb{R}$, $f(x) := e^x$. Let $S \subset \mathbb{R}$ be a totally bounded subset. For any $r > 0$, S can be covered by finitely many open balls $B_i(r) \subset \mathbb{R}$ of radius r, $i = 1, \ldots, n$, which are bounded subsets of $\mathbb{R}$ and so they are all totally bounded. Being continuous on $\mathbb{R}$, by Theorem 4.1.7, f is uniformly continuous on every compact subset of $\mathbb{R}$. Since every bounded subset of $\mathbb{R}$ is contained in a compact subset of $\mathbb{R}$, f is uniformly continuous on every bounded subset of $\mathbb{R}$. So in particular, f is uniformly continuous on each $B_i(r)$. By Exercise 4.1, Part B, Problem #21, $f(B_i(r))$ is totally bounded for each $i = 1, \ldots, n$. Therefore, $f(S) = \bigcup_{i=1}^{n} f(B_i(r))$ is totally bounded. Hence f preserves total boundedness. However, we have seen in Exercise 4.1, Part A, #4 that it is not uniformly continuous on $\mathbb{R}$.

23. Let $f : X \to Y$ be any function and $g : Y \to Z$ be $1-1$ and continuous. Suppose $h := g \circ f : X \to Z$ is uniformly continuous.
    (a) If Y is compact, show that f is also uniformly continuous.
    (b) If g is not $1-1$, is the assertion of (a) still valid?
    (c) If X and Z are compact but Y is not, is the assertion of (a) still valid?

Solution.

(a) Since g is $1-1$ and continuous and Y is compact, by Theorem 2.3.16, $g^{-1} : g(Y) \to Y$ exists and is continuous. Since Y is compact and g is continuous, $g(Y)$ is also compact and so by Theorem 4.1.7, g^{-1} is uniformly continuous. Hence $f = g^{-1} \circ h$ is also uniformly continuous by Theorem 4.1.8.

(b) The assertion of (a) will no longer be valid.

Example: Let $Y \neq$ singleton be any compact space, $f : X \to Y$ be any discontinuous function and $g : Y \to Z$ be a constant function. Being a constant function, g is continuous but not

$1-1$. Then $h = g \circ f$ is a constant function and so it is uniformly continuous. However, f is not uniformly continuous.

(c) The assertion of (a) will no longer be valid.

Example: Let $X := [0,2]$, $Y := [0,1] \cup (2,3]$, $Z := [0,2]$. Then X and Z are compact but Y is not. Consider $f : X \to Y$ and $g : Y \to Z$ given by

$$f(x) := \begin{cases} x & x \in [0,1] \\ x+1 & x \in (1,2], \end{cases}$$

$$[12pt]g(y) := \begin{cases} y & y \in [0,1] \\ y-1 & y \in (2,3]\ . \end{cases}$$

Then g is $1-1$ and continuous on Y, $h(x) = g \circ f(x) = x$ is uniformly continuous on X, but f is not continous on X.

24. Let $D \subset X$ be a dense subset of X, Y be a complete metric space, and $f : D \to Y$ be uniformly continuous. Show that f can be extended to a uniformly continuous function $F : X \to Y$ in the sense that $F|_D = f$, and that such an extension is unique.

Proof. Since D is dense in X, for any $x \in X$, we have a sequence $\{x_n\}_{n\in\mathbb{N}}$ in D which converges to x as $n \to \infty$. In particular, $\{x_n\}_{n\in\mathbb{N}}$ is Cauchy and so by Exercise 4.1, Part B, Problem #18, $\{f(x_n)\}_{n\in\mathbb{N}}$ is a Cauchy sequence in the complete metric space Y. Hence $\{f(x_n)\} \to$ some $y \in Y$ as $n \to \infty$. Define $F(x) := y$.

We claim that $F : X \to Y$ such defined is actually well-defined, that is, for any other sequence $\{t_n\}_{n\in\mathbb{N}}$ in D which converges to x as $n \to \infty$, we must have $\{f(t_n)\} \to y$ as $n \to \infty$.

In fact, for any $\varepsilon > 0$, by the uniform continuity of f on D, there exists $\delta > 0$ such that $d(f(p), f(q)) < \varepsilon$ for any p, $q \in D$ with $d(p,q) < \delta$. Since $\{x_n\} \to x$ and $\{t_n\} \to x$ as $n \to \infty$, there exists $N \in \mathbb{N}$ such that $d(x_n, x) < \delta/2$ and $d(t_n, x) < \delta/2$ for all $n \geq N$. By triangle inequality, we have $d(x_n, t_n) < \delta$ for all $n \geq N$. Hence $d(f(x_n), f(t_n)) < \varepsilon$ for all $n \geq N$. As it is known that $\{f(x_n)\} \to y$ as $n \to \infty$, this forces $\{f(t_n)\} \to y$ as $n \to \infty$ (why?) This proves the claim and $F : X \to Y$ is well-defined.

Next, for any $x \in D$, the constant sequence $\{x_n := x\}_{n\in\mathbb{N}}$ clearly converges to x as $n \to \infty$, hence $F(x) = f(x)$ and so $F|_D = f$.

We move onto show the uniform continuity of F on X. For any $\varepsilon > 0$, choose $\delta > 0$ as above. For any p, $q \in X$ with $d(p, q) < \delta/3$, there exist sequences $\{p_n\}_{n\in\mathbb{N}}$ and $\{q_n\}_{n\in\mathbb{N}}$ in D such that $\{p_n\} \to p$ and $\{q_n\} \to q$ as $n \to \infty$. Hence there exists $N \in \mathbb{N}$ such that $d(p_n, p) < \delta/3$ and $d(q_n, q) < \delta/3$ for all $n \geq N$. By triangle inequality,

$$d(p_n, q_n) \leq d(p_n, p) + d(p, q) + d(q, q_n) < \delta \quad \text{for all } n \geq N .$$

Hence $d(f(p_n), f(q_n)) < \varepsilon$ for all $n \geq N$. On the other hand, by the definition of F, we have $F(p) = \lim_{n\to\infty} f(p_n)$ and $F(q) = \lim_{n\to\infty} f(q_n)$. Hence there exists $M \in \mathbb{N}$ such that $d(F(p), f(p_n)) < \varepsilon$ and $d(F(q), f(q_n)) < \varepsilon$ for all $n \geq M$. Therefore,

$$\begin{aligned}
&d(F(p), F(q)) \\
&\leq d(F(p), F(p_n)) + d(F(p_n), F(q_n)) + d(F(q_n), F(q)) \\
&= d(F(p), f(p_n)) + d(f(p_n), f(q_n)) + d(f(q_n), F(q)) \\
&< 3\varepsilon
\end{aligned}$$

for all $n \geq \max\{N, M\}$. Hence F is uniformly continuous on X.

Finally, uniqueness. If G is another extension of f to X such that G is (uniformly) continuous, then by construction, $F|_D = G|_D$. For any $x \in X \setminus D$, there is a sequence $\{x_n\}_{n\in\mathbb{N}}$ in D which converges to x as $n \to \infty$. Then as both F and G are continuous, we have $G(x) = \lim_{n\to\infty} G(x_n) = \lim_{n\to\infty} F(x_n) = F(x)$ and so $G = F$. Hence uniqueness.

25. Let $f : [0, \infty) \to \mathbb{R}$ be a function such that $\lim_{n\to\infty} f(n+x) = 0$ for every $x \in [0, 1)$.
    (a) If f is uniformly continuous, prove that $\lim_{x\to\infty} f(x) = 0$.
    (b) Determine whether (a) remain valid if the continuity of f is only pointwise but not uniform.

Solution:

(a) Since f is uniformly continuous, for all $\varepsilon > 0$, there exists $\delta > 0$ such that

$$|f(x) - f(y)| < \frac{\varepsilon}{2}$$

for any $x, y \in [0, \infty)$ with $|x - y| < \delta$.

Let $K \in \mathbb{N}$ be large enough such that $K > 1/\delta$. Partition the interval $[0, 1]$ into $0 = x_0 < \ldots < x_K = 1$ such that $0 < x_i - x_{i-1} < \delta$ for all $1 \le i \le K$. For each fixed $i \in \{0, \ldots, K-1\}$, since $f(n + x_i) \to 0$ as $n \to \infty$, there exists $N_i \in \mathbb{N}$ such that

$$\big|f(n + x_i)\big| < \frac{\varepsilon}{2} \quad \text{whenever } n \ge N_i \, .$$

For any $x > 0$, if as usual we denote by $[x]$ the largest integer that does not exceed x, then $x - [x] \in [0, 1)$ and so there exists $i_0 \in \{0, \ldots, K-1\}$ such that $x - [x] \in [x_{i_0}, x_{i_0+1})$. Hence $0 \le x - ([x] + x_{i_0}) < \delta$ and

$$\Big|f(x) - f\left([x] + x_{i_0}\right)\Big| < \frac{\varepsilon}{2} \, .$$

In particular, for any $x \ge N := \max\{N_i : i = 0, \ldots, K-1\}$, we have $[x] \ge N_{i_0}$ and so

$$|f(x)| \le \big|f([x] + x_{i_0})\big| + \big|f(x) - f([x] + x_{i_0})\big| < \frac{\varepsilon}{2} + \frac{\varepsilon}{2} = \epsilon \, .$$

That is, $\lim_{x\to\infty} f(x) = 0$.

(b) *Answer*: No, (a) will no longer hold.

Example: Consider the function $f : [0, \infty) \to \mathbb{R}$ defined by

$$f(x) := \begin{cases} n(x - n) & x \in \left[n, n + \frac{1}{n}\right] \ , \ n \ge 3 \\ 2 - n(x - n) & x \in \left[n + \frac{1}{n}, n + \frac{2}{n}\right] \ , \ n \ge 3 \\ 0 & \text{otherwise.} \end{cases}$$

It is readily seen that f is continuous on $[0, \infty)$. For any $x \in (0, 1)$ and any $n > \frac{2}{x}$, since $n + x \in (n + \frac{2}{n}, n+1)$, we have

$f(x+n) = 0$. Besides, when $x = 0$, $f(x+n) = f(n) = 0$ for all $n \in \mathbb{N}$. Hence for any $x \in [0,1)$, $f(n+x) \to 0$ as $n \to \infty$. However, since $f(n+\frac{1}{n}) = 1$ for all $n \geq 3$, $\lim_{x\to\infty} f(x)$ fails to exist.

26. Suppose that X and Y are metric spaces and $f : X \to Y$ is a function. Show that the following two statements are equivalent:
   (a) f is uniformly continuous.
   (b) For all sequences $\{u_n\}_{n\in\mathbb{N}}$, $\{x_n\}_{n\in\mathbb{N}}$ in X, if $\{d_X(u_n, x_n)\} \to 0$ as $n \to \infty$, then $\{d_Y(f(u_n), f(x_n))\} \to 0$ as $n \to \infty$.

Proof. (a)$\Rightarrow$(b): Suppose f is uniformly continuous, then for any $\varepsilon > 0$, there exists $\delta = \delta(\varepsilon) > 0$ such that

$$d_X(u,x) < \delta \quad \Rightarrow \quad d_Y(f(u), f(x)) < \varepsilon \ .$$

Let $\{u_n\}_{n\in\mathbb{N}}$, $\{x_n\}_{n\in\mathbb{N}}$ be two sequences in X such that $\{d_X(u_n, x_n)\} \to 0$ as $n \to \infty$. Then there exists $N \in \mathbb{N}$ such that

$$d_X(u_n, x_n) < \delta \quad \text{for all } n \geq N \ .$$

Hence

$$d_Y(f(u_n), f(x_n)) < \varepsilon \quad \text{for all } n \geq N \ ,$$

which implies

$$\{d_Y(f(u_n), f(x_n))\} \to 0 \quad \text{as } n \to \infty \ .$$

(b)$\Rightarrow$(a): Assume to the contrary that f is not uniformly continuous. Then there exists $\varepsilon > 0$ such that for any $\delta > 0$, there exists u, $x \in X$ with $d_X(u,x) < \delta$ but $d_Y(f(u), f(x)) \geq \varepsilon$. So for any $n \in \mathbb{N}$, we can find u_n, $x_n \in X$ such that

$$d_X(u_n, x_n) < \frac{1}{n} \quad \text{and} \quad d_Y(f(u_n), f(x_n)) > \varepsilon \ .$$

Notice that that means we have two sequences $\{u_n\}_{n\in\mathbb{N}}$, $\{x_n\}_{n\in\mathbb{N}}$ in X with $\{d_X(u_n, x_n)\} \to 0$ as $n \to \infty$ but $\{d_Y(f(u_n), f(x_n))\} \not\to 0$ as $n \to \infty$, a contradiction.

4.2 Contraction and Banach's Fixed Point Theorem

A special example of uniformly continuous functions of a metric space X into itself is called a *contraction*. It is a distance-decreasing function in such a way that the distance between any two distinct points will decrease at least by a factor strictly less than 1. Definition 4.2.1 below will make it precise. It turns out that such functions play an important role in the existence and uniqueness problem of solutions to many differential and integral equations.

Definition 4.2.1. *A function $f : X \to X$ is said to be a contraction of X if there exists a positive number $0 < \alpha < 1$ called a contraction constant such that*

$$d\big(f(x), f(y)\big) \leq \alpha\, d(x,y) \quad \textit{for all } x, y \in X \ .$$

Observe that the contraction constant of any contraction is not unique. In fact, if $0 < \alpha < 1$ is a contraction constant of a contraction f, then any number $\beta \in [\alpha, 1)$ is also a contraction constant of f.

It is also clear by definition that every contraction of X is uniformly continuous on X.

Note that implicitly imposed in Definition 4.2.1, a contraction of X must be a self-map on X, that is, a function which maps X *into itself*, although it does not have to be surjective.

Examples 4.2.2:

(i) $f : \mathbb{R} \to \mathbb{R}$ defined by $f(x) = \frac{x}{2}$ is a contraction.
Proof. As $|f(x) - f(y)| = |\frac{x}{2} - \frac{y}{2}| = \frac{1}{2}\,|x - y|$ for any x, $y \in \mathbb{R}$, f is a contraction with contraction constant $\frac{1}{2}$.

(ii) $f : (2, \infty) \to (1, \infty)$ defined by $f(x) = \frac{x}{2}$ is *not* a contraction.
Proof. Observe that although we do have $|f(x) - f(y)| = |\frac{x}{2} - \frac{y}{2}| = \frac{1}{2}\,|x - y|$ for any x, $y \in (2, \infty)$, since the image of $(2, \infty)$ under f is $(1, \infty) \not\subset (2, \infty)$, according to Definition 4.2.1, f is not a contraction.

(iii) $f : (2, \infty) \to (2, \infty)$ defined by $f(x) = 1 + \frac{x}{2}$ is a contraction.
Proof. Observe first that $f\big((2, \infty)\big) = (2, \infty)$. Similar to (i), we have

$$|f(x) - f(y)| = \left|\left(1 + \frac{x}{2}\right) - \left(1 + \frac{y}{2}\right)\right| = \frac{1}{2}\left|x - y\right|$$

for any x, $y \in (2, \infty)$ and so f is a contraction with contraction constant $\frac{1}{2}$.

Remark. Every contraction is *distance-decreasing*, that is,

$$d(f(x), f(y)) < d(x, y) \qquad \text{for all } x \neq y \in X \ .$$

(In some books, a distance-decreasing function is known as a *contractive mapping*). However, the converse is not true. In fact, a function is a contraction means the ratio

$$\frac{d(f(x), f(y))}{d(x, y)} \leq \alpha < 1 \qquad \text{for all } x \neq y \ ,$$

while a distance-decreasing function is one with ratio

$$\frac{d(f(x), f(y))}{d(x, y)} < 1 \qquad \text{for all } x \neq y \ ,$$

but the ratio is not "bounded away from" 1, that is, the ratio could get arbitrarily close to 1 instead of "staying away from" 1.

Example 4.2.3: $f : \mathbb{R} \to \mathbb{R}$ defined by $f(x) = \sqrt{x^2 + 1}$ is not a contraction although it is distance-decreasing. In fact, it is easy to see that

$$\frac{d\big(f(x), f(y)\big)}{d(x, y)} = \frac{x + y}{\sqrt{x^2 + 1} + \sqrt{y^2 + 1}} < 1$$

for all $x \neq y$ in $\mathbb{R}$ and so f is distance-decreasing. However, for any $y \in \mathbb{R}$, we have

$$\frac{d\big(f(x), f(y)\big)}{d(x, y)} = \frac{x + y}{\sqrt{x^2 + 1} + \sqrt{y^2 + 1}} \to 1 \quad \text{as } x \to \infty \ .$$

Thus the contraction constant doesn't exist.

Definition 4.2.4. *Let $f : X \to X$ be a function. A point $p \in X$ is called a fixed point of f or a point fixed by f if $f(p) = p$.*

Examples 4.2.5.

(i) In Example 4.2.2 (i), the contraction f of the complete metric space $\mathbb{R}$ has a unique fixed point 0.

(ii) In Example 4.2.2 (iii), the contraction f of the space $(2, \infty)$ does not have any fixed point. In fact, solving $f(x) = x$ we get $x = 2$, which is not within $(2, \infty)$. This suggests that we consider the same function f on $[2, \infty)$. It is easily verified that $f : [2, \infty) \to [2, \infty)$ and in this case f has 2 as a unique fixed point in $[2, \infty)$. Note that $[2, \infty)$ is complete.

(iii) In Example 4.2.3, the function f is only distance-decreasing but not a contraction. Furthermore, as $f(x) \neq x$ for any $x \in \mathbb{R}$, f does not have any fixed point in the complete metric space $\mathbb{R}$.

Theorem 4.2.6 (Banach's Fixed point theorem). *Every contraction of a complete metric space has a unique fixed point.*

Proof. Let $f : X \to X$ be a contraction with contraction constant $0 < \alpha < 1$. Fix $a \in X$ and define a sequence $\{p_n\}_{n \in \mathbb{N}}$ in X by

$$p_0 := a\,, \quad p_{n+1} := f(p_n)\,, \quad n \geq 0\,.$$

Then for all $n \geq 1$,

$$d(p_{n+1}, p_n) = d\big(f(p_n), f(p_{n-1})\big) \leq \alpha d(p_n, p_{n-1})$$

and so by induction,

$$d(p_{n+1}, p_n) \leq \alpha^n d(p_1, p_0) = c\alpha^n\,,$$

where we denote by $c := d(p_1, p_0)$. Hence for any $m > n$,

$$d(p_m, p_n) \leq \sum_{k=n}^{m-1} d(p_{k+1}, p_k) \leq c \sum_{k=n}^{m-1} \alpha^k \leq \frac{c}{1 - \alpha} \alpha^n\,.$$

Since $0 < \alpha < 1$, $\alpha^n \to 0$ as $n \to \infty$ and thus $\{p_n\}_{n\in\mathbb{N}}$ is a Cauchy sequence. Since X is complete, $\{p_n\} \to$ some $p \in X$ as $n \to \infty$. By the continuity of f, we have

$$f(p) = \lim_{n\to\infty} f(p_n) = \lim_{n\to\infty} p_{n+1} = p \ .$$

Hence p is a fixed point of f. Finally, if $q \in X$ is another fixed point of f, then

$$d(p,q) = d\big(f(p), f(q)\big) \leq \alpha d(p,q) \ .$$

Since $\alpha < 1$, this can happen only if $d(p,q) = 0$ or $p = q$. Hence the uniqueness. □

At first sight, Banach's Fixed Point Theorem looks so theoretical that it may not be readily applicable. However, it turns out that it is not the case. It has very effective applications in the existence and uniqueness problem of solutions to many differential and integral equations. Furthermore, the proof of Banach's Fixed Point Theorem also leads to an iterative method of approximating the unique solution of such equations.

Example 4.2.7. Consider the *Fredholm Equation*

$$f(x) = \lambda \int_a^b K(x,y) f(y) dy + \varphi(x) \ ,$$

where $K \in C([a,b] \times [a,b])$, $\varphi \in C[a,b]$, and $\lambda \in \mathbb{R}$ are given. Show that the Fredholm equation has a unique solution $f \in C[a,b]$ for $|\lambda| \ll$.

Proof. It is clear that if $a = b$ or $\lambda = 0$ or $K \equiv 0$, then $f = \varphi$ is the unique solution. So we assume that $a < b$, $\lambda \neq 0$, and $K \not\equiv 0$.

Consider the metric space $(C[a,b], d)$, where

$$d(g,h) := \sup\Big\{\big|g(x) - h(x)\big| : x \in [a,b]\Big\} \ .$$

It will be shown later in Corollary 5.1.13 that it is complete. So for the time being we will just take it for granted. Consider the operator

$$F : C[a,b] \to C[a,b]$$

defined by

$$F(f)(x) := \lambda \int_a^b K(x,y)f(y)dy + \varphi(x)\ , \quad f \in C[a,b]\ .$$

Note that $f \in C[a,b]$ solves the Fredholm Equation if and only if it is a fixed point of the operator F. Hence the problem of determining the solutions to the Fredholm Equation reduces to that of finding the fixed points of F. In order to do the latter, we apply Banach's Fixed Point Theorem. Modulo the proof that $(C[a,b],d)$ is complete, we only need to show that F is a contraction.

For any f_1, $f_2 \in C[a,b]$,

$$\begin{aligned} d(F(f_1),F(f_2)) &= \sup_{x\in[a,b]} \Big|F(f_1)(x) - F(f_2)(x)\Big| \\ &= \sup_{x\in[a,b]} \Big|\lambda \int_a^b K(x,y)\big(f_1(y) - f_2(y)\big)\,dy\Big| \\ &\le \sup_{x\in[a,b]} |\lambda| \int_a^b \Big|K(x,y)\Big|\,\Big|f_1(y) - f_2(y)\Big|\,dy \\ &\le |\lambda| \int_a^b M\ d(f_1,f_2)\,dy \\ &\le |\lambda|\,M(b-a)\,d(f_1,f_2) \end{aligned}$$

where $M := \sup\big\{|K(x,y)| : (x,y) \in [a,b]\times[a,b]\big\} > 0$. Note that since K is continuous on the compact set $[a,b]\times[a,b]$, it is bounded and hence $M < \infty$. Write $\alpha := |\lambda| M(b-a)$. Whenever $|\lambda| < \frac{1}{M(b-a)}$, we have $0 < \alpha < 1$ and

$$d\big(F(f_1),F(f_2)\big) \le \alpha\, d(f_1,f_2) \quad \text{for all } f_1,f_2 \in C[a,b]\ .$$

Thus F is a contraction. By Banach's Fixed Point Theorem, F has a unique fixed point $f \in C[a,b]$ which is also the unique solution to the Fredholm Equation.

Example 4.2.8. Let's work out a concrete example of the following Fredholm Equation

$$f(x) = \frac{1}{2}\int_0^1 xyf(y)\,dy + x, \qquad x \in [0,1], \tag{*}$$

and estimate its solution with accuracy less than 0.01 at each point in $[0,1]$.

Same as Example 4.2.7, we consider the complete metric space $(C[0,1], d)$ with

$$d(g,h) := \sup\{|g(x) - h(x)| : x \in [0,1]\},$$

and the operator

$$\begin{aligned} F : C[0,1] &\to C[0,1] \\ f &\mapsto F(f) \end{aligned}$$

defined by

$$F(f)(x) := \frac{1}{2}\int_0^1 xyf(y)\,dy + x, \qquad x \in [0,1].$$

Note that $\sup\{|xy| : (x,y) \in [0,1]\times[0,1]\} = 1$. Hence for any f_1, $f_2 \in C[0,1]$,

$$\begin{aligned} d(F(f_1), F(f_2)) &= \sup_{x\in[0,1]} \left|F(f_1)(x) - F(f_2)(x)\right| \\ &\le \sup_{x\in[0,1]} \frac{1}{2}\int_0^1 |xy|\,|f_1(y) - f_2(y)|\,dy \\ &\le \frac{1}{2}\cdot 1 \int_0^1 |f_1(y) - f_2(y)|\,dy \\ &\le \frac{1}{2}\int_0^1 d(f_1,f_2)\,dy \\ &= \frac{1}{2} d(f_1,f_2) \end{aligned}$$

and so F is a contraction with contraction constant $\alpha = \frac{1}{2}$. By Banach's Fixed Point Theorem, it has a unique fixed point $f \in C[0,1]$ and so the given integral equation $(*)$ has this as a unique solution.

To approximate this unique solution, we follow the proof of Banach's Fixed Point Theorem and pick as the starting point the simple function

$$f_0 := 0 \ .$$

Then we have

$$\begin{aligned} f_1 = \ & F(f_0) &= x \ , \\ f_2 = \ & F(f_1) &= \frac{7}{6}x \ , \\ f_3 = \ & F(f_2) &= \frac{43}{36}x \ , \\ & \cdots\cdots \end{aligned}$$

We could go on and compute f_n for $n \geq 4$. By the proof of Banach's Fixed Point Theorem, the sequence $\{f_n\}_{n\in\mathbb{N}}$ should converge to the unique fixed point f of F, which is the unique solution to the Fredholm Equation $(*)$. Hence by choosing large enough $n \in \mathbb{N}$, we can use f_n to approximate, with respect to the metric d, the unique solution f to $(*)$ up to any prescribed accuracy. In fact, from the proof of Banach's Fixed Point Theorem, we have

$$d(f_n, f_m) \leq \frac{d(f_1, f_0)}{1-\alpha}\alpha^n \qquad \text{for all } m > n \ .$$

So the error by using f_n to estimate the unique solution f to (*) with respect to d is

$$d(f_n, f) = \lim_{m\to\infty} d(f_n, f_m) \leq \frac{d(f_1, f_0)}{1-\alpha}\alpha^n \ .$$

Now it is easy to compute

$$\begin{aligned} d(f_1, f_0) &= \sup\{|f_1(x) - f_0(x)| : x \in [0,1]\} \\ &= \sup\{|x| : x \in [0,1]\} = 1 \ , \end{aligned}$$

and so the error by using f_n to estimate the unique solution f to (*) is

$$d(f_n, f) \leq \frac{1}{1-\alpha}\alpha^n = \frac{1}{1-\frac{1}{2}} \cdot \frac{1}{2^n} = \frac{1}{2^{n-1}} \ .$$

So for example, if we want to approximate f within an accuracy of 0.01 at any point in $[0, 1]$, we need

$$\frac{1}{2^{n-1}} < 0.01 \ .$$

Solving this, we get $n \geq 8$. So f_8 approximates f with an accuracy of 0.01 at any point in $[0, 1]$.

Remark. For this simple example, it is rather obvious by seeing the first few f_n's that all f_n's are scalar multiples of x and it is easy to speculate that this is also the case for its limit f. Hence we may simply put $f(x) := kx$ into $(*)$ and solve for k. In fact, solving

$$kx = \frac{1}{2}\int_0^1 xy(ky)\, dy + x \ ,$$

we have $k = \dfrac{6}{5}$ and so the unique solution to $(*)$ is simply $y = \dfrac{6}{5}x$.

Exercise 4.2

Part A: True or False Questions

For each of the following statements, determine if it is true or false. If it is true, prove it. If it is false, give a counterexample or provide proper justification.

1. Every uniformly continuous function $f : X \to X$ on a complete metric space X has a fixed point.

   Answer: False.
   Example: The function $f : \mathbb{R} \to \mathbb{R}$ given by $f(x) := x + 1$, $x \in \mathbb{R}$, is clearly uniformly continuous on the complete metric space $\mathbb{R}$. But f has no fixed point in $\mathbb{R}$.

2. $f : (0, \infty) \to \mathbb{R}$ defined by $f(x) := \ln x$ is a contraction.

   Answer: False.
   Justification. By Exercise 4.1, Part A, Problem #5, f is not uniformly continuous. Hence it is not a contraction.

3. $f : \mathbb{R} \to \mathbb{R}$ defined by $f(x) := \cos(\cos x)$ is a contraction.

   Answer: True.
   Proof. By Mean Value Theorem,

   $$\Big|\cos(\cos x) - \cos(\cos y)\Big| = \Big|x - y\Big| \, \Big|\sin(\cos c) \sin c\Big|$$

   for some c between x and y. As $\cos c \in [-1, 1]$, we have $|\sin(\cos c) \sin c| \leq \sin 1$. Hence

   $$\Big|\cos(\cos x) - \cos(\cos y)\Big| \leq (\sin 1) \Big|x - y\Big| \quad \text{for all } x, y \in \mathbb{R}$$

   and so f is a contraction with contraction constant $0 < \sin 1 < 1$.

4. $f : \mathbb{R} \to \mathbb{R}$ defined by $f(x) := \ln(1 + e^x)$ is a contraction.

   Answer: False.

Justification. Since $x = \ln(1 + e^x)$ has no solution, f has no fixed point. As $\mathbb{R}$ is complete, by Banach's Fixed Point Theorem, f cannot be a contraction.

5. The function $f : [1, \infty) \to \left[\frac{1}{2}, \infty\right)$ given by $f(x) = \frac{x}{2}$ is a contraction of $[1, \infty)$.

   Answer: False.
   Justification. The function f is not a self map, as $f(1) = \frac{1}{2} \notin [1, \infty)$. Hence by definition, it is not a contraction of $[1, \infty)$.

6. The function $f : (-1, \infty) \to \mathbb{R}$ defined by $f(x) := \dfrac{1}{1+x}$ is a contraction of $[-\frac{1}{2}, 0]$.

   Answer: False.
   Justification. Note that f is strictly decreasing, $f(-\frac{1}{2}) = 2$, and $f(0) = 1$. Hence $f([-\frac{1}{2}, 0]) = [1, 2] \not\subset [-\frac{1}{2}, 0]$ and so f is not a contraction of $[-\frac{1}{2}, 0]$.

7. The function $f : (-1, \infty) \to \mathbb{R}$ defined by $f(x) := \dfrac{1}{1+x}$ is a contraction of $[0, 1]$.

   Answer: False. We have $f([0, 1]) = [\frac{1}{2}, 1] \subset [0, 1]$. However, it's not a contraction. In fact, it is easy to see that

$$\frac{|f(x) - f(y)|}{|x - y|} = \frac{1}{(1+x)(1+y)} \quad \text{for any } 0 \leq x < y \leq 1\,.$$

   By taking $x = 0$ and letting y approaches 0^+, the right hand side approaches 1 indefinitely and so the contraction constant does not exist.

8. The function $f : (-1, \infty) \to \mathbb{R}$ defined by $f(x) := \dfrac{1}{1+x}$ has a fixed point in $[0, 1]$.

   Answer: True.

Proof. Solving the equation $f(x) = x$, one has a unique solution $x = \frac{\sqrt{5}-1}{2} \in [0,1]$.

9. Let X be complete. Then every uniformly continuous function $f : X \to X$ having a unique fixed point in X must be a contraction of X.

   Answer: False.

   Example: The identity function $f : \mathbb{R} \to \mathbb{R}$, $f(x) := x$, $x \in \mathbb{R}$, is clearly uniformly continuous, with a unique fixed point $x = 0$, and $\mathbb{R}$ is complete. However, f is not a contraction.

Part B: Problems

1. Show that the function $f : [0, \infty) \to \mathbb{R}$ defined by

$$f(x) := \ln\left(1 + e^{-x}\right), \quad x \in [0, \infty),$$

has a unique fixed point. Find explicitly the fixed point.

Solution. Note that f is continuous on $[0, \infty)$ and differentiable in $(0, \infty)$ with $f'(x) = \dfrac{-e^{-x}}{1+e^{-x}}$. In particular, $|f'(x)| \leq \dfrac{1}{2}$ for all $x \in (0, \infty)$. Hence for any x, $y \in [0, \infty)$ with $x \neq y$, by Mean Value Theorem,

$$|f(x) - f(y)| = |f'(c)|\,|x - y|$$

for some c between x and y. This gives

$$|f(x) - f(y)| \leq \frac{1}{2}\,|x - y|$$

and hence f is a contraction, with contraction constant $\frac{1}{2}$. By Banach's Fixed Point Theorem, f has a unique fixed point p.

To find p, we can either imitate the proof of Banach's Fixed Point Theorem, starting with any point $x_0 \in [0, \infty)$, compute explicitly the sequence $\{x_n\}_{n\in\mathbb{N}}$ with $x_n := f(x_{n-1})$, $n \in \mathbb{N}$, and we will have $p = \lim_{n\to\infty} x_n$. For example, if we pick $x_0 = 0$, then

$$\begin{aligned} x_1 &= f(x_0) = f(0) = \ln 2 \ , \\ x_2 &= f(x_1) = f(\ln 2) = \ln \frac{3}{2} \ , \\ x_3 &= f(x_2) = f\left(\ln \frac{3}{2}\right) = \ln \frac{5}{3} \ , \\ x_4 &= f(x_3) = f\left(\ln \frac{5}{3}\right) = \ln \frac{8}{5} \\ &\cdots \end{aligned}$$

Inductively, it is easy to verify that $e^{x_{n+1}} = 1 + \dfrac{1}{e^{x_n}}$ for all $n \in \mathbb{N}$. As we know that $\{x_n\}_{n\in\mathbb{N}}$ converges to p, we have

$$e^p = \lim_{n\to\infty} e^{x_{n+1}} = \lim_{n\to\infty}\left(1 + \frac{1}{e^{x_n}}\right) = 1 + \frac{1}{e^p} \ .$$

Solving this, we get $e^p = \dfrac{1+\sqrt{5}}{2}$ and so $p = \ln\left(\dfrac{1+\sqrt{5}}{2}\right)$.

Alternatively, we can simply solve the equation $p = f(p)$ explicitly and get $p = \ln\left(\dfrac{1+\sqrt{5}}{2}\right)$.

2. For any $k \in C([0,1]\times[0,1])$ with $|k| \le M < 1$ on $[0,1]\times[0,1]$, show that the integral equation

$$f(x) + \int_0^1 k(x,y)\,f(y)\,dy = e^{x^2} \qquad x \in [0,1]$$

has a unique solution $f \in C[0,1]$.

Proof. First observe that if $k \equiv 0$ on $-0,1]$, then the given integral equation has a unique solution $f(x) = e^{x^2}$. so we assume that $k \not\equiv 0$ on $[0,1]$ and so $M > 0$. As usual, we take for granted that $(C([0,1]),d)$ with metric d defined by

$$d(f,g) := \sup\{|f(x) - g(x)| : x \in [0,1]\}$$

is complete. Define an operator $T : (C[0,1], d) \to C([0,1], d)$ by

$$T(f)(x) := e^{x^2} - \int_0^1 k(x,y)\, f(y)\, dy$$

for and $f \in C[0,1]$ and $x \in [0,1]$. Then f is a solution to the given integral equation if and only if it is a fixed point of T. Now for any f, $g \in C[0,1]$, we have

$$\begin{aligned}\left|T(f)(x) - T(g)(x)\right| &= \left|\int_0^1 k(x,y)\,(f(y) - g(y))\, dy\right| \\ &\leq \int_0^1 \left|k(x,y)\right| \left|f(y) - g(y)\right| dy \\ &\leq Md(f,g)\ .\end{aligned}$$

So T is a contraction of $C[0,1]$ with contraction constant $0 < M < 1$ and so by Banach's Fixed Point Theorem, T has a unique fixed point in $C[0,1]$, which is in turn the unique solution to the given integral equation.

3. Let $X := (0,\infty)$ with the usual Euclidean metric and $C^1(X;X)$ be the collection of all continuously differentiable positive functions on $(0,\infty)$. If $f \in C^1(X,X)$ satisfies $x|f'(x)| \leq kf(x)$ for all $x \in X$, where $k \in (0,1)$ is a constant, show that f has a unique fixed point.

Proof. For any $0 < x < u$, we have

$$\begin{aligned}\left|\ln f(u) - \ln f(x)\right| &= \left|\int_x^u \frac{f'(s)}{f(s)}\, ds\right| \leq \int_x^u \frac{|f'(s)|}{f(s)}\, ds \\ &\leq \int_x^u \frac{k}{s}\, ds = k\,|\ln u - \ln x|\ .\end{aligned}$$

On X, we consider another metric

$$\tilde{d}_X(u,x) := \left|\ln u - \ln x\right| \quad \text{for all } u, x \in X\ .$$

It is not hard to verify that $(X, \tilde{d}_X)$ is a complete metric space and from the above estimate, we have

$$\tilde{d}(f(u), f(x)) \leq k\, \tilde{d}(u, x) \quad \text{for all } u, x \in X$$

and so $f : (X, \tilde{d}_X) \to (X, \tilde{d}_X)$ is a contraction with contraction constant $k \in (0, 1)$. By Banach's Fixed Point Theorem, f has a unique fixed point.

4. Let $f : X \to X$ be a distance-decreasing function (or a contractive function) of X, that is, $d(f(x), f(y)) < d(x, y)$ for any $x \neq y \in X$.
   (a) Prove that f has at most one fixed point.
   (b) If X is complete, is it necessarily true that f must have a fixed point?
   (c) If X is compact, prove that f has exactly one fixed point.
   (d) If X is compact, is it necessarily true that f must be a contraction?

Proof.

(a) Suppose p and q are distinct fixed points of f, then

$$d(p, q) = d(f(p), f(q)) < d(p, q)$$

which is impossible. So f has at most one fixed point.

(b) *Answer*: No.

Example: Let $X := (1, \infty)$ equipped with the Euclidean metric, and $f : X \to X$ be defined by $f(x) := \sqrt{x}$, $x \in X$. Then

$$|f(x) - f(y)| = |\sqrt{x} - \sqrt{y}| < |\sqrt{x} - \sqrt{y}| \cdot |\sqrt{x} + \sqrt{y}| = |x - y| \,.$$

Hence f is distance-decreasing. However, as $f(x) < x$ for all $x \in X$, f has no fixed point in X.

Alternative Example: Let $f : \mathbb{R} \to \mathbb{R}$ be defined by $f(x) := \sqrt{x^2+1}$, $x \in \mathbb{R}$. As shown in Example 4.2.3, f is distance-decreasing. However, it is easily seen that $f(x) = x$ has no solution and so f has no fixed point.

(c) Define $g : X \to \mathbb{R}$ by $g(x) := d\big(x, f(x)\big)$ for $x \in X$. It is evident that g is continuous on X. If X is compact, g attains its minimum in X at some point, say, $a \in X$. If $f(a) \neq a$, since f is distance-decreasing, we have

$$g(f(a)) = d\big(f(a), f(f(a))\big) < d\big(a, f(a)\big) = g(a) ,$$

which contradicts to the assumption that g attains its minimum at a. Therefore, $f(a) = a$. That is, a is a fixed point of f. Finally, the uniqueness of the fixed point is guaranteed by (a).

(d) *Answer*: No.

Example: Let $X := [0, \frac{1}{2}]$ equipped with the Euclidean metric, and $f(x) := x^2$, $x \in X$. Then X is compact and we have

$$|f(x) - f(y)| = |x^2 - y^2| = |x - y|\,|x + y| \leq |x - y|$$

for all $x, y \in [0, \frac{1}{2}]$. Note that the equality holds if and only if $|x + y| = 1$, which forces $x = y = \frac{1}{2}$. Hence for any $x \neq y \in X$, we have $|f(x) - f(y)| < |x - y|$ and so f is distance-decreasing.

But then for $x \neq y \in X$, $\dfrac{|f(x) - f(y)|}{|x - y|} = |x + y|$ can be arbitrarily close to 1 and so there is no contraction constant. Hence f is not a contraction of X.

5. Let X be a complete metric space and $f : X \to X$. Suppose there exists $n \in \mathbb{N}$ such that $f^{(n)} := f \circ f \circ \cdots \circ f$ (n iterates) is a contraction of X.

   (a) Show that f has a unique fixed point in X.

   (b) Is f necessarily a contraction?

Proof.

(a) By Banach's Fixed Point Theorem, $f^{(n)}$ has a unique fixed point, and we denote it as x_0. As $f^{(n)}(f(x_0)) = f^{(n+1)}(x_0) = f(f^{(n)}(x_0)) = f(x_0)$, we see that $f(x_0)$ is also a fixed point of $f^{(n)}$. The uniqueness implies that $f(x_0) = x_0$, which in turn implies that x_0 is a fixed point of f.
Finally, it is evident that every fixed point of f must also be a fixed point of $f^{(n)}$. Hence x_0 is the only fixed point of f.

(b) *Answer*: No.
Example: Consider the function $f : \mathbb{R} \to \mathbb{R}$ defined by

$$f(x) := \begin{cases} -x & \text{if } x \geq 0 \\ \frac{-x}{2} & \text{if } x < 0 \,. \end{cases}$$

Then $f^2(x) = \frac{x}{2}$ is a contraction on the complete metric space $\mathbb{R}$, but as $|f(1) - f(2)| = 1 \not< |1 - 2|$, f is not a contraction.

6. Let X be a complete metric space and $f : X \to X$. For any $n \in \mathbb{N}$, let

$$a_n := \sup_{x_1 \neq x_2} \frac{d\left(f^{(n)}(x_1), f^{(n)}(x_2)\right)}{d(x_1, x_2)} \,.$$

If $\sum_{n=1}^{\infty} a_n < \infty$, show that f has a unique fixed point in X.
Proof. As $a_n \geq 0$ for all n and $\sum_{n=1}^{\infty} a_n < \infty$, there exists $N \in \mathbb{N}$ such that $a_N < \frac{1}{2}$. Hence for any $x_1 \neq x_2$, we have

$$\frac{d\left(f^{(N)}(x_1), f^{(N)}(x_2)\right)}{d(x_1, x_2)} \leq a_N < \frac{1}{2}$$

and so $f^{(N)}$ is a contraction of X with contraction constant $\frac{1}{2}$. The assertion now follows from Exercise 4.2, Part B, Problem # 5.

7. Let (X, d) be a complete metric space and $T : X \to X$. If there exists $0 < K < \frac{1}{2}$ such that

$$d\big(T(x), T(y)\big) \leq K\big[d\big(x, T(x)\big) + d\big(y, T(y)\big)\big]$$

for all x, $y \in X$, show that T has a unique fixed point in X.

Proof. Fix $x_0 \in X$. Define a sequence $\{x_n\}_{n\in\mathbb{N}}$ in X by $x_n := T(x_{n-1})$ for all $n \in \mathbb{N}$. By assumption, for each $n \in \mathbb{N}$,

$$\begin{aligned}
d(x_{n+1}, x_n) &= d\big(T(x_n), T(x_{n-1})\big) \\
&\leq K\big[d\big(x_n, T(x_n)\big) + d\big(x_{n-1}, T(x_{n-1})\big)\big] \\
&= K\big[d(x_n, x_{n+1}) + d(x_{n-1}, x_n)\big]
\end{aligned}$$

and so

$$d(x_{n+1}, x_n) \leq \frac{K}{1-K}\, d(x_{n-1}, x_n)\,.$$

Inductively, we have

$$d(x_{n+1}, x_n) \leq \Big(\frac{K}{1-K}\Big)^n d(x_1, x_0)\,.$$

Since $0 < K < \frac{1}{2}$, we have $\frac{K}{1-K} < 1$. Thus for any n, $p \geq 1$,

$$\begin{aligned}
d(x_{n+p}, x_n) &\leq d(x_{n+p}, x_{n+p-1}) + \cdots + d(x_{n+1}, x_n) \\
&\leq \sum_{i=n}^{n+p-1} \Big(\frac{K}{1-K}\Big)^i d(x_1, x_0) \\
&\leq \Big(\frac{K}{1-K}\Big)^n \left(\frac{1}{1-\frac{K}{1-K}}\right) d(x_1, x_0) \\
&= \Big(\frac{K}{1-K}\Big)^n \Big(\frac{1-K}{1-2K}\Big) d(x_1, x_0)\,.
\end{aligned}$$

It is then evident that $\{x_n\}_{n\in\mathbb{N}}$ is a Cauchy sequence in the complete metric space X, hence it is convergent to some element, say $x \in X$. Note that

$$\begin{aligned}
d\big(x, T(x)\big) &\leq d(x, x_n) + d\big(x_n, T(x_n)\big) + d\big(T(x_n), T(x)\big) \\
&\leq d(x, x_n) + d\big(x_n, T(x_n)\big) \\
&\quad + K\big[d\big(x_n, T(x_n)\big) + d\big(x, T(x)\big)\big]
\end{aligned}$$

and so

$$d(x, T(x)) \le \frac{1}{1-K}\, d(x, x_n) + \frac{K+1}{1-K}\, d\big(x_n, T(x_n)\big) \ .$$

Letting $n \to \infty$, the right hand side vanishes and so we have $x = T(x)$, that is, x is a fixed point of T. Finally, the uniqueness follows from the observation that if z is a fixed point of T, then

$$d(z, x) = d(T(z), T(x)) \le K\big[d\big(z, T(z)\big) + d\big(x, T(x)\big)\big] = 0 \ .$$

8. Let $\varphi : (0, \infty) \to (0, \infty)$ be a monotonic increasing function with $\lim_{n\to\infty} \varphi^{(n)}(t) = 0$ for every $t \in (0, \infty)$. Suppose X is complete and $f : X \to X$ satisfies

$$d\big(f(x), f(y)\big) \le \varphi\big(d(x, y)\big) \quad \text{for all } x, y \in X \ .$$

   (a) Show that $\varphi(\varepsilon) < \varepsilon$ for all $\varepsilon > 0$.
   (b) Show that f is uniformly continuous on X.
   (c) Fix $a \in X$. For any $n \in \mathbb{N}$, let $x_n := f^{(n)}(a)$. Show that $\{d(x_{n+1}, x_n)\} \to 0$ as $n \to \infty$.
   (d) Show that $\{x_n\}_{n\in\mathbb{N}}$ is a Cauchy sequence.
   (e) Show that f has a unique fixed point in X.

Proof.

   (a) Fix $\varepsilon > 0$. If $\varphi(\varepsilon) \ge \varepsilon$, since φ is increasing, we have $\varphi^{(2)}(\varepsilon) \ge \varphi(\varepsilon) \ge \varepsilon$, and so inductively, $\varphi^{(n)}(\varepsilon) \ge \varepsilon$ for all n, which contradicts to the assumption that $\lim_{n\to\infty} \varphi^{(n)}(t) = 0$ for all $t > 0$. Hence $\varphi(\varepsilon) < \varepsilon$.
   (b) For any $\varepsilon > 0$, take $\delta := \varepsilon$. Then for any x, $y \in X$ with $d(x, y) < \delta = \varepsilon$, by the monotonicity of φ and (a), we have

$$d\big(f(x), f(y)\big) \le \varphi\big(d(x, y)\big) \le \varphi(\varepsilon) < \varepsilon \ .$$

   Thus f is uniformly continuous on X.

(c) If $f(a) = a$, then $x_n = a$ for all n and there is nothing to prove. So assume that $f(a) \neq a$. So $d(x_1, a) > 0$ and we have

$$\begin{aligned} d(x_{n+1}, x_n) &= d\big(f(x_n), f(x_{n-1})\big) \leq \varphi\big(d(x_n, x_{n-1})\big) \\ &\leq \varphi^{(2)}\big(d(x_{n-1}, x_{n-2})\big) \\ &\leq \cdots \leq \varphi^{(n)}\big(d(x_1, x)\big) \end{aligned}$$

and so $\lim_{n\to\infty} d(x_{n+1}, x_n) \leq \lim_{n\to\infty} \varphi^{(n)}\big(d(x_1, x)\big) = 0$.

(d) For any $\varepsilon > 0$, by (a), $\varepsilon - \varphi(\varepsilon) > 0$ and so by (c), there exists $N \in \mathbb{N}$ such that

$$d(x_{n+1}, x_n) < \frac{\varepsilon - \varphi(\varepsilon)}{2} \quad \text{for all } n \geq N .$$

Since φ is increasing, for any $z \in B(x_N, \varepsilon)$, we have

$$\begin{aligned} d\left(f(z), x_N\right) &\leq d\big(f(z), f(x_N)\big) + d\big(f(x_N), x_N\big) \\ &\leq \varphi\big(d(z, x_N)\big) + d(x_{N+1}, x_N) \\ &\leq \varphi(\varepsilon) + \frac{\varepsilon - \varphi(\varepsilon)}{2} \\ &< \varepsilon , \end{aligned}$$

which implies that

$$f\big(B(x_N, \varepsilon)\big) \subset B(x_N, \varepsilon).$$

Hence for all $m > N$, $x_m = f^{(m-N)}(x_N) \in B(x_N, \varepsilon)$ and thus for all $m, n \geq N$,

$$d(x_m, x_n) \leq d(x_m, x_N) + d(x_n, x_N) < 2\varepsilon .$$

Hence $\{x_n\}_{n\in\mathbb{N}}$ is a Cauchy sequence.

(e) Since X is complete, by (d), the sequence $\{x_n\}$ converges in X to some point x. As f is continuous, we have

$$x = \lim_{n\to\infty} x_{n+1} = \lim_{n\to\infty} f(x_n) = f(x) ,$$

which implies that x is a fixed point of f.

Finally, if there exists $y \in X \setminus \{x\}$ such that $f(y) = y$, then

$$d(x,y) = d\big(f(x), f(y)\big) \le \varphi\big(d(x,y)\big) < d(x,y) ,$$

which is absurd. Hence the assertion.

9. On $C[a,b]$, taking for granted that the metric

$$d(g,h) := \sup\{|g(x) - h(x)| : x \in [a,b]\} , \quad g,h \in C[a,b]$$

makes $(C[a,b], d)$ a complete metric space, show that the Volterra Equation

$$f(x) = \lambda \int_a^x K(x,y) f(y)\,dy + \varphi(x) , \quad x \in [a,b] ,$$

where $K \in C([a,b] \times [a,b])$, $\varphi \in C[a,b]$, and $\lambda \in \mathbb{R}$ are given, has a unique solution $f \in C[a,b]$.

Proof. If $\lambda = 0$ or $K \equiv 0$ on $[a,b] \times [a,b]$, it is obvious that $f = \varphi$ is the unique solution to the Volterra Equation. Hence assume that $\lambda \neq 0$ and $K \not\equiv 0$. Define an operator

$$\begin{aligned} F : C[a,b] &\to C[a,b] \\ f &\mapsto F(f) \end{aligned}$$

by

$$F(f)(x) := \lambda \int_a^x K(x,y) f(y)\,dy + \varphi(x) , \quad x \in [a,b] .$$

Then $f \in C[a,b]$ is a solution to the Volterra Equation if and only if it is a fixed point of F in $C[a,b]$.

For any $f, g \in C[a,b]$ and $x \in [a,b]$, we have

$$\begin{aligned} \big|F(f)(x) - F(g)(x)\big| &= |\lambda| \left| \int_a^x K(x,y)\big(f(y) - g(y)\big)\,dy \right| \\ &\le |\lambda|\, M(x-a)\, d(f,g) , \end{aligned}$$

where $M > 0$ is an upper bound for $|K|$ in $[a,b] \times [a,b]$. Next, we have

$$\begin{aligned}
&\left|F^2(f)(x) - F^2(g)(x)\right| \\
&= |\lambda| \left|\int_a^x K(x,y)\big(F(f)(y) - F(g)(y)\big)\,dy\right| \\
&\le |\lambda| \int_a^x |K(x,y)| \left|F(f)(y) - F(g)(y)\right| dy \\
&\le |\lambda|^2 M^2 \, d(f,g) \int_a^x (y-a)\,dy \\
&= |\lambda|^2 M^2 \frac{(x-a)^2}{2}\, d(f,g)
\end{aligned}$$

and so by induction,

$$|F^n(f)(x) - F^n(g)(x)| \le |\lambda|^n M^n \frac{(x-a)^n}{n!}\, d(f,g)\,.$$

Hence

$$d(F^n(f), F^n(g)) \le |\lambda|^n M^n \frac{(b-a)^n}{n!}\, d(f,g)\,.$$

In particular, whenever n is large enough, or to be precise, whenever $|\lambda|^n M^n \frac{(b-a)^n}{n!} < 1$, F^n is a contraction. By Exercise 4.2, Part B, Problem # 5, F has a unique fixed point $f \in C[a,b]$, which is in turn the unique solution to the Volterra Equation.

10. Estimate the solution in $C[0,1]$ to the integral equation

$$f(x) = \frac{1}{3}\int_0^1 (x^2 + y^2) f(y)\,dy + x$$

within an error of 0.01 everywhere in $[0,1]$.

Solution. We refer the readers to Example 4.2.7, and consider the operator $F : C[0,1] \to [0,1]$ defined by

$$F(f)(x) := \frac{1}{3}\int_0^1 (x^2 + y^2) f(y)\,dy + x\,, \qquad x \in [0,1]\,.$$

Using the same notations as those in Example 4.2.7, we have $\lambda = \frac{1}{3}$, $M = \sup\{x^2 + y^2 : (x, y) \in [0,1] \times [0,1]\} = 2$, $b = 1$, $a = 0$. So $\lambda M(b-a) = \frac{2}{3} < 1$ and so F is a contraction, with contraction constant $\alpha = \lambda M(b-a) = \frac{2}{3}$. By Banach's Fixed Point Theorem, F has a unique fixed point f which is also the unique solution to the given integral equation. To approximate the unique solution, we pick as the starting point the simple function

$$f_0 := 0 \ .$$

Then we have

$$\begin{aligned} f_1 &= F(f_0) = x \ , \\ f_2 &= F(f_1) = \frac{1}{3}\int_0^1 (x^2 + y^2)ydy + x = \frac{1}{6}x^2 + x + \frac{1}{12} \ , \\ f_3 &= F(f_2) = \cdots \ . \end{aligned}$$

Note from the proof of Banach's Fixed Point Theorem, we have

$$d(f_n, f_m) \leq \frac{d(f_1, f_0)}{1-\alpha}\alpha^n \quad \text{for all } m > n \ .$$

So the error by using f_n to estimate the unique solution f is

$$d(f_n, f) = \lim_{m\to\infty} d(f_n, f_m) \leq \frac{d(f_1, f_0)}{1-\alpha}\alpha^n \ .$$

Now it is easy to compute

$$d(f_1, f_0) = \sup\{|f_1(x) - f_0(x)| : x \in [0,1]\} = 1 \ ,$$

and so the error by using f_n to estimate the unique solution f is

$$d(f_n, f) \leq \frac{1}{1-\alpha}\alpha^n = \frac{1}{1-\frac{2}{3}}\left(\frac{2}{3}\right)^n = \frac{2^n}{3^{n-1}} \ .$$

In particular, to approximate f within an error of 0.01 in d, we solve

$$\frac{2^n}{3^{n-1}} < 0.01$$

and get $n \geq 15$, hence we could use f_{15} as an approximation.

11. Let $F : D \subset \mathbb{R}^2 \to \mathbb{R}$ be a continuous function satisfying a *Lipschitz condition* in y, that is, there exists $L > 0$ such that

$$\left|F(x,y_1) - F(x,y_2)\right| \le L|y_1 - y_2|$$

for all (x,y_1), $(x,y_2) \in D$. Show that for any $(x_0,y_0) \in D$, the *initial value problem*

$$\begin{cases} \dfrac{dy}{dx}(x) = F\big(x,y(x)\big) & \text{for all } x \in I \\ y(x_0) = y_0 \end{cases} \tag{*}$$

has a unique solution $y = y(x)$ on the interval $I = [x_0 - \delta, x_0 + \delta]$ for suitably chosen $\delta > 0$.

Proof. Observe first that $y = y(x)$ is a solution to the given initial value problem (*) if and only if

$$y(x) = y_0 + \int_{x_0}^{x} F\big(t,y(t)\big)\,dt\,, \quad x \in I\,. \tag{**}$$

Consider the metric space $\big(C(I), d\big)$ with

$$d(f,g) := \sup\left\{\left|f(x) - g(x)\right| : x \in I\right\}.$$

We again take it for granted that $(C(I), d)$ is complete. Define

$$\begin{aligned} T : C(I) &\to C(I) \\ h &\mapsto T(h) \end{aligned}$$

by

$$(Th)(x) := y_0 + \int_{x_0}^{x} F\big(t,h(t)\big)\,dt\,, \quad x \in I\,.$$

Then $y = y(x)$ solves (**) if and only if it is a fixed point of T. Thus, by Banach's Fixed Point Theorem, the problem reduces to showing that T is a contraction on $I = [x_0 - \delta, x_0 + \delta]$ with a suitably chosen δ, which is by now rather straight forward. In fact, by

$$\begin{aligned} d(Tf, Tg) &= \sup\left\{\left|(Tf)(x) - (Tg)(x)\right| : x \in I\right\} \\ &\le \sup\left\{\left|\int_{x_0}^{x} |F(t, f(t)) - F(t, g(t))| dt\right| : x \in I\right\} \\ &\le \sup\left\{\left|L \int_{x_0}^{x} \left|f(t) - g(t)\right| dt\right| : x \in I\right\} \\ &\le \sup\left\{\left|L \cdot d(f,g) \cdot (x - x_0)\right| : x \in I\right\} \\ &\le \delta L\, d(f,g) , \end{aligned}$$

we see that T is a contraction with contraction constant δL whenever $\delta < \frac{1}{L}$. Hence the integral equation (**) and thus in turn the initial value problem (*) has a unique solution in $I = [x_0 - \delta, x_0 + \delta]$.

12. Approximate the solution to the nonlinear initial value problem

$$\begin{cases} \dfrac{dy}{dx} = x - y^2 \\ y(0) = \dfrac{1}{2} \end{cases} \quad \text{on} \quad -\frac{1}{4} \le x \le \frac{1}{4}$$

within an error of 0.05 everywhere in $\left[\frac{1}{4}, \frac{1}{4}\right]$.

Proof. Comparing with Exercise 4.2, Part B, Problem #11, we write

$$F(x, y) := x - y^2 .$$

Then

$$\left|F(x, y_1) - F(x, y_2)\right| = |y_1^2 - y_2^2| = |y_1 + y_2|\, |y_1 - y_2| .$$

This suggests that we consider F on the domain

$$D := \left\{(x, y) \in \mathbb{R}^2 : |x| \le \frac{1}{4} , \left|y - \frac{1}{2}\right| \le \frac{1}{2}\right\} ,$$

on which we have

$$\left|F(x_1, y_1) - F(x_2, y_2)\right| \le 2|y_1 - y_2| .$$

From Exercise 4.2, Part B, Problem #11, the initial value problem has a unique solution $y = y(x)$ on the interval $\left[-\frac{1}{4}, \frac{1}{4}\right]$. Now observe that the initial value problem is equivalent to the integral equation

$$y = \frac{1}{2} + \int_0^x (t - y^2)\, dt$$

whose unique solution is precisely the unique fixed point of the operator $T : C\left[-\frac{1}{4}, \frac{1}{4}\right] \to C\left[-\frac{1}{4}, \frac{1}{4}\right]$ given by

$$(Ty)(x) := \frac{1}{2} + \int_0^x (t - y^2)\, dt , \quad x \in \left[-\frac{1}{4}, \frac{1}{4}\right] ,$$

which is a contraction with contraction constant $\alpha := 2 \cdot \frac{1}{4} = \frac{1}{2} < 1$. From the proof of Banach's Fixed Point Theorem, the unique solution to the initial value problem is the limit of the sequence $\{T^n y_0\}_{n \in \mathbb{N}}$ for any initial function $y_0(x)$. So we set

$$\begin{aligned}
y_0(x) &:= \frac{1}{2} , \\
y_1(x) &:= (Ty_0)(x) = \frac{1}{2} + \int_0^x \left[t - \left(\frac{1}{2}\right)^2\right] dt \\
&= \frac{1}{2} - \frac{x}{4} + \frac{x^2}{2} , \\
y_2(x) &:= (Ty_1)(x) = \frac{1}{2} + \int_0^x \left[t - \left(\frac{1}{2} - \frac{t}{4} + \frac{t^2}{2}\right)^2\right] dt \\
&= \frac{1}{2} - \frac{x}{4} + \frac{5}{8}x^2 - \frac{3}{16}x^3 + \frac{1}{16}x^4 - \frac{1}{20}x^5 , \\
&\cdots\cdots
\end{aligned}$$

Denote by $Y(x)$ the unique solution. From the proof of Banach's Fixed Point Theorem, we have

$$\begin{aligned}
d(y_n, Y) &\le \frac{\alpha^n}{1 - \alpha}\, d(y_1, y_0) \\
&= \frac{\left(\frac{1}{2}\right)^n}{1 - \frac{1}{2}} \sup\left\{\left|\frac{x}{4} - \frac{x^2}{2}\right| : x \in \left[-\frac{1}{4}, \frac{1}{4}\right]\right\} \\
&\le \frac{1}{2^{n-1}}\left(\frac{3}{32}\right) .
\end{aligned}$$

In particular,

$$d(y_1, Y) \le 0.09375 \ ,$$
$$d(y_2, Y) \le 0.046875 \ .$$

Hence the function y_2 is a good approximation to the unique solution Y, with an error < 0.05 in the sup norm, i.e., with error < 0.05 everywhere on $\left[-\frac{1}{4}, \frac{1}{4}\right]$.

Chapter 5

Uniform Convergence

This chapter is devoted to sequences and series of real-valued functions on a metric space X. In general, similar to the case of a sequence or a series of real numbers, we are interested in the convergence of a sequence or a series of functions. In particular, we are interested to understand which property shared by each term of a sequence or series of functions could be passed on to its limit function. The complication here is that unlike the case of sequences or series of real numbers for which there is a simple and unique sense of convergence, there are different senses of convergence of a sequence or a series of functions. It turns out that some properties shared by each term of a sequence or series of functions can be passed on to its limit function under certain senses of convergence but not under other senses. That makes the study of sequences and series of functions more interesting.

For the sake of simplicity, throughout this Chapter, we shall assume that all our functions are real-valued functions defined on a nonempty subset $S \subset \mathbb{R}$. The readers should be able to tell that most results in this Chapter are also valid in the context where S is a nonempty subset of a general metric space X.

5.1 Sequence of Functions

Let $\{f_n\}_{n\in\mathbb{N}}$ be a sequence of functions on S and f be a function on S. Observe that for every $x \in S$, $\{f_n(x)\}_{n\in\mathbb{N}}$ is a sequence of real numbers for which we can talk about its convergence. In case the sequence of real numbers $\{f_n(x)\}_{n\in\mathbb{N}}$ is convergent to $f(x)$ at every

$x \in S$, that is, in case

$$\lim_{n\to\infty} f_n(x) = f(x) \quad \text{for every } x \in S,$$

the sequence $\{f_n\}_{n\in\mathbb{N}}$ is said to *converge pointwisely on S to the function f*, and is denoted as "$\{f_n\} \to f$ *pointwisely on* S", or "$\lim_{n\to\infty} f_n = f$ *pointwisely on* S", or, if no confusions may arise, simply "$\lim_{n\to\infty} f_n = f$ *on* S". Equivalently, $\{f_n\}_{n\in\mathbb{N}}$ converges pointwisely on S to the function f if for every $\varepsilon > 0$ and every $x \in S$, there exists $N \in \mathbb{N}$ such that

$$|f_n(x) - f(x)| < \varepsilon \quad \text{for every } n \geq N\ .$$

Note that naturally, the natural number N is dependent on the given positive number ε and also on the point $x \in S$ under consideration. So in general, when $\varepsilon > 0$ or $x \in S$ varies, the number N will also vary accordingly. Therefore, in general, if we want to make this observation clearer, we would say that for any $\varepsilon > 0$ and any $x \in S$, there exists $N = N(\varepsilon, x) \in \mathbb{N}$ such that

$$|f_n(x) - f(x)| < \varepsilon \quad \text{for every } n \geq N(\varepsilon, x)\ .$$

Now a natural question is: what common properties of the f_n's can be passed on to their *pointwise limit* f? For instance, we pose the following questions:

Question 1: If f_n is continuous for each $n \in \mathbb{N}$ and $\{f_n\} \to f$ pointwisely, must f be continuous?

Observation: Let $c \in S$ be any point in S and we are to determine whether f is continuous at c. By definition, we are to check whether it is true that $\lim_{x\to c} f(x)$ equals $f(c)$. Now by assumption, $\{f_n\} \to f$ pointwisely on S and so $\lim_{n\to\infty} f_n(x) = f(x)$ for every $x \in S$. Hence in particular,

$$\lim_{x\to c} f(x) = \lim_{x\to c}\left(\lim_{n\to\infty} f_n(x)\right)\ .$$

On the other hand, for each $n \in \mathbb{N}$, f_n is continuous at c. So we have $\lim_{x \to c} f_n(x) = f_n(c)$. Moreover, we have $\lim_{n \to \infty} f_n(c) = f(c)$ and so

$$\lim_{n \to \infty} \left(\lim_{x \to c} f_n(x) \right) = \lim_{n \to \infty} f_n(c) = f(c) \ .$$

Combining, the question reduces to determining whether we have the equality

$$\lim_{x \to c} \left(\lim_{n \to \infty} f_n(x) \right) \stackrel{?}{=} \lim_{n \to \infty} \left(\lim_{x \to c} f_n(x) \right) \ .$$

In other words, Question 1 is to determine whether the two limits $\lim_{x \to c}$ and $\lim_{n \to \infty}$ commute.

Question 2: If $\{f_n\} \to f$ pointwisely on an interval $[a, b] \subset \mathbb{R}$, f_n is integrable over $[a, b]$ for every $n \in \mathbb{N}$, and f is integrable over $[a, b]$, is it necessarily true that

$$\lim_{n \to \infty} \int_a^b f_n(x)dx \stackrel{?}{=} \int_a^b f(x)dx \ .$$

Observation: As $\lim_{n \to \infty} f_n(x) = f(x)$ for every $x \in S$, the problem reduces to determining whether it is true that

$$\lim_{n \to \infty} \int_a^b f_n(x)dx \stackrel{?}{=} \int_a^b \lim_{n \to \infty} f(x)dx \ .$$

That is, the problem is to determine whether $\lim_{n \to \infty}$ can be taken out from the integral sign, or equivalently, whether $\lim_{n \to \infty}$ commutes with integration.

Question 3: If $\{f_n\} \to f$ pointwisely on an interval $(a, b) \subset \mathbb{R}$ and f_n is differentiable in (a, b) for every $n \in \mathbb{N}$, is it necessarily true that f is differentiable in (a, b) with $f' = \lim_{n \to \infty} f_n'$?

Observation: As $\lim_{n \to \infty} f_n(x) = f(x)$ for every $x \in S$, the problem reduces to determining whether it is true that $\lim_{n \to \infty} f_n$ is differentiable, and, if $\lim_{n\to\infty} f'_n$ exists, whether it is true that

$$\lim_{n\to\infty} f'_n(x) \stackrel{?}{=} \big(\lim_{n\to\infty} f_n\big)'(x) , \quad x \in (a,b) .$$

That is, the problem is to determine whether $\lim_{n\to\infty}$ commutes with differentiation.

The following simple Examples show that in general, the answers to all these three questions are negative.

Example 5.1.1. For any $n \in \mathbb{N}$, let $f_n : [0,1] \to \mathbb{R}$ be defined by $f_n(x) := x^n$, $x \in [0,1]$. Then

$$f_n(x) \to f(x) = \begin{cases} 0 & \text{if } x \in [0,1) \\ 1 & \text{if } x = 1 \end{cases}$$

pointwisely on $[0,1]$. Note that all f_n's are continuous on [0,1] but the pointwise limit f is not.

Example 5.1.2. For any $n \in \mathbb{N}$, let $f_n : [0,1] \to \mathbb{R}$ be defined by $f_n(x) := n^2(1-x)x^n$, $x \in [0,1]$. Then

$$f_n(x) \to f(x) :\equiv 0$$

pointwisely on $[0,1]$. Observe that

$$\int_0^1 f_n(x)dx = \frac{n^2}{(n+1)(n+2)} \quad \text{for all } n \in \mathbb{N}$$

and so

$$\lim_{n\to\infty} \int_0^1 f_n(x)dx = \lim_{n\to\infty} \frac{n^2}{(n+1)(n+2)} = 1 \neq 0 = \int_0^1 f(x)dx .$$

Hence in general, limit and integration do not commute.

Example 5.1.3. (i). For any $n \in \mathbb{N}$, let $f_n : (-1,1) \to \mathbb{R}$ be defined by

$$f_n(x) := \sqrt{x^2 + \frac{1}{n}}\,, \quad x \in (-1,1)\,.$$

It is clear that each f_n is differentiable in $(-1,1)$ and

$$f_n(x) \to f(x) := |x|$$

pointwisely in $(-1,1)$ but f is not differentiable at 0. So the pointwise limit of a sequence of differentiable functions may not be differentiable.

(ii). Let $\{f_n\}_{n\in\mathbb{N}}$ be the sequence of differentiable functions on $\mathbb{R}$ defined by

$$f_n(x) := \frac{1}{\sqrt{n}} \sin nx\,, \quad x \in \mathbb{R}\,, \quad n \in \mathbb{N}\,.$$

Observe that for any $x \in \mathbb{R}$,

$$|f_n(x)| = \left|\frac{1}{\sqrt{n}} \sin nx\right| \le \left|\frac{1}{\sqrt{n}}\right| \to 0 \quad \text{as } n \to \infty\,,$$

thus $\{f_n\} \to f :\equiv 0$ pointwisely on $\mathbb{R}$. Note that f is differentiable in $\mathbb{R}$ with $f' \equiv 0$. However,

$$f_n'(x) = \sqrt{n}\cos nx\,, \quad x \in \mathbb{R}$$

does not converge at any $x \in \mathbb{R}$. Hence in particular, limit and differentiation do not commute.

Although these examples are rather disappointing in the sense that they gave negative answers to the three questions posed at the beginning of the section, a careful analysis shows that they share a common feature that while the sequences of functions $\{f_n\}_{n\in\mathbb{N}}$ are all pointwisely convergent, the *speeds* of the convergence at various points are very different. In other words, for the same $\varepsilon > 0$, at different points $x \in S$, we need very different corresponding integer $N \in \mathbb{N}$. That is, there is no single $N \in \mathbb{N}$ that is good for all points $x \in S$. This important observation leads to the following definition.

Definition 5.1.4. *A sequence of functions $\{f_n\}_{n\in\mathbb{N}}$ on S is said to converge uniformly on S to a function f, denoted by $\{f_n\} \to f$ uniformly on S, if for any $\varepsilon > 0$, there exists $N \in \mathbb{N}$ such that*

$$\big|f_n(x) - f(x)\big| < \varepsilon \quad \textit{for all } n \geq N \textit{ and all } x \in S\ .$$

In this case, we write $f = \lim_{n\to\infty} f_n$ uniformly on S and call f the uniform limit of $\{f_n\}_{n\in\mathbb{N}}$ on S.

Note that in Definition 5.1.4, the integer $N \in \mathbb{N}$ depends only on ε but not on $x \in S$. That is, the same $N \in \mathbb{N}$ is good for all points $x \in S$, or equivalently, we have a *uniform* N for all $x \in S$.

Remark. It is obvious by definition that uniform convergence implies pointwise convergence.

Theorem 5.1.5. *Let $\{f_n\}_{n\in\mathbb{N}}$ be a sequence of functions on S. The following statements are all equivalent:*

(a) *$\{f_n\} \to f$ uniformly on S.*

(b) *For any $\varepsilon > 0$, there exists $N \in \mathbb{N}$ such that*

$$\sup\big\{\big|f_n(x) - f(x)\big| : x \in S\big\} < \varepsilon \quad \textit{for all } n \geq N\ .$$

(c) $\sup\{|f_n(x) - f(x)| : x \in S\} \to 0$ *as $n \to \infty$.*

Proof. (a) $\Rightarrow$ (b): Suppose $\{f_n\} \to f$ uniformly on S. For any $\varepsilon > 0$, there exists $N \in \mathbb{N}$ such that

$$\big|f_n(x) - f(x)\big| < \frac{\varepsilon}{2} \quad \text{for all } n \geq N \text{ and all } x \in S\ .$$

Hence

$$\sup\big\{\big|f_n(x) - f(x)\big| : x \in S\big\} \leq \frac{\varepsilon}{2} < \varepsilon \quad \text{for all } n \geq N\ .$$

(b) $\Rightarrow$ (a): It follows from definition of uniform convergence on S.

(b) $\Leftrightarrow$ (c): Observe that $\left\{\sup\left\{\left|f_n(x) - f(x)\right| : x \in S\right\}\right\}_{n\in\mathbb{N}}$ is a sequence of real numbers. The equivalence of (b) and (c) now follows immediately from definition of limit of a sequence of real numbers. □

Back to the three Questions posed at the beginning of this Section. With the stronger condition that the convergence of the sequence of functions $\{f_n\}_{n\in\mathbb{N}}$ is uniform, we have an affirmative answer to Question 1:

Theorem 5.1.6. *The uniform limit of a sequence of continuous functions is continuous. More precisely, let $\{f_n\}_{n\in\mathbb{N}}$ be a sequence of functions on S such that $\{f_n\} \to f$ uniformly on S. If f_n is continuous at $c \in S$ for every $n \in \mathbb{N}$, then so is f. In particular,*

$$\lim_{x\to c}\lim_{n\to\infty} f_n(x) = \lim_{x\to c} f(x) = f(c) = \lim_{n\to\infty} f_n(c) = \lim_{n\to\infty}\lim_{x\to c} f_n(x) \ .$$

Proof. For any $\varepsilon > 0$, since $\{f_n\} \to f$ uniformly on S, there exists $N \in \mathbb{N}$ such that

$$\left|f_N(x) - f(x)\right| < \frac{\varepsilon}{3} \quad \text{for all } x \in S \ .$$

Since f_N is continuous at c, there exists $\delta > 0$ such that

$$\left|f_N(x) - f_N(c)\right| < \frac{\varepsilon}{3} \quad \text{for all } x \in S \text{ with } |x - c| < \delta \ .$$

Hence for any $x \in S$ with $|x - c| < \delta$,

$$\begin{aligned}\left|f(x) - f(c)\right| &\le \left|f(x) - f_N(x)\right| + \left|f_N(x) - f_N(c)\right| + \left|f_N(c) - f(c)\right| \\ &< \frac{\varepsilon}{3} + \frac{\varepsilon}{3} + \frac{\varepsilon}{3} = \varepsilon \ .\end{aligned}$$

□

Corollary 5.1.7. *The uniform limit of a sequence of uniformly continuous functions is uniformly continuous.*

Proof. The proof is analogous to that of Theorem 5.1.6, noting that here c can be an arbitrary point in S. In fact, for any $\varepsilon > 0$, since $\{f_n\} \to f$ uniformly on S, there exists $N \in \mathbb{N}$ such that

$$\left|f_N(x) - f(x)\right| < \frac{\varepsilon}{3} \quad \text{for all } x \in S \ .$$

Since f_N is uniformly continuous in S, there exists $\delta > 0$ such that

$$\left|f_N(x) - f_N(y)\right| < \frac{\varepsilon}{3} \quad \text{for all } x, y \in S \text{ with } |x - y| < \delta \ .$$

Hence for any x, $y \in S$ with $|x - y| < \delta$,

$$\begin{aligned}\left|f(x) - f(y)\right| &\leq \left|f(x) - f_N(x)\right| + \left|f_N(x) - f_N(y)\right| + \left|f_N(y) - f(y)\right| \\ &< \frac{\varepsilon}{3} + \frac{\varepsilon}{3} + \frac{\varepsilon}{3} = \varepsilon \ .\end{aligned}$$

□

Under the condition of uniform convergence, we also have an affirmative answer to Question 2:

Theorem 5.1.8. *Let $\{f_n\}_{n\in\mathbb{N}}$ be a sequence of continuous functions on $[a, b]$ which converges uniformly to f on $[a, b]$. Then*

$$\int_a^b \lim_{n\to\infty} f_n = \int_a^b f = \lim_{n\to\infty} \int_a^b f_n \ .$$

Proof. By Theorem 5.1.6, being the uniform limit of the sequence of continuous functions $\{f_n\}_{n\in\mathbb{N}}$, f is also continuous. Hence all f_n's and f are integrable over $[a, b]$. Now as $\{f_n\} \to f$ uniformly on $[a, b]$, for any $\varepsilon > 0$, there exists $N \in \mathbb{N}$ such that

$$\left|f_n(x) - f(x)\right| < \frac{\varepsilon}{b - a} \quad \text{for all } n \geq N \text{ and all } x \in [a, b] \ .$$

Hence

$$\left| \int_a^b f_n - \int_a^b f \right| = \left| \int_a^b f_n - f \right| \le \int_a^b |f_n - f| < \varepsilon$$

for all $n \ge N$. Hence the assertion. □

Remark. In general, the condition that $\{f_n\} \to f$ uniformly guarantees that $\left\{ \int_a^b f_n \right\} \to \int_a^b f$ as $n \to \infty$, but this is not a necessary condition. To be a little more precise, the conditions that $\{f_n\} \to f$ pointwisely and $\left\{ \int_a^b f_n \right\} \to \int_a^b f$ as $n \to \infty$ together do not necessarily imply that $\{f_n\} \to f$ uniformly. For example, consider $f_n(x) := x^n$ on $[0, 1]$, $n \in \mathbb{N}$. Then

$$\{f_n(x)\} \to f(x) = \begin{cases} 0 & x \in [0,1) \\ 1 & x = 1 \end{cases} \qquad \text{as } n \to \infty$$

and

$$\int_0^1 f = 0 = \lim_{n\to\infty} \frac{1}{n+1} = \lim_{n\to\infty} \int_0^1 f_n \ ,$$

but then we have seen that $\{f_n\} \not\twoheadrightarrow f$ uniformly on $[0, 1]$.

We need a bit more preparation before we could tackle the remaining question, Question 3. Recall that on $C[a,b]$,

$$d(f,g) := \sup\left\{ \left|f(x) - g(x)\right| : x \in [a,b] \right\}$$

is a well defined metric (again, will see in a few moments that it makes $C[a,b]$ a complete metric space). So every $f \in C[a,b]$ now has two identities. One, it is an element in the abstract metric space $\big(C[a,b], d\big)$, and two, it is a continuous function $f : [a,b] \to \mathbb{R}$. Likewisely, a sequence $\{f_n\}_{n\in\mathbb{N}}$ in $C[a,b]$ also has two identities. One, it is a sequence of points in the abstract metric space $\big(C[a,b], d\big)$, and two, it is a sequence of continuous functions $f_n : [a,b] \to \mathbb{R}$, $n \in \mathbb{N}$. Therefore, we can talk about its convergence in the metric d, its pointwise convergence as a sequence of continuous functions on $[a,b]$, and its uniform convergence as a sequence of continuous functions on $[a,b]$.

Clearly, uniform convergence implies pointwise convergence but not vice versa. On the other hand, the following important Theorem gives the relation between uniform convergence (when $\{f_n\}_{n\in\mathbb{N}}$ is considered as a sequence of continuous functions on $[a,b]$) and convergence in the metric d (when $\{f_n\}_{n\in\mathbb{N}}$ is considered as a sequence of points in the abstract metric space $(C[a,b],d)$).

Theorem 5.1.9. *Let $\{f_n\}_{n\in\mathbb{N}}$ be a sequence in $\big(C[a,b],d\big)$, where*

$$d(f,g) := \sup\big\{\big|f(x)-g(x)\big| : x\in[a,b]\big\}.$$

Then $\{f_n\}\to f$ uniformly on $[a,b]$ if and only if $\{f_n\}\to f$ in d, that is, if and only if $\{d(f_n,f)\}\to 0$ as $n\to\infty$. Hence for a sequence of functions in $C\big([a,b],d\big)$, uniform convergence is equivalent to convergence in the "sup metric".

Proof. By Theorem 5.1.5,

$$\begin{aligned}
&\{f_n\}\to f \quad \text{uniformly on } [a,b]\\
\iff &\text{for any } \varepsilon>0, \text{ there exists } N\in\mathbb{N} \text{ s.t.}\\
&\quad d(f_n,f) = \sup_{x\in[a,b]}\big\{\big|f_n(x)-f(x)\big|\big\} < \varepsilon \text{ for all } n\geq N\\
\iff &\big\{d(f_n,f)\big\}\to 0 \text{ as } n\to\infty.
\end{aligned}$$

□

Recall that in a general metric space, a sequence is convergent implies it is Cauchy but the converse is in general not true unless the metric space under consideration is complete. However, for a sequence of functions on S, uniform convergence is equivalent to "uniform Cauchyness".

Definition 5.1.10. *A sequence of functions $\{f_n\}_{n\in\mathbb{N}}$ on S is said to be uniformly Cauchy in S if for any $\varepsilon>0$, there exists $N\in\mathbb{N}$ such that*

$$\big|f_n(x)-f_m(x)\big| < \varepsilon \quad \textit{for all } n,m\geq N \textit{ and all } x\in S\,.$$

Note that in Definition 5.1.10, similar to the definition of uniform convergence, the integer $N \in \mathbb{N}$ depends only on ε but not on $x \in S$. That is, the same $N \in \mathbb{N}$ is good for all points $x \in S$, or equivalently, we have a *uniform* N for all $x \in S$.

Theorem 5.1.11 (Cauchy's criterion for uniform convergence of sequences). *Let $\{f_n\}_{n\in\mathbb{N}}$ be a sequence of real-valued functions on S. Then $\{f_n\} \to f$ uniformly on S for some function f if and only if $\{f_n\}_{n\in\mathbb{N}}$ is uniformly Cauchy in S.*

Proof. ($\Rightarrow$) Obvious by triangle inequality.
($\Leftarrow$) Since $\{f_n\}_{n\in\mathbb{N}}$ is uniformly Cauchy in S, for any $x \in S$, $\{f_n(x)\}_{n\in\mathbb{N}}$ is a Cauchy sequence of real numbers. Since $\mathbb{R}$ is complete, this Cauchy sequence is convergent in $\mathbb{R}$. Naturally the limit of this sequence depends on the point x and so we can call it $f(x)$. Do this for all $x \in S$ and we obtain a well-defined function $f : S \to \mathbb{R}$ given by $f(x) := \lim_{n\to\infty} f_n(x)$ for every $x \in S$. Now as $\{f_n\}_{n\in\mathbb{N}}$ is uniformly Cauchy in S, for any $\varepsilon > 0$, there exists $N \in \mathbb{N}$ such that

$$\left|f_n(x) - f_m(x)\right| < \varepsilon \quad \text{for all } n, m \geq N \text{ and any } x \in S \ .$$

Hence for any $n \geq N$ and any $x \in S$, we have

$$\left|f_n(x) - f(x)\right| = \lim_{m\to\infty} \left|f_n(x) - f_m(x)\right| \leq \varepsilon \ .$$

Therefore,

$$d(f_n, f) = \sup\Big\{|f_n(x) - f(x)| : x \in S\Big\} \leq \varepsilon$$

for all $n \geq N$. Hence $\{f_n\}_{n\in\mathbb{N}}$ is uniformly convergent on S. □

Remark. At first sight, Theorem 5.1.11 is rather astonishing, as for a general metric space, Cauchy sequences may not converge unless the space is complete. Here for Theorem 5.1.11, it seems that no completeness is required. But then the careful readers would be able to tell that certain completeness condition has already been implicitly imposed, namely, the completeness of the target space $\mathbb{R}$. In fact, it is easy to see that the completeness of $\mathbb{R}$ is critically used in the proof and so the result is not really peculiar at all.

Remark. It's time to do a little recap here. For any sequence of real-valued functions $\{f_n\}_{n\in\mathbb{N}}$ on $[a,b] \in \mathbb{R}$, the following conditions are all equivalent:

(i) $\{f_n\} \to f$ uniformly on $[a,b]$.

(ii) For any $\varepsilon > 0$, there exists $N \in \mathbb{N}$ such that

$$|f_n(x) - f(x)| < \varepsilon \quad \text{for all } n \geq N \text{ and all } x \in [a,b]\,.$$

(iii) For any $\varepsilon > 0$, there exists $N \in \mathbb{N}$ such that

$$\sup\left\{|f_n(x) - f(x)| : x \in [a,b]\right\} < \varepsilon \quad \text{for all } n \geq N\,.$$

(iv) $\sup\left\{|f_n(x) - f(x)| : x \in [a,b]\right\} \to 0$ as $n \to \infty$.

(v) $\{d(f_n, f)\} \to 0$ as $n \to \infty$, where d is the "sup" metric on $C[a,b]$ defined by

$$d(f,g) := \sup\left\{|f_n(x) - f(x)| : x \in [a,b]\right\} \text{ for } f,g \in C[a,b]\,.$$

(vi) $\{f_n\}_{n\in\mathbb{N}}$ is uniformly Cauchy.

So to prove that $\{f_n\} \to f$ uniformly on $[a,b]$, it suffices to prove any of the preceding equivalent statements. Specifically, if the f_n's are given with explicit formula and the pointwise limit of $\{f_n\}_{n\in\mathbb{N}}$ exists and is known or easily computed, in general, it would be easier to check the uniform convergence by checking conditions (ii), (iii), (iv), or (v). On the other hand, if the explicit formulae of f_n's are not given or are very complicated, and the pointwise limit of $\{f_n\}_{n\in\mathbb{N}}$ is not known, then we may have to resort to condition (vi).

To the contrary, in order to prove that $\{f_n\} \not\to f$ uniformly on $[a, b]$, it suffices to show any of the following equivalent statements which are precisely the negation of the preceding equivalent statements:

(i)′ $\{f_n\} \not\to f$ uniformly on $[a, b]$.

(ii)′ There exists $\varepsilon > 0$ such that for any $N \in \mathbb{N}$,

$$|f_n(x) - f(x)| \geq \varepsilon \quad \text{for some } n \geq N \text{ and some } x \in [a, b] \,.$$

(iii)′ There exists $\varepsilon > 0$ such that for any $N \in \mathbb{N}$,

$$\sup\left\{|f_n(x) - f(x)| : x \in [a, b]\right\} \geq \varepsilon \quad \text{for some } n \geq N \,.$$

(iv)′ $\sup\left\{|f_n(x) - f(x)| : x \in [a, b]\right\} \not\to 0 \quad$ as $n \to \infty$.

(v)′ $\{d(f_n, f)\} \not\to 0$ as $n \to \infty$, where d is the "sup" metric on $C[a, b]$ defined by

$$d(f, g) := \sup\left\{|f_n(x) - f(x)| : x \in [a, b]\right\} \text{ for } f, g \in C[a, b] \,.$$

(vi)′ $\{f_n\}_{n\in\mathbb{N}}$ is not uniformly Cauchy, that is, there exists $\varepsilon > 0$ such that for any $N \in \mathbb{N}$, we can find n, $m \geq N$ and $x \in S$ such that

$$|f_n(x) - f_m(x)| \geq \varepsilon \,.$$

Example 5.1.12. We investigate the convergence of the sequence $\{f_n\}_{n\in\mathbb{N}}$ on $[0, 1]$ defined by $f_n(x) := x^n$, $x \in [0, 1]$. As shown in Example 5.1.1,

$$f_n(x) \to f(x) = \begin{cases} 0 & \text{if } x \in [0, 1) \\ 1 & \text{if } x = 1 \end{cases} \tag{*}$$

pointwisely on $[0, 1]$ as $n \to \infty$. Since all f_n's are continuous on $[0, 1]$ but the pointwise limit f is not, by Theorem 5.1.6, the convergence is not uniform on $[0, 1]$.

Instead of proving it indirectly as above, it is more illuminating and helps us better understand the behavior of the f_n's if we prove it by first principle: Observe that for any fixed $n \in \mathbb{N}$, no matter how large or small it is, by choosing $t \in [0, 1)$ close enough to 1, the value of $f_n(t) = t^n$ can be made as close to 1 as we wish. Hence we pick $\varepsilon := \frac{1}{2}$, then for every $n \in \mathbb{N}$, we can find $t \in [0, 1)$ such that $|f_n(t) - f(t)| = |f_n(t)| > \frac{1}{2} = \varepsilon$. This shows statement (ii)$'$ in the preceding Remark and so the convergence $\{f_n\} \to f$ is not uniform on $[0, 1]$.

Alternatively, by the same observation, we have

$$\sup\left\{\left|f_n(x) - f(x)\right| : x \in [0, 1]\right\} > \frac{1}{2} \quad \text{for any } n \in \mathbb{N}$$

which is statement (iii)$'$ in the preceding Remark and so the convergence $\{f_n\} \to f$ is only pointwise but not uniform on $[0, 1]$.

The interested readers can easily prove that the convergence is not uniform on $[0, 1]$ by establishing (iv)$'$, (v)$'$, or (vi)$'$ in the preceding remark.

Although the convergence of $\{f_n\} \to f$ on $[0, 1]$ is only pointwise but not uniform, it is interesting to observe that it becomes uniform when restricted to a slightly smaller domain $[0, a]$ for any $0 \le a < 1$. In fact, for any $\varepsilon > 0$, let $N \in \mathbb{N}$ be large enough such that $a^N < \varepsilon$. Then

$$|f_n(x) - f(x)| = x^n \le a^n \le a^N < \varepsilon \text{ for all } n \ge N \text{ and all } x \in [0, a] .$$

Hence $\{f_n\} \to f$ uniformly on $[0, a]$.

Alternatively, for any $\varepsilon > 0$, let $N \in \mathbb{N}$ be large enough such that $2a^N < \varepsilon$. Then

$$\begin{aligned} |f_n(x) - f_m(x)| &= |x^n - x^m| = x^m \left|x^{n-m} - 1\right| \\ &\le x^N (|x|^{n-m} + 1) \le 2a^N < \varepsilon \end{aligned}$$

for all $n > m \ge N$. Thus $\{f_n\}_{n \in \mathbb{N}}$ is uniformly Cauchy in $[0, a]$ and so it is uniformly convergent there. Since the pointwise limit of $\{f_n\}_{n \in \mathbb{N}}$ is f, we conclude that $\{f_n\} \to f$ uniformly on $[0, a]$.

The interested readers are encouraged to check whether the convergence is uniform in $[0, 1)$.

With Cauchy's Criterion for uniform convergence at hand, we are now in a position to prove the following long overdue result.

Corollary 5.1.13. *For any $f, g \in C[a, b]$, define*

$$d(f, g) := \sup \{|f(x) - g(x)| : x \in [a, b]\}.$$

Then $(C[a, b], d)$ is a complete metric space.

Proof. The fact that d is a well-defined metric on $C[a, b]$ has been shown before. So it remains to show completeness. Let $\{f_n\}_{n\in\mathbb{N}}$ be a Cauchy sequence in $C[a, b]$. Then for any $\varepsilon > 0$, there exists $N \in \mathbb{N}$ such that

$$d(f_n, f_m) < \varepsilon \quad \text{for all } m, n \geq N .$$

Hence

$$|f_n(x) - f_m(x)| \leq d(f_n, f_m) < \varepsilon \quad \text{for all } m, n \geq N \text{ and all } x \in [a, b] .$$

That is, $\{f_n\}_{n\in\mathbb{N}}$ is uniformly Cauchy in $[a, b]$ and so by Theorem 5.1.11, it is uniformly convergent on $[a, b]$ to some function $f \in C[a, b]$. By Theorem 5.1.6, we have $f \in C[a, b]$. Finally, by Theorem 5.1.9, we have $\{d(f_n, f)\} \to 0$ as $n \to \infty$, that is, $\{f_n\} \to f$ in $(C[a, b], d)$ and so $(C[a, b], d)$ is complete. □

Finally, we shall address Question 3. Theorems 5.1.6 and 5.1.8 naturally lead us speculate that Question 3 would also have an affirmative answer in case the convergence of the sequence of functions $\{f_n\}_{n\in\mathbb{N}}$ is uniform. However, the following Example shows that unfortunately, it is not the case.

Example 5.1.14. Let us revisit Example 5.1.3 again.

(i) Observe that the sequence of differentiable functions

$$f_n(x) := \sqrt{x^2 + \frac{1}{n}}\,, \quad x \in (-1,1), \quad n \in \mathbb{N}$$

actually converges uniformly to $f(x) := |x|$ in $(-1,1)$. In fact, for all $x \in (-1,1)$ we have

$$\begin{aligned} \big|f_n(x) - f(x)\big| &= \left|\sqrt{x^2 + \frac{1}{n}} - |x|\right| \\ &= \left|\frac{\left(x^2 + \frac{1}{n}\right) - |x|^2}{\sqrt{x^2 + \frac{1}{n}} + |x|}\right| \\ &\leq \frac{\frac{1}{n}}{\sqrt{\frac{1}{n}}} = \frac{1}{\sqrt{n}} \to 0 \quad \text{as } n \to \infty\,. \end{aligned}$$

Hence $\{f_n\} \to f$ uniformly in $(-1,1)$. However, it is clear that f is not differentiable at 0. Hence the uniform limit of a sequence of differentiable functions may not be differentiable.

(ii) Observe that the sequence of differentiable functions

$$f_n(x) := \frac{1}{\sqrt{n}}\sin nx\,, \quad x \in \mathbb{R}\,, \quad n \in \mathbb{N}$$

actually converges uniformly to $f \equiv 0$ on $\mathbb{R}$. In fact,

$$\begin{aligned} &\sup\{|f_n(x) - 0| : x \in \mathbb{R}\} \\ &= \sup\left\{\left|\frac{1}{\sqrt{n}}\sin nx\right| : x \in \mathbb{R}\right\} = \frac{1}{\sqrt{n}} \to 0 \text{ as } n \to \infty\,. \end{aligned}$$

Hence $\{f_n\} \to f$ uniformly on $\mathbb{R}$. Clearly, $f' \equiv 0$ on $\mathbb{R}$ but

$$f_n'(x) = \sqrt{n}\cos nx\,, \quad x \in \mathbb{R}\,,$$

does not converge at any point in $\mathbb{R}$. Hence the derivative of the uniform limit of a sequence of differentiable functions may not be equal to the limit of the sequence of derivatives (the latter may not even exist.)

Example 5.1.14 shows that in general, the uniform limit of a sequence of differentiable functions may not necessarily be differentiable, and even if it is differentiable, its derivative may not be equal to the limit of the sequence of derivatives. While that is a bit disappointing, we do have an interesting result which says that if a sequence of functions is convergent at one point and if the sequence of derivatives is uniformly convergent, then the sequence of functions is uniformly convergent to a differentiable function, with derivative equals to the uniform limit of the sequence of derivatives. More precisely, we have

Theorem 5.1.15. *Let $\{f_n\}_{n\in\mathbb{N}}$ be a sequence of differentiable functions on a bounded open set $S := (a, b) \subset \mathbb{R}$. Suppose*

(i) *$\{f_n'\} \to$ some function g uniformly on S, and*

(ii) *there exists $x_0 \in S$ such that $\{f_n(x_0)\}_{n\in\mathbb{N}}$ converges,*

then

(a) *$\{f_n\} \to$ some function f uniformly on S, and*

(b) *f is differentiable, with $f' = g = (\lim_{n\to\infty} f_n)'$ on S.*

Proof. For every $n \in \mathbb{N}$, by our experience in elementary calculus, we consider the difference quotient of the differentiable function f_n as follows. For any fixed $c \in S$, we define

$$g_n(x) := \begin{cases} \dfrac{f_n(x) - f_n(c)}{x - c} & \text{if } x \neq c \\ f_n'(c) & \text{if } x = c\ . \end{cases}$$

Then by construction, g_n is continuous at c. On the other hand, by Mean Value Theorem, for any $x \neq c$,

$$\begin{aligned} \left|g_n(x) - g_m(x)\right| &= \left|\frac{\left[f_n(x) - f_m(x)\right] - \left[f_n(c) - f_m(c)\right]}{x - c}\right| \\ &= \left|f_n'(t) - f_m'(t)\right| \end{aligned}$$

for some $t \in S$ between x and c. Since $\{f_n'\}_{n\in\mathbb{N}}$ converges uniformly on S, it is uniformly Cauchy in S and so the same is true for $\{g_n\}_{n\in\mathbb{N}}$.

(a) In order to show that $\{f_n\}_{n\in\mathbb{N}}$ is uniformly convergent, in view of Theorem 5.1.11, it suffices to show that it is uniformly Cauchy. By the construction above, we have

$$f_n(x) - f_n(c) = g_n(x)(x-c) \quad \text{for all} \quad x \in S$$

and so

$$f_n(x) - f_m(x) = f_n(c) - f_m(c) + (x-c)\big[g_n(x) - g_m(x)\big]$$

for all $x \in S$. Note that this is true for any fixed $c \in S$. In particular, we take $c := x_0$. As $\big\{f_n(x_0)\big\}_{n\in\mathbb{N}}$ converges, there exists $N_1 \in \mathbb{N}$ such that

$$\big|f_n(x_0) - f_m(x_0)\big| < \frac{\varepsilon}{2} \quad \text{for all } n, m \geq N_1 \ .$$

On the other hand, since $\{g_n\}_{n\in\mathbb{N}}$ is uniformly Cauchy in S, there exists $N_2 \in \mathbb{N}$ such that

$$\big|g_n(x) - g_m(x)\big| < \frac{\varepsilon}{2(b-a)} \quad \text{for all } n, m \geq N_2 \text{ and all } x \in S \ .$$

Thus

$$\begin{aligned} \big|f_n(x) - f_m(x)\big| &\leq \big|f_n(x_0) - f_m(x_0)\big| + |x - x_0|\big|g_n(x) - g_m(x)\big| \\ &< \frac{\varepsilon}{2} + (b-a) \cdot \frac{\varepsilon}{2(b-a)} \\ &= \varepsilon \end{aligned}$$

for all n, $m \geq \max\{N_1, N_2\}$ and all $x \in S$. Hence $\{f_n\}_{n\in\mathbb{N}}$ is uniformly Cauchy in S. By Theorem 5.1.11, $\{f_n\}_{n\in\mathbb{N}}$ is uniformly convergent in S to some function f.

(b) For any $c \in S$,

$$\begin{aligned} f'(c) &= \lim_{x\to c} \frac{f(x)-f(c)}{x-c} \\ &= \lim_{x\to c}\lim_{n\to\infty} \frac{f_n(x)-f_n(c)}{x-c} \\ &= \lim_{x\to c}\lim_{n\to\infty} g_n(x) \ . \end{aligned}$$

Since $\{g_n\}_{n\in\mathbb{N}}$ is uniformly Cauchy, it is uniformly convergent and so by Theorem 5.1.6, we have

$$\begin{aligned} f'(c) &= \lim_{x\to c}\lim_{n\to\infty} g_n(x) \\ &= \lim_{n\to\infty}\lim_{x\to c} g_n(x) \\ &= \lim_{n\to\infty} g_n(c) \\ &= \lim_{n\to\infty} f_n'(c) \\ &= g(c) \ . \end{aligned}$$

□

Condition (ii) of Theorem 5.1.15 requires the convergence of the sequence of functions $\{f_n\}_{n\in\mathbb{N}}$ at one single point $x_0 \in S$. It is a bit mysterious in that on the one hand it is rather weak as we do not even require pointwise convergence on S, and on the other hand, the requirement of convergence of the sequence $\{f_n\}_{n\in\mathbb{N}}$ at one single point seems a bit too arbitrary. The following example addresses on these.

Example 5.1.16. (Condition (ii) of Theorem 5.1.15 is necessary.) Let $S := (0,1) \subset \mathbb{R}$ and for any $n \in \mathbb{N}$, define $f_n(x) := \ln nx$, $x \in S$. Then $\{f_n\}_{n\in\mathbb{N}}$ is a sequence of differentiable functions on S. Note that $f_n'(x) = \frac{1}{x}$ for all $n \in \mathbb{N}$ and so

$$\{f_n'(x)\} \to g(x) := \frac{1}{x} \quad \text{uniformly on } S \ ,$$

but $\{f_n\}_{n\in\mathbb{N}}$ does not converge at any $x \in S$. Hence condition (ii) of Theorem 5.1.15 fails to hold and certainly $\{f_n\}_{n\in\mathbb{N}}$ is not uniformly convergent on S.

Exercise 5.1

Part A: True or False Questions

For each of the following statements, determine if it is true or false. If it is true, prove it. If it is false, give a counterexample or provide proper justification.

1. Let $\{f_n\}_{n\in\mathbb{N}}$, $\{g_n\}_{n\in\mathbb{N}}$ be sequences of real-valued functions on $S \subset \mathbb{R}$. If $\{f_n\} \to f$ and $\{g_n\} \to g$ uniformly on S, then $\{f_n + g_n\} \to f + g$ uniformly on S.

   Answer: True. Since $\{f_n\} \to f$ and $\{g_n\} \to g$ uniformly on S, for any $\varepsilon > 0$, there exists $N \in \mathbb{N}$ such that $|f_n(x) - f(x)| < \frac{\varepsilon}{2}$ and $|g_n(x) - g(x)| < \frac{\varepsilon}{2}$ for all $x \in S$ and all $n \geq N$. Hence

   $$\left|(f_n+g_n)(x)-(f(x)+g(x))\right| \leq \left|f_n(x)-f(x)\right|+\left|g_n(x)-g(x)\right| < \varepsilon$$

   for all $x \in S$ and all $n \geq N$. That is, $\{f_n + g_n\} \to f + g$ uniformly on S.

2. The sequence of functions $\{f_n\}_{n\in\mathbb{N}}$ on $[0,\infty)$ defined by $f_n(x) := \frac{1}{1+nx}$, $x \in [0,\infty)$, is uniformly convergent on $[0,\infty)$.

   Answer: False.

   Justification. It is evident that $\{f_n(x)\} \to f(x) := \begin{cases} 1 & \text{if } x = 0 \\ 0 & \text{if } x \neq 0 \end{cases}$ pointwisely on $[0,\infty)$. Since f is not continuous on $[0,\infty)$, the convergence is not uniform on $[0,\infty)$.

3. The sequence of functions $\{f_n\}_{n\in\mathbb{N}}$ on $[0,\infty)$ defined by $f_n(x) := \frac{1}{1+nx}$, $x \in [0,\infty)$, is uniformly convergent on $(0,\infty)$.

   Answer: False.

   Justification. It is evident that $\{f_n(x)\} \to f(x) :\equiv 0$ pointwisely on $(0,\infty)$. Although f is continuous, the convergence is still not uniform. In fact, take $\varepsilon = \frac{1}{2}$. Then for every $n \in \mathbb{N}$, the point $x_n := \frac{1}{n}$ would give

   $$|f_n(x) - f(x)| = \frac{1}{2} \geq \varepsilon\,.$$

4. The sequence of functions $\{f_n\}_{n\in\mathbb{N}}$ on $[0,1]$ defined by $f_n(x) := \sum_{i=0}^{n} x^i(1-x)$, $x \in [0,1]$, is uniformly convergent on $[0,1]$.

   Answer: False.

   Justification. It is evident that

$$\{f_n(x)\} \to f(x) := \begin{cases} 1 & \text{if } x \in [0,1) \\ 0 & \text{if } x = 1 \end{cases}$$

   pointwisely on $[0,1]$. Since f is not continuous on $[0,1]$, the convergence is not uniform on $[0,1]$.

5. The sequence of functions $\{f_n\}_{n\in\mathbb{N}}$ on $[0,\infty)$ defined by $f_n(x) := \frac{x}{1+nx}$, $x \in [0,\infty)$, is uniformly convergent on $[0,\infty)$.

   Answer: True.

   Proof. It is clear that $f_n(0) = 0$ for all $n \in \mathbb{N}$ and so $\{f_n(0)\} \to 0$ as $n \to \infty$. On the other hand, for $x > 0$, we have $0 < f_n(x) \leq \frac{x}{nx} = \frac{1}{n} \to 0$ as $n \to \infty$. Hence $\{f_n\} \to f :\equiv 0$ pointwisely on $[0,\infty)$. It is not hard to see that the convergence is actually uniform. In fact, for any $\varepsilon > 0$, let $N \in \mathbb{N}$ be large enough such that $\frac{1}{N} < \varepsilon$. Then $|f_n(x) - 0| \leq \frac{1}{n} \leq \frac{1}{N} < \varepsilon$ for all $n \geq N$ and all $x \in (0,\infty)$. Thus $\{f_n\} \to 0$ uniformly on $(0,\infty)$. Combining, we have $\{f_n\} \to 0$ uniformly on $[0,\infty)$.

6. The sequence of functions $\{f_n\}_{n\in\mathbb{N}}$ on $[0,1]$ defined by $f_n(x) := x(1-x)^n$, $x \in [0,1]$, is uniformly convergent on $[0,1]$.

   Answer: True.

   Proof. It is clear that $f_n(0) = 0, f_n(1) = 0$ for all $n \in \mathbb{N}$ and so $\{f_n(0)\} \to 0$ and $\{f_n(1)\} \to 0$ as $n \to \infty$. On the other hand, for any $0 < x < 1$, we have $0 < f_n(x) \leq (1-x)^n \to 0$ as $n \to \infty$. Hence $\{f_n\} \to f :\equiv 0$ pointwisely on $[0,1]$. To show that the convergence is actually uniform, observe first that by elementary calculus, the maximum of the function $f_n(x)$ on $[0,1]$ is attained at the point $\frac{1}{n+1}$, and so for any $x \in [0,1]$, we have

$$|f_n(x)-f(x)| = x(1-x)^n \leq \frac{1}{n+1}\left[1-\frac{1}{n+1}\right]^n = \frac{n^n}{(n+1)^{n+1}} .$$

This suggests that for any $\varepsilon > 0$, we take $N \in \mathbb{N}$ large enough such that $\frac{N^N}{(N+1)^{N+1}} < \varepsilon$. This is possible since $\lim_{N\to\infty} \frac{N^N}{(N+1)^{N+1}} = 0$. Then

$$|f_n(x) - f(x)| \leq \frac{n^n}{(n+1)^{n+1}} \leq \frac{N^N}{(N+1)^{N+1}} < \epsilon$$

for any $n \geq N$ and any $x \in [0,1]$. Thus the convergence $\{f_n\} \to f$ is uniform on $[0,1]$.

7. The sequence of functions $\{f_n\}_{n\in\mathbb{N}}$ on $[0,\infty)$ defined by $f_n(x) := \tan^{-1}(nx)$, $x \geq 0$, is uniformly convergent on $[0,\infty)$.

   Answer: False.

   Justification. It is clear that

$$\{f_n\} \to f := \begin{cases} 0 & \text{if } x = 0 \\ \frac{\pi}{2} & \text{if } x > 0 \end{cases}$$

   pointwisely on $[0,\infty)$. Since f is not continuous on $[0,\infty)$, the convergence is not uniform.

8. The sequence of functions $\{f_n\}_{n\in\mathbb{N}}$ on $[0,\infty)$ defined by $f_n(x) := \tan^{-1}(nx)$, $x \geq 0$, is uniformly convergent on $[a,\infty)$ for any $a > 0$.

   Answer: True.

   Proof. It is clear that $\{f_n\} \to f :\equiv \frac{\pi}{2}$ pointwisely on $[a,\infty)$. To show that the convergence is actually uniform, we first observe that

$$|f_n(x) - f(x)| = \left|\tan^{-1}(nx) - \frac{\pi}{2}\right| = \cot^{-1}(nx) .$$

   Hence we need to show that for $n \in \mathbb{N}$ large enough, $\cot^{-1}(nx)$ will be smaller than a prescribed positive number $\varepsilon > 0$ for all $x \geq a$.

Now by the fact that $\cot^{-1}$ is monotonically decreasing on $[0, \infty)$, we have

$$\begin{aligned} & \cot^{-1}(nx) < \varepsilon && \text{for all } x \geq a \\ \iff \quad & nx > \cot\varepsilon && \text{for all } x \geq a \\ \iff \quad & n > \frac{\cot\varepsilon}{x} && \text{for all } x \geq a \end{aligned}$$

and so this suggests that we pick $N \in \mathbb{N}$ such that $N > \frac{\cot\varepsilon}{a}$. Then for any $n \geq N$ and any $x \geq a$, we have

$$n \geq N > \frac{\cot\varepsilon}{a} \geq \frac{\cot\varepsilon}{x}$$

and so

$$|f_n(x) - f(x)| = \cot^{-1}(nx) < \varepsilon\ .$$

Thus the convergence is uniform.

9. The sequence of functions $\{f_n\}_{n\in\mathbb{N}}$ on $[0, \infty)$ defined by $f_n(x) := \tan^{-1}(nx)$, $x \geq 0$, is uniformly convergent on $(0, \infty)$.

   Answer: False.

   Justification. Using the analysis of Exercise 5.1, Part A, Problem #8, we see that the convergence is uniform on $(0, \infty)$ if and only if for any given $\varepsilon > 0$, there exists $N \in \mathbb{N}$ large enough such that

   $$|f_n(x) - f(x)| = \cot^{-1}(nx) < \varepsilon \text{ for all } x \in (0, \infty) \text{ and all } n \geq N\ ,$$

   or

   $$n > \frac{\cot\varepsilon}{x} \quad \text{for all } x \in (0, \infty) \text{ and all } n \geq N\ .$$

   This requires $N > \dfrac{\cot\varepsilon}{x}$ for all $x \in (0, \infty)$, which is clearly impossible. Thus the convergence is not uniform.

10. Let $\{f_n\}_{n\in\mathbb{N}}$ be a sequence of functions uniformly convergent to f on $\mathbb{R}$. If all f_n's are discontinuous at 0, then f is discontinuous at 0.

Answer: False.

Example: For any $n \in \mathbb{N}$, let $f_n : \mathbb{R} \to \mathbb{R}$ be defined by

$$f_n(x) := \begin{cases} \frac{1}{n} & \text{if } x \in \mathbb{Q} \\ 0 & \text{otherwise} . \end{cases}$$

Then none of the f_n's is continuous at 0 but $\{f_n\} \to f :\equiv 0$ uniformly on $\mathbb{R}$ and f is continuous everywhere.

11. Let $\{f_n\}_{n\in\mathbb{N}}$ be a sequence of functions on $S \subset \mathbb{R}$. If $\{f_n\}_{n\in\mathbb{N}}$ is uniformly convergent on every subset of S which is closed in $\mathbb{R}$, then $\{f_n\}$ is uniformly convergent on S.

    Answer: False.

    Example: Consider $S := (0,1) \subset \mathbb{R}$ and for every $n \in \mathbb{N}$, $f_n(x) := x^n$, $x \in S$. By Example 5.1.12, $\{f_n\}_{n\in\mathbb{N}}$ is uniformly convergent to $f :\equiv 0$ on every subset $[a, b] \subset S$. Now let $C \subset S$ be a subset of S which is closed in $\mathbb{R}$. It is evident that C is contained in some $[a, b] \subset S$. Hence $\{f_n\}_{n\in\mathbb{N}}$ is uniformly convergent to $f \equiv 0$ on C. But then by Example 5.1.12 again, $\{f_n\}_{n\in\mathbb{N}}$ is not uniformly convergent to $f \equiv 0$ on S.

12. The pointwise limit of a sequence of bounded functions is bounded.

    Answer: False.

    Example: Consider the sequence of functions $\{f_n\}_{n\in\mathbb{N}}$ on $(0,1)$ given by $f_n(x) := \frac{n}{nx+1}$, $x \in (0,1)$. Clearly, $\{f_n(x)\} \to f(x) := \frac{1}{x}$ pointwisely on $(0,1)$. Furthermore, for each $n \in \mathbb{N}$,

    $$\frac{n}{n+1} \leq f_n \leq n \quad \text{on } (0,1) ,$$

    hence each f_n is bounded on $(0,1)$ but the pointwise limit $f(x) = \frac{1}{x}$ is not.

13. The uniform limit of a sequence of bounded functions is bounded.

Answer: True.

Proof. Let $\{f_n\}_{n\in\mathbb{N}}$ be a sequence of functions on $S \subset \mathbb{R}$ such that each f_n is bounded on S and $\{f_n\} \to f$ uniformly on S. So there exists $N \in \mathbb{N}$ such that

$$|f_n - f| < 1 \quad \text{on } S \quad \text{for all } n \geq N \, .$$

Since f_N is bounded on S, there exists $M_N > 0$ such that $|f_N| \leq M_N$ on S. Hence

$$|f| \leq |f_N| + |f_N - f| < M_N + 1 \quad \text{on } S \, .$$

That is, the uniform limit f is bounded on S.

14. The uniform limit of a sequence of contractions $\{f_n\}_{n\in\mathbb{N}}$ is also a contraction.

Answer: False.

Example: For any $n \in \mathbb{N}$, let $f_n : [0,1] \to [0,1]$ be defined by $f_n(x) := (1 - \frac{1}{n})\,x$, $x \in [0,1]$. Then for every $n \in \mathbb{N}$, f_n is a contraction on $[0,1]$ with contraction constant $0 < 1 - \frac{1}{n} < 1$. Furthermore, it is not hard to see that $\{f_n\}_{n\in\mathbb{N}}$ converges uniformly to the function $f(x) := x$ on $[0,1]$. But f is not a contraction. Details are left to the readers.

15. If a sequence of functions $\{f_n\}_{n\in\mathbb{N}}$ converges pointwisely to a continuous function f, then the convergence is uniform.

Answer: False.

Example: For any $n \in \mathbb{N}$, let $f_n : \mathbb{R} \to \mathbb{R}$ be defined by $f_n(x) := \dfrac{1}{n^3(x - \frac{1}{n})^2 + 1}$. Then $\{f_n\} \to f :\equiv 0$ pointwisely on $\mathbb{R}$. However, for any $n \in \mathbb{N}$, we have

$$\left| f_n\left(\frac{1}{n}\right) - f\left(\frac{1}{n}\right) \right| = \left| f_n\left(\frac{1}{n}\right) - 0 \right| = 1 \, ,$$

thus the convergence is not uniform.

16. If a sequence of continuous functions $\{f_n\}_{n\in\mathbb{N}}$ converges uniformly to f and a sequence of uniformly continuous functions $\{g_n\}_{n\in\mathbb{N}}$ converges uniformly to g, then $\{f_n \circ g_n\}_{n\in\mathbb{N}}$ converges uniformly to $f \circ g$.

    Answer: False.

    Example: For any $n \in \mathbb{N}$, let $f_n, g_n : \mathbb{R} \to \mathbb{R}$ be defined by $f_n(x) := x^2 + \frac{1}{n}$ and $g_n(x) := x + \frac{1}{n}$, respectively. Then each f_n is continuous on $\mathbb{R}$ and each g_n is uniformly continuous on $\mathbb{R}$. Furthermore, $\{f_n\}_{n\in\mathbb{N}}$ converges uniformly in $\mathbb{R}$ to the function $f(x) := x^2$ and $\{g_n\}_{n\in\mathbb{N}}$ converges uniformly in $\mathbb{R}$ to the function $g(x) := x$. Observe also that $f_n \circ g_n(x) = (x + \frac{1}{n})^2 + \frac{1}{n}$ and $f \circ g(x) = x^2$, $x \in \mathbb{R}$. Now for all $n \in \mathbb{N}$, we have

$$\Big|f_n \circ g_n(n) - f \circ g(n)\Big| = \left|\left[\left(n + \frac{1}{n}\right)^2 + \frac{1}{n}\right] - n^2\right|$$
$$= 2 + \frac{1}{n^2} + \frac{1}{n} \geq 2\,,$$

    thus $\{f_n \circ g_n\}_{n\in\mathbb{N}}$ does not converge uniformly to $f \circ g$.

17. If $\{f_n\}_{n\in\mathbb{N}}$ converges uniformly to f in $\mathbb{R}$, then $\{g \circ f_n\}_{n\in\mathbb{N}}$ converges uniformly to $g \circ f$ in $\mathbb{R}$ for any continuous function g on $\mathbb{R}$.

    Answer: False.

    Example: For any $n \in \mathbb{N}$, let $f_n : \mathbb{R} \to \mathbb{R}$ be defined by $f_n(x) := x + \frac{1}{n}$, $x \in \mathbb{R}$. It is easy to check that $\{f_n\}_{n\in\mathbb{N}}$ converges uniformly to the function $f(x) := x$ on $\mathbb{R}$. Let $g : \mathbb{R} \to \mathbb{R}$ be defined by $g(x) := x^2$, $x \in \mathbb{R}$. Then g is continuous on $\mathbb{R}$. Since

$$g \circ f_n(x) = \left(x + \frac{1}{n}\right)^2 = x^2 + \frac{2x}{n} + \frac{1}{n^2}$$

    for all $x \in \mathbb{R}$ and all $n \in \mathbb{N}$, we have $\{g \circ f_n(x)\} \to g(x) = x^2$ pointwisely on $\mathbb{R}$. However, as for every $n \in \mathbb{N}$ we have

$$\Big|g \circ f_n(n) - g \circ f(n)\Big| = \left|\left(n^2 + \frac{2n}{n} + \frac{1}{n^2}\right) - n^2\right| = 2 + \frac{1}{n^2} \geq 2\,,$$

    the convergence is not uniform.

18. If $\{f_n\}_{n\in\mathbb{N}}$ converges uniformly to f in $\mathbb{R}$, then for any function g on $\mathbb{R}$, $\{f_n \circ g\}_{n\in\mathbb{N}}$ converges uniformly to $f \circ g$ in $\mathbb{R}$.

    Answer: True.

    Proof. By the uniform convergence $\{f_n\} \to f$ in $\mathbb{R}$, for any $\varepsilon > 0$, there exists $N \in \mathbb{N}$ such that

    $$\big|f_n(y) - f(y)\big| < \varepsilon \quad \text{for any } n \geq N \text{ and any } y \in \mathbb{R} \,.$$

    In particular,

    $$\Big|f_n\big(g(x)\big) - f\big(g(x)\big)\Big| < \varepsilon \quad \text{for any } n \geq N \text{ and any } x \in \mathbb{R} \,.$$

    Thus $\{f_n \circ g\} \to f \circ g$ uniformly in $\mathbb{R}$.

19. If $\{f_n\}_{n\in\mathbb{N}}$ is a sequence of continuous functions on $[0,1]$ which is convergent uniformly to f on $[0,1]$, then for any continuous function g on $\mathbb{R}$, $\{g \circ f_n\}_{n\in\mathbb{N}}$ converges uniformly to $g \circ f$ on $[0,1]$.

    Answer: True.

    Proof. Since continuous functions on compact sets are bounded, each f_n is bounded. Since $\{f_n\} \to f$ uniformly on $[0,1]$, there exists $N \in \mathbb{N}$ such that

    $$\big|f_n(x) - f_m(x)\big| < 1 \quad \text{for any } n, m \geq N \text{ and any } x \in [0,1] \,.$$

    In particular, if $|f_N| \leq M$ on $[0,1]$, then $|f_n| \leq M+1$ on $[0,1]$ for all $n \geq N$ and so without loss of generality, we simply assume that $|f_n| \leq M+1$ for all $n \in \mathbb{N}$. Since g is continuous, it is uniformly continuous on the compact set $S := [-M-1, M+1]$, which is a superset of the range of all f_n's. So for any $\varepsilon > 0$, there exists $\delta > 0$ such that

    $$\big|g(y_1) - g(y_2)\big| < \varepsilon \quad \text{for any } y_1, y_2 \in S \text{ with } |y_1 - y_2| < \delta \,.$$

Since $\{f_n\} \to f$ uniformly on $[0,1]$, there exists $K \in \mathbb{N}$ such that

$$\big|f_n(x) - f(x)\big| < \delta \quad \text{for all } x \in [0,1] \text{ and all } n \geq K\ .$$

Therefore,

$$\Big|g\big(f_n(x)\big) - g\big(f(x)\big)\Big| < \varepsilon \quad \text{for all } x \in [0,1] \text{ and all } n \geq K\ .$$

Thus $\{g \circ f_n\} \to g \circ f$ uniformly on $[0,1]$.

20. If a sequence of uniformly continuous functions $\{f_n\}_{n\in\mathbb{N}}$ converges uniformly to f and a sequence of functions $\{g_n\}_{n\in\mathbb{N}}$ converges uniformly to g, then $\{f_n \circ g_n\}$ converges uniformly to $f \circ g$.

Answer: True.

Proof. Let $\{g_n\}_{n\in\mathbb{N}}$ be a sequence of functions on $S \subset \mathbb{R}$ such that $\{g_n\} \to g$ uniformly on S, and $\{f_n\}_{n\in\mathbb{N}}$ be a sequence of uniformly continuous functions on $T \subset \mathbb{R}$ such that $\{f_n\} \to f$ uniformly on T. In order to make sense, we assume that $T \supset \bigcup_{n\in\mathbb{N}} g_n(S)$. By Corollary 5.1.7, being the uniform limit of a sequence of uniformly continuous functions, f is uniformly continuous on T. Hence for any $\varepsilon > 0$, there exists $\delta > 0$ such that

$$\big|f(y_1) - f(y_2)\big| < \frac{\varepsilon}{2} \quad \text{for any } y_1, y_2 \in T \text{ with } \big|y_1 - y_2\big| < \delta\ .$$

Since $\{g_n\} \to g$ uniformly on S, there exists $N_1 \in \mathbb{N}$ such that

$$\big|g_n(x) - g(x)\big| < \delta \quad \text{for any } n \geq N_1 \text{ and any } x \in S\ .$$

Hence

$$\Big|f\big(g_n(x)\big) - f\big(g(x)\big)\Big| < \frac{\varepsilon}{2} \quad \text{for any } n \geq N_1 \text{ and any } x \in S\ .$$

On the other hand, as $\{f_n\} \to f$ uniformly on S, there exists $N_2 \in \mathbb{N}$ such that

$$\big|f_n(y) - f(y)\big| < \frac{\varepsilon}{2} \quad \text{for any } n \geq N_2 \text{ and all } y \in T\ .$$

In particular,

$$\left|f_n(g_n(x)) - f(g_n(x))\right| < \frac{\varepsilon}{2} \quad \text{for any } n \geq N_2 \text{ and all } x \in S\,.$$

Therefore,

$$\begin{aligned}
&\left|f_n(g_n(x)) - f(g(x))\right| \\
&\leq \left|f_n(g_n(x)) - f(g_n(x))\right| + \left|f(g_n(x)) - f(g(x))\right| \\
&< \frac{\varepsilon}{2} + \frac{\varepsilon}{2} = \varepsilon
\end{aligned}$$

for any $n \geq \max\{N_1, N_2\}$ and any $x \in S$, that is, $\{f_n \circ g_n\} \to f \circ g$ uniformly on S.

Part B: Problems

1. (a) Find the pointwise limit of the sequence of the functions $\{e^{-nx}\}_{n\in\mathbb{N}}$ on $[0, 1]$.

   (b) Find the limit of the sequence $\{e^{-nx}\}_{n\in\mathbb{N}}$ in $(C[0, 1], d)$, where

   $$d(f, g) := \left(\int_0^1 |f(x) - g(x)|^p\, dx\right)^{1/p}$$

   with $p \geq 1$.

   (c) What is your observation?

   Solution.

   (a) It is elementary to see that

   $$\{e^{-nx}\} \to f(x) := \begin{cases} 0 & \text{if } x \in (0, 1] \\ 1 & \text{if } x = 0\,. \end{cases}$$

   (b) It is easy to see that

   $$d(e^{-nx}, 0) = \left(\int_0^1 e^{-pnx}\, dx\right)^{1/p} = \left(\frac{1 - e^{-pn}}{pn}\right)^{1/p} \to 0$$

as $n \to \infty$, thus $\{e^{-nx}\} \to 0$ in $(C[0,1], d)$.

(c) Convergence in different contexts can be very different.

2. Consider the sequence of functions $\{f_n\}_{n\in\mathbb{N}}$ on $[0,1]$ defined by $f_n(x) := \sin[x^n(1-x^n)]$, $x \in [0,1]$.
   (a) Does $\{f_n(x)\}_{n\in\mathbb{N}}$ converge pointwisely on $[0,1]$?
   (b) Does $\{f_n(x)\}_{n\in\mathbb{N}}$ converge uniformly on $[0,1]$?
   (c) Does $\lim_{n\to\infty}\int_0^1 f_n(x)\,dx$ exist?

Solution.

(a) *Answer*: Yes.

Proof. Clearly $f_n(0) = f_n(1) = 0$ for all $n \in \mathbb{N}$. For any $x \in (0,1)$, since $\sin y \le y$ for all $y \in (0,1)$, we have $|f_n(x)| \le x^n(1-x^n) \le x^n \to 0$ as $n \to \infty$. Therefore, $\{f_n\} \to 0$ pointwisely on $[0,1]$.

(b) *Answer*: No.

Justification. Since $f_n\left(\sqrt[n]{\frac{1}{2}}\right) = \sin\frac{1}{4}$, we have

$$\sup\{|f_n(x) - 0| : x \in [0,1]\} \ge \frac{1}{4} \quad \text{for all } n \in \mathbb{N}$$

and so the convergence cannot be uniform.

(c) *Answer*: Yes, and the limit is 0. Hence in particular,

$$\lim_{n\to\infty}\int_0^1 f_n(x)\,dx = 0 = \int_0^1 \lim_{n\to\infty} f_n(x)\,dx\,.$$

Justification. Similar to (a), we have

$$\left|\int_0^1 f_n(x)\,dx\right| \le \int_0^1 x^n(1-x^n)\,dx = \frac{n}{(n+1)(2n+1)} \to 0$$

as $n \to \infty$.

3. Consider the sequence of functions $\{f_n\}_{n\in\mathbb{N}}$ on $\mathbb{R}$ defined by $f_n(x) := \dfrac{1}{n}e^{-n^2x^2}$, $x \in \mathbb{R}$.

(a) Show that $\{f_n(x)\}_{n\in\mathbb{N}}$ converges pointwisely on $\mathbb{R}$ to some function f. Find f.

(b) Determine whether $\{f_n\}_{n\in\mathbb{N}}$ converges uniformly on $\mathbb{R}$.

(c) Show that $\lim_{n\to\infty} f_n' = f'$ on $\mathbb{R}$ (that is, $\{f_n'\} \to f'$ pointwisely on $\mathbb{R}$).

(d) Show that the convergence $\{f_n'\} \to f'$ cannot be uniform on any non-degenerate interval containing 0.

Solution.

(a) Since $|f_n(x)| \le \frac{1}{n}$ for all $x \in \mathbb{R}$, $\{f_n\} \to f :\equiv 0$ pointwisely on $\mathbb{R}$.

(b) For the same reason as that in (a), we have

$$\sup\left\{|f_n(x) - f(x)| : x \in \mathbb{R}\right\} \le \frac{1}{n} \to 0 \quad \text{as } n \to \infty .$$

Hence $\{f_n\} \to f \equiv 0$ uniformly on $\mathbb{R}$.

(c) We have $f_n'(x) = -2nxe^{-n^2x^2}$ for all $x \in \mathbb{R}$, $n \in \mathbb{R}$. Clearly, $f_n'(0) = 0$ for all $n \in \mathbb{N}$. On the other hand, for any $x \ne 0$, by elementary analysis, we have $\{ne^{-n^2x^2}\} \to 0$ as $n \to \infty$. Hence $f_n' \to 0 \equiv f'$ pointwisely on $\mathbb{R}$.

(d) Let $I \subset \mathbb{R}$ be a non-degenerate interval with $0 \in I$. For any $N \in \mathbb{N}$, there exists $n \ge N$ such that $\frac{1}{n} \in I$ or $-\frac{1}{n} \in I$. Without loss of generality, assume that $\frac{1}{n} \in I$. Then

$$\sup\left\{|f_n'(x) - 0| : x \in I\right\} \ge \left|f_n'\left(\frac{1}{n}\right)\right| = \frac{2}{e} ,$$

thus the convergence $\{f_n'\} \to f'$ cannot be uniform on I.

4. For each n, define $g_n : \mathbb{R} \to \mathbb{R}$ by

$$g_n(x) := \begin{cases} 0 & \text{if } x = 0 \text{ or } x \notin \mathbb{Q} , \\ b + \frac{1}{n} & \text{if } x = \frac{a}{b},\ a, b \in \mathbb{Z}, a \ne 0, b > 0, (a, b) = 1 . \end{cases}$$

Show that $\{g_n\}_{n\in\mathbb{N}}$ is uniformly convergent on $\mathbb{R}$.

Proof. Note that for any $n, m \in \mathbb{N}$,

$$g_n(x)-g_m(x) = \begin{cases} 0 & \text{if } x = 0 \text{ or } x \notin \mathbb{Q}\,, \\ \frac{1}{n} - \frac{1}{m} & \text{if } x = \frac{a}{b}\,, a, b \in \mathbb{Z}, b > 0, (a,b) = 1\,. \end{cases}$$

Hence

$$|g_n(x) - g_m(x)| \le \left|\frac{1}{n} - \frac{1}{m}\right| \quad \text{for all } x \in \mathbb{R}\,.$$

As $\{\frac{1}{n}\}_{n\in\mathbb{N}}$ is a Cauchy sequence, $\{g_n\}_{n\in\mathbb{N}}$ is uniformly Cauchy. By Cauchy's criterion, $\{g_n\}_{n\in\mathbb{N}}$ is uniformly convergent on $\mathbb{R}$.

5. Let $\{f_n\}_{n\in\mathbb{N}}$ be a sequence of uniformly continuous functions on $S \subset \mathbb{R}$, and $\{x_n\}_{n\in\mathbb{N}}$ a sequence in S with $\{x_n\} \to x \in S$ as $n \to \infty$.
   (a) If $\{f_n\}_{n\in\mathbb{N}}$ converges uniformly to some function f on S, show that $\{f_n(x_n)\} \to f(x)$ as $n \to \infty$.
   (b) If the convergence $\{f_n\} \to f$ is only pointwise, would the assertion in (a) still be true?

Proof.

(a) As $\{f_n\} \to f$ uniformly on S, for any $\varepsilon > 0$, there exists $N_1 \in \mathbb{N}$ such that

$$|f_n(x) - f(x)| < \frac{\varepsilon}{2} \quad \text{for all } n \ge N_1 \text{ and all } x \in S\,.$$

By Corollary 5.1.7, being the uniform limit of a sequence of uniformly continuous functions, f is uniformly continuous on S. Hence there exists $\delta > 0$ such that

$$|f(p) - f(q)| < \frac{\varepsilon}{2} \quad \text{for any } p, q \in S \text{ with } |p - q| < \delta\,.$$

Since $\{x_n\} \to x$ as $n \to \infty$, there exists $N_2 \in \mathbb{N}$ such that

$$|x_n - x| < \delta \quad \text{for all } n \ge N_2\,.$$

Combining, we have

$$|f_n(x_n) - f(x)| \le |f_n(x_n) - f(x_n)| + |f(x_n) - f(x)| < \varepsilon$$

for any $n \ge \max\{N_1, N_2\}$. The assertion follows.

(b) In case the convergence $\{f_n\} \to f$ is only pointwise, the assertion in (a) fails to hold.

Example: Consider the sequence of functions $\{f_n\}_{n\in\mathbb{N}}$ on $[0,1]$ defined by $f_n(x) := x^n$, $x \in [0,1]$. We have seen that

$$\{f_n(x)\} \to f(x) := \begin{cases} 0 & \text{if } x \in [0,1) \\ 1 & \text{if } x = 1 \end{cases}$$

pointwisely but not uniformly on $[0,1]$. Let $x_n := 1 - \frac{1}{n} \in [0,1]$ for all $n \in \mathbb{N}$. We have $\{x_n\} \to 1$ as $n \to \infty$. But then

$$f_n(x_n) = \left(1 - \frac{1}{n}\right)^n \to \frac{1}{e} \neq 1 = f(1) \text{ as } n \to \infty\,.$$

6. Show that every uniformly convergent sequence of bounded functions is *uniformly bounded*. More precisely, let $\{f_n\}_{n\in\mathbb{N}}$ be a sequence of functions on $S \subset \mathbb{R}$ which is uniformly convergent. If each f_n is bounded on S, i.e., $\sup_{x\in S} |f_n(x)| < \infty$ for each $n \in \mathbb{N}$, then $\{f_n\}_{n\in\mathbb{N}}$ is *uniformly bounded*, that is, $\sup\{|f_n(x)| : x \in S, n \in \mathbb{N}\} < \infty$.

Proof. As $\{f_n\}_{n\in\mathbb{N}}$ is uniformly convergent, it must be uniformly Cauchy. So there exists $N \in \mathbb{N}$ such that

$$|f_n(x) - f_N(x)| < 1 \quad \text{for all } n \geq N \text{ and all } x \in S\,.$$

It follows that

$$|f_n(x)| \leq |f_N(x)| + 1 \quad \text{for all } n \geq N \text{ and all } x \in S\,.$$

For any $n \in \mathbb{N}$, write $M_n := \sup_{x\in S} |f_n(x)|$. If we set

$$M := \max\{M_1, M_2, \ldots, M_{N-1}, 1 + M_N\} > 0\,,$$

it is easy to see that

$$\sup_{x\in S} |f_n(x)| \leq M \quad \text{for all } n \in \mathbb{N}\,.$$

That is, $\{f_n\}_{n\in\mathbb{N}}$ is uniformly bounded.

7. (Dini's Theorem) Let $\{f_n\}_{n\in\mathbb{N}}$ be a sequence of functions on $S \subset \mathbb{R}$. Suppose
   (i) S is compact,
   (ii) f_n is continuous for every $n \in \mathbb{N}$, and
   (iii) for each $x \in S$, $\{f_n(x)\}_{n\in\mathbb{N}} \downarrow 0$ as $n \to \infty$,

   show that $\{f_n\} \to 0$ uniformly on S.

   Proof. Let $\varepsilon > 0$ be given. For any $n \in \mathbb{N}$, since f_n is continuous, the subset $O_n := f_n^{-1}\big((-\infty, \varepsilon)\big) = \{x \in S : f_n(x) < \varepsilon\} \subset S$ is open. Furthermore, for any $x \in O(n)$, since $\{f_n(x)\}_{n\in\mathbb{N}}$ is decreasing, we have

   $$f_{n+1}(x) \leq f_n(x) < \varepsilon$$

   and so $x \in O_{n+1}$. Hence $\{O_n\}_{n\in\mathbb{N}}$ is increasing. On the other hand, for each $x \in S$, since $\{f_n(x)\} \downarrow 0$ as $n \to \infty$, there exists $n_x \in \mathbb{N}$ such that $f_{n_x}(x) < \varepsilon$ and so $x \in O_{n_x}$. Thus $\{O_n : n \in \mathbb{N}\}$ is an open cover of S. By the compactness of S, there exists $N \in \mathbb{N}$ such that $S = \bigcup_{n=1}^{N} O_n$. Since $\{O_n\}_{n\in\mathbb{N}}$ is increasing, we have $S = O_N$. In particular, for any $n \geq N$ and any $x \in S = O_N$,

   $$0 \leq f_n(x) \leq f_N(x) < \varepsilon \ .$$

   This shows the $\{f_n\} \to 0$ uniformly on S.

8. In Exercise 5.1, Part A, Problem #7, is condition (i), i.e., the compactness of S, necessary?

   Answer: Yes, compactness is necessary.
   Example: For any $n \in \mathbb{N}$, consider $f_n : \mathbb{R} \to \mathbb{R}$ given by

   $$f_n(x) := \frac{x}{n} \ .$$

   Clearly, f_n is continuous for every $n \in \mathbb{N}$ and $f_n(x) \downarrow 0$ as $n \to \infty$ for every $x \in \mathbb{R}$, but $\mathbb{R}$ is not compact. Note that since for every

$n \in \mathbb{N}$,

$$\sup\left\{\left|f_n(x) - f(x)\right| : x \in \mathbb{R}\right\} = \sup\left\{\frac{|x|}{n} : x \in \mathbb{R}\right\} = \infty\,,$$

we have

$$\sup\left\{\left|f_n(x) - f(x)\right| : x \in \mathbb{R}\right\} \to \infty \quad \text{as } n \to \infty$$

and so the convergence $\{f_n\} \to 0$ is not uniform on $\mathbb{R}$.

9. In Exercise 5.1, Part A, Problem #7, is condition (iii), i.e., the monotonicity of $\{f_n\}_{n\in\mathbb{N}}$, necessary?

   Answer: Yes, monotonicity is necessary.

   Example: Consider the sequence of functions $\{f_n\}_{n\in\mathbb{N}}$ on $[0,1]$ given by

$$f_n(x) := \begin{cases} n^2 x & 0 \le x \le \frac{1}{n} \\ -n^2 x + 2n & \frac{1}{n} \le x \le \frac{2}{n} \\ 0 & \frac{2}{n} \le x \le 1\,. \end{cases}$$

   Clearly f_n is continuous on X for each n, $[0,1]$ is compact, and $f_n \to f \equiv 0$ pointwisely on X. However, the sequence $\{f_n(x)\}_{n\in\mathbb{N}}$ is not monotone. Observe that

$$\sup\left\{\left|f_n(x) - f(x)\right| : x \in X\right\} = n \to \infty \quad \text{as } n \to \infty\,,$$

   hence the convergence is not uniform.

10. In Exercise 5.1, Part A, Problem #7, is condition (ii), i.e., f_n is continuous for all n necessary?

    Answer: Yes, save for finitely many n's, the continuity of f_n's is necessary.

    Example: Consider the sequence of functions $\{f_n\}_{n\in\mathbb{N}}$ on $[0,1]$ given by

$$f_n(x) := \begin{cases} 0 & \text{if } x = 0 \\ 1 & \text{if } 0 < x < 1/n \\ 0 & \text{if } \frac{1}{n} \le x \le 1\,. \end{cases}$$

Clearly, $[0,1]$ is compact and $\{f_n\} \downarrow 0$ pointwisely on $[0,1]$, but none of the f_n's is continuous on $[0,1]$. As $\left|f_n(\frac{1}{2n}) - 0\right| = 1$ for all $n \in \mathbb{N}$, the convergence is not uniform.

11. Let $\{f_n\}_{n\in\mathbb{N}}$ be a sequence of functions defined on $S \subset \mathbb{R}$. Suppose
    (i) S is compact,
    (ii) f_n is continuous for every $n \in \mathbb{N}$, and
    (iii) there is a continuous function $f : S \to \mathbb{R}$ such that $\{f_n(x)\}_{n\in\mathbb{N}} \downarrow f(x)$ as $n \to \infty$ for every $x \in S$,

    show that $\{f_n\} \to f$ uniformly on S. What if the pointwise limit f is not continuous on S?

    Solution. For any $n \in \mathbb{N}$, define $g_n := f_n - f$. As $\{f_n\} \downarrow f$ in S, $\{g_n\} \downarrow 0$ in S. As all f_n's and f are continuous, all g_n's are continuous. Hence all conditions in Exercise 5.1, Part A, Problem #7 are satisfied and so $\{g_n\} \to 0$ uniformly on S. Hence $\{f_n\} \to f$ uniformly on S.
    However, in case the pointwise limit f is not continuous, we may not have uniform convergence.
    Example: Consider the sequence of functions $\{f_n\}_{n\in\mathbb{N}}$ on $[0,1]$ given by $f_n(x) := \frac{1}{1+nx}$, $x \in [0,1]$. Clearly, $[0,1]$ is compact, f_n is continuous for all $n \in \mathbb{N}$, and $\{f_n(x)\} \downarrow f(x) := \begin{cases} 1 & \text{if } x = 0 \\ 0 & \text{if } x \neq 0 \end{cases}$.
    Note that f is not continuous. Observe that as shown in Exercise 5.1, Part A, Problem #2, the convergence $\{f_n\} \to f$ is not uniform.

12. Let $\{f_n\}_{n\in\mathbb{N}}$ be a sequence of continuous real-valued functions defined and uniformly bounded (cf. Exercise 5.1, Part B, Problem #6) on $S \subset \mathbb{R}$. Suppose for each $x \in S$, the sequence of real numbers $\{f_n(x)\}_{n\in\mathbb{N}}$ is monotonically decreasing. Determine whether $\{f_n\}$ is uniformly convergent on S.

    Answer: No.

Example: Let $S := [0, 1]$ and for all $n \in \mathbb{N}$,

$$f_n(x) := \begin{cases} -nx & 0 \leq x \leq \frac{1}{n} \\ -1 & \frac{1}{n} < x \leq 1 \,. \end{cases}$$

Clearly, each f_n is continuous on S and for every $x \in S$, $\{f_n(x)\}_{n\in\mathbb{N}}$ is monotonically decreasing. On the other hand, it is readily seen that $|f_n(x)| \leq 1$ for all $n \in \mathbb{N}$ and all $x \in S$ and so $\{f_n\}_{n\in\mathbb{N}}$ is uniformly bounded on S. Furthermore, it is easy to see that

$$\{f_n(x)\} \to f(x) := \begin{cases} 0 & x = 0 \\ -1 & x \in (0, 1] \,. \end{cases}$$

But then as the pointwise limit f is not continuous on S, the convergence $\{f_n\} \to f$ cannot be uniform.

13. Let $\{g_n\}_{n\in\mathbb{N}}$ be a sequence of continuous real-valued functions defined and uniformly bounded on a compact set $S \subset \mathbb{R}$. If $\{g_n\}_{n\in\mathbb{N}}$ converges pointwisely on S, determine whether the convergence must be uniform.

    Answer: No.

    Example: Let $S := [0, 1]$ equipped with the Euclidean metric. For any $n \in \mathbb{N}$, let $g_n : S \to \mathbb{R}$ be defined by $g_n(x) := x^n$, $x \in S$. Clearly, g_n is continuous and $|g_n| \leq 1$ on S for each $n \in \mathbb{N}$. So $\{g_n\}_{n\in\mathbb{N}}$ is uniformly bounded. Moreover, it is easy to see that

$$g_n(x) \to g(x) := \begin{cases} 0 & x \in [0, 1) \\ 1 & x = 1 \end{cases} \qquad \text{as } n \to \infty \,.$$

    As the pointwise limit g of $\{g_n\}_{n\in\mathbb{N}}$ is not continuous on S, the convergence cannot be uniform.

14. For each n, define $f_n : \mathbb{R} \to \mathbb{R}$ by $f_n(x) := (1 + \frac{1}{n})x$, $x \in \mathbb{R}$. Determine whether $\{f_n\}_{n\in\mathbb{N}}$ is uniformly convergent on every bounded non-degenerate interval $I \subset \mathbb{R}$.

Solution.

Answer: Yes. (Note, however, that $\{f_n\} \not\to f$ uniformly on $\mathbb{R}$ – cf. the solution to Exercise 5.1, Part B, Problem #8. Details are left with the interested readers.)

Proof. It is clear from the definition of the f_n's that $\{f_n(x)\}_{n\in\mathbb{N}}$ converges pointwisely to the function $f(x) := x$ on $\mathbb{R}$. Let I be any bounded interval. Then there exists $M > 0$ such that $I \subset [-M, M]$, that is, $|x| \leq M$ for all $x \in I$. For any $\varepsilon > 0$, let $N \in \mathbb{N}$ be large enough such that $N > \frac{M}{\varepsilon}$. Then

$$|f_n(x) - f(x)| = \frac{|x|}{n} \leq \frac{M}{n} \leq \frac{M}{N} < \varepsilon$$

for all $n \geq M$ and all $x \in I$. Hence $\{f_n\} \to f$ uniformly on I.

Alternatively, it can be proved by showing that the sequence $\{f_n\}_{n\in\mathbb{N}}$ is uniformly Cauchy without first finding the pointwise limit of the sequence: For any $\varepsilon > 0$, since $\{\frac{1}{n}\}_{n\in\mathbb{N}}$ is Cauchy, there is $N \in \mathbb{N}$ such that $|\frac{1}{n} - \frac{1}{m}| < \frac{\varepsilon}{M}$ whenever $n, m \geq N$. As a result,

$$|f_n(x) - f_m(x)| = \left|\frac{1}{n} - \frac{1}{m}\right| |x| \leq \left|\frac{1}{n} - \frac{1}{m}\right| M < \frac{\varepsilon}{M} \cdot M = \varepsilon$$

for all $n, m \geq N$ and all $x \in I$. So $\{f_n\}_{n\in\mathbb{N}}$ is uniformly Cauchy and hence uniformly convergent on I.

15. Let $\{f_n\}_{n\in\mathbb{N}}$ and $\{g_n\}_{n\in\mathbb{N}}$ be sequences of functions on $S \subset \mathbb{R}$. If $\{f_n\} \to f$ and $\{g_n\} \to g$ uniformly on S, is it necessarily true that $\{f_n g_n\} \to fg$ uniformly on S?

    Answer: No.

    Example: Consider the sequences of functions $\{f_n\}_{n\in\mathbb{N}}$ and $\{g_n\}_{n\in\mathbb{N}}$ on $\mathbb{R}$ defined by $f_n(x) := x$ and $g_n(x) := \frac{1}{n}$, $n \in \mathbb{N}$, $x \in \mathbb{R}$. Clearly we have $\{f_n\} \to x$ and $\{g_n\} \to 0$ uniformly on $\mathbb{R}$, but as shown in the solution to Exercise 5.1, Part B, Problem #7, $f_n g_n = \frac{x}{n} \not\to 0$ uniformly on $\mathbb{R}$.

16. In Exercise 5.1, Part B, Problem #15, if in addition, all f_n's and g_n's are bounded on S, is it necessarily true that $\{f_n g_n\} \to fg$ uniformly on S?

Answer: Yes.

Proof. By Exercise 5.1, Part B, Problem #6, both $\{f_n\}_{n\in\mathbb{N}}$ and $\{g_n\}_{n\in\mathbb{N}}$ are uniformly bounded on S. Hence there exist $K,\ L \in \mathbb{R}$ such that $|f_n| \le K$ and $|g_n| \le L$ on S for all $n \in \mathbb{N}$. Let $M = \max\{K, L\}$. Then

$$|f_n|, |g_n| \le M \qquad \text{on } S \quad \text{for all } n \in \mathbb{N}$$

and so

$$|f|, |g| \le M \qquad \text{on } S\ .$$

By the uniform convergence of $\{f_n\}$ and $\{g_n\}$ to f and g, respectively, for any $\varepsilon > 0$, there exist $N_1,\ N_2 \in \mathbb{N}$ such that

$$\begin{aligned} |f_n(x) - f(x)| &< \frac{\varepsilon}{2M} \quad \text{for all } x \in S \text{ and all } n \ge N_1\ , \\ |g_n(x) - g(x)| &< \frac{\varepsilon}{2M} \quad \text{for all } x \in S \text{ and all } n \ge N_2\ . \end{aligned}$$

Hence for all $x \in S$ and all $n \ge \max\{N_1, N_2\}$,

$$\begin{aligned} &|f_n g_n(x) - f(x)g(x)| \\ &\le |f_n(x)\big(g_n(x) - g(x)\big)| + |g(x)\big(f_n(x) - f(x)\big)| \\ &< M \cdot \frac{\varepsilon}{2M} + M \cdot \frac{\varepsilon}{2M} \\ &= \varepsilon \end{aligned}$$

and so $\{f_n g_n\} \to fg$ uniformly on S.

17. Let $f : \mathbb{R} \to \mathbb{R}$ be a function. For any $a \in \mathbb{R}$, let $f_a : \mathbb{R} \to \mathbb{R}$ be the shifted function $f_a(x) := f(x - a)$, $x \in \mathbb{R}$. Show that f is uniformly continuous if and only if whenever $\{a_n\}_{n\in\mathbb{N}}$ is a sequence of real numbers which converges to zero, the sequence shifted functions $\{f_{a_n}\}_{n\in\mathbb{N}}$ converges uniformly to f.

Proof. ($\Rightarrow$): Let $f : \mathbb{R} \to \mathbb{R}$ be uniformly continuous. For any $\varepsilon > 0$, there exists $\delta > 0$ such that

$$|f(x_1) - f(x_2)| < \varepsilon \quad \text{for any } x_1, x_2 \in \mathbb{R} \text{ with } |x_1 - x_2| < \delta\ .$$

Let $\{a_n\}_{n\in\mathbb{R}}$ be any sequence with $\{a_n\} \to 0$ as $n \to \infty$. Then there exists $N \in \mathbb{N}$ such that

$$|a_n| < \delta \quad \text{for all } n \geq N\ .$$

Hence for any $x \in \mathbb{R}$ and any $n \geq N$, we have $|(x - a_n) - x| = |a_n| < \delta$ and so

$$|f_{a_n}(x) - f(x)| = |f(x - a_n) - f(x)| < \varepsilon\ .$$

Thus $\{f_{a_n}\} \to f$ uniformly on $\mathbb{R}$.

($\Leftarrow$): It is more convenient to use contrapositive arguments. So suppose $\{f_{a_n}\} \to f$ uniformly whenever $\{a_n\} \to 0$ as $n \to \infty$, but f is not uniformly continuous on $\mathbb{R}$. Then there exists $\varepsilon > 0$ such that for each $n \in \mathbb{N}$, we can find x_n, $y_n \in \mathbb{R}$ with $|x_n - y_n| < 1/n$ but $|f(x_n) - f(y_n)| \geq \varepsilon$. For any $n \in \mathbb{N}$, write $a_n := x_n - y_n$. By construction, $\{a_n\} \to 0$ as $n \to \infty$. So by assumption, $\{f_{a_n}\} \to f$ uniformly on $\mathbb{R}$. Hence there exists $N \in \mathbb{N}$ such that

$$|f_{a_n}(x) - f(x)| < \varepsilon \quad \text{for all } n \geq N \text{ and all } x \in \mathbb{R}\ .$$

In particular, we have

$$\begin{aligned} |f(y_N) - f(x_N)| &= |f(x_N - a_N) - f(x_N)| \\ &= |f_{a_N}(x_N) - f(x_N)| < \varepsilon\ , \end{aligned}$$

which is a contradiction. Therefore, f is uniformly continuous.

5.2 Series of Functions

From our experience with elementary calculus, associated to an infinite sequence of real numbers there is a corresponding formal infinite series of real numbers with the same convergence as the sequence. Conversely, associated to any formal infinite series of real numbers, there is a corresponding infinite sequence of real numbers with the same convergence as the series. To be more precise, starting with an infinite sequence of real numbers $\{a_n\}_{n\in\mathbb{N}}$, we define

$$\begin{cases} b_1 & := a_1 \\ b_{k+1} & := a_{k+1} - a_k \ , \quad k \in \mathbb{N} \ , \end{cases}$$

and form the formal infinite series $\sum_{k=1}^{\infty} b_k$. Observe that for every $n \in \mathbb{N}$, we have

$$\sum_{k=1}^{n} b_k = a_n \ .$$

Furthermore, the infinite sequence $\{a_n\}_{n\in\mathbb{N}}$ is convergent if and only if the formal infinite series $\sum_{k=1}^{\infty} b_k$ is convergent, and in this case we write

$$\lim_{n\to\infty} a_n = \lim_{n\to\infty} \sum_{k=1}^{n} b_k = \sum_{k=1}^{\infty} b_k \ .$$

Conversely, starting with a formal infinite series $\sum_{k=1}^{\infty} a_k$, we define, for any $n \in \mathbb{N}$, the n-th partial sum

$$s_n := \sum_{k=1}^{n} a_k \ ,$$

and form the infinite sequence of partial sums $\{s_n\}_{n\in\mathbb{N}}$. Then the formal infinite series $\sum_{k=1}^{\infty} a_k$ is convergent if and only if the infinite sequence of partial sums $\{s_n\}_{n\in\mathbb{N}}$ is convergent, and in this case

$$\lim_{n\to\infty} s_n = \lim_{n\to\infty} \sum_{k=1}^{n} a_k = \sum_{k=1}^{\infty} a_k \ .$$

So there is a natural $1-1$ correspondence between infinite sequences of real numbers and formal infinite series of real numbers. In certain situations, it is easier to deal with sequences, while for other situations, since we have the mechanism of addition, it could be easier to deal with infinite series.

Analogously, there is a natural $1-1$ correspondence between infinite sequences of functions and formal infinite series of functions. Namely, starting with an infinite sequence of functions $\{f_n\}_{n\in\mathbb{N}}$ on $S \subset \mathbb{R}$, we define

$$\begin{cases} g_1 & := f_1 \\ g_{k+1} & := f_{k+1} - f_k \ , \quad k \in \mathbb{N} \ , \end{cases}$$

and form the formal infinite series of functions $\sum_{k=1}^{\infty} g_k$. Then for every $n \in \mathbb{N}$, we have

$$\sum_{k=1}^{n} g_k = f_n \ .$$

Furthermore, the infinite sequence of functions $\{f_n\}_{n\in\mathbb{N}}$ is convergent pointwisely/uniformly on S if and only if the formal infinite series of functions $\sum_{k=1}^{\infty} g_k$ is convergent pointwisely/uniformly on S, and in this case

$$\lim_{n\to\infty} f_n = \lim_{n\to\infty} \sum_{k=1}^{n} g_k = \sum_{k=1}^{\infty} g_k \ .$$

Conversely, starting with a formal infinite series of functions $\sum_{k=1}^{\infty} f_k$ on S, we define, for any $n \in \mathbb{N}$, the n-th partial sum

$$s_n := \sum_{k=1}^{n} f_k \ ,$$

and form the infinite sequence of partial sums $\{s_n\}_{n\in\mathbb{N}}$. Then the formal infinite series $\sum_{k=1}^{\infty} f_k$ is convergent pointwisely/uniformly on S if and only if the infinite sequence of partial sums $\{s_n\}_{n\in\mathbb{N}}$ is convergent pointwisely/uniformly on S, and in this case

$$\lim_{n\to\infty} s_n = \lim_{n\to\infty} \sum_{k=1}^{n} f_k = \sum_{k=1}^{\infty} f_k .$$

Again, in certain situations, it would be easier to deal with series of functions instead of sequences of functions. Hence it will be handy to display the series version of the results in the preceding section.

Definition 5.2.1. *Let $\sum_{k=1}^{\infty} f_k$ be a formal series of functions on S. For any $n \in \mathbb{N}$, The n-th partial sum of the series is defined as*

$$s_n(x) := \sum_{k=1}^{n} f_k(x) , \quad x \in S .$$

The series $\sum_{k=1}^{\infty} f_k$ is said to converge uniformly on S if the sequence of partial sums $\{s_n\}_{n\in\mathbb{N}}$ is uniformly convergent on S, that is, there is a function $f : S \to \mathbb{R}$ such that $\{s_n\} \to f$ uniformly on S. In this case f is said to be the uniform limit of the series and we write

$$\sum_{k=1}^{\infty} f_k(x) = f(x) \quad \text{uniformly on } S .$$

Theorem 5.2.2 (Cauchy's criterion for uniform convergence of infinite series). *$\sum_{n=1}^{\infty} f_n$ converges uniformly on S if and only if for every $\varepsilon > 0$, there exists $N \in \mathbb{N}$ such that*

$$\Big| \sum_{k=n+1}^{n+p} f_k(x) \Big| < \varepsilon \quad \text{for all } n \geq N,\ p \in \mathbb{N}, \text{ and all } x \in S . \qquad (*)$$

Proof. Consider the sequence of partial sums $s_n := \sum_{k=1}^{n} f_k$. In view of Theorem 5.1.11, $\sum_{n=1}^{\infty} f_n$ is uniformly convergent on S if and only if $\{s_n\}_{n\in\mathbb{N}}$ is uniformly convergent on S if and only if $\{s_n\}_{n\in\mathbb{N}}$ is uniformly Cauchy on S if and only if for any $\varepsilon > 0$, there exists $N \in \mathbb{N}$ such that

$$\left|\sum_{k=n+1}^{n+p} f_k(x)\right| = \left|s_{n+p}(x) - s_n(x)\right| < \varepsilon$$

for all $n \geq N$, $p \in \mathbb{N}$ and all $x \in S$. $\square$

Corollary 5.2.3. *If $f_n : S \to \mathbb{R}$ is continuous at $c \in S$ for each $n \in \mathbb{N}$ and $\sum_{n=1}^{\infty} f_n(x) = f(x)$ uniformly on S, then f is also continuous at c.*

Proof. Consider the sequence of partial sums $s_n := \sum_{k=1}^{n} f_k$. Then by assumption, $\{s_n\} \to f$ uniformly on S. Since each f_k is continuous on S, the same is true for each s_n. By Theorem 5.1.6, being the uniform limit of $\{s_n\}_{n\in\mathbb{N}}$, f is continuous on S. $\square$

A very useful tool to check uniform convergence of series of functions is the following

Theorem 5.2.4 (Weierstrass M-Test). *Let $\sum_{n=1}^{\infty} f_n$ be a formal series of functions on S satisfying*

(i) *for each $n \in \mathbb{N}$, $|f_n| \leq a_n$ on S for some $a_n \in \mathbb{R}$, and*
(ii) *$\sum_{n=1}^{\infty} a_n$ converges,*

then $\sum_{n=1}^{\infty} f_n$ converges uniformly on S.

Proof. Since $\sum_{n=1}^{\infty} a_n$ is convergent, for any $\varepsilon > 0$, there exists $N \in \mathbb{N}$ such that $\sum_{k=N+1}^{\infty} a_k < \varepsilon$. Thus for all $n \geq N$, $p \in \mathbb{N}$, and all $x \in S$,

$$\left|\sum_{k=n+1}^{n+p} f_k(x)\right| \leq \sum_{k=n+1}^{n+p} \left|f_k(x)\right| \leq \sum_{k=n+1}^{n+p} a_k \leq \sum_{k=N+1}^{\infty} a_k < \varepsilon\ .$$

The result now follows from Theorem 5.2.2. $\square$

Example 5.2.5. We have seen in Example 5.1.12 that for every $0 \leq a < 1$, the sequence of functions $\{f_n\}_{n\in\mathbb{N}}$ on $[0, a]$ defined by $f_n(x) := x^n$ is uniformly convergent on $[0, a]$. It would be constructive to see how this can be shown by first converting the sequence of functions to its corresponding formal series of functions and then proving the uniform convergence of the latter by using Weierstrass M-Test, as follows.

Following the prescription at the beginning of this Section, we define

$$\begin{cases} g_1 & := f_1 \\ g_n & := f_n - f_{n-1} \ , \quad n \geq 2 \ . \end{cases}$$

Then the sequence of functions $\{f_n\}_{n\in\mathbb{N}}$ converges uniformly on $[0, a]$ if and only if the series of functions $\sum_{n=1}^{\infty} g_n$ converges uniformly on $[0, a]$, and in this case they have the same limit. So the problem boils down to showing the uniform convergence of $\sum_{n=1}^{\infty} g_n$ on $[0, a]$. Now for any $x \in [0, a]$, we have

$$|g_1(x)| = |f_1(x)| = x \leq a \ ,$$
$$|g_n(x)| \leq |f_n(x)| + |f_{n-1}(x)| = x^n + x^{n-1} \leq a^n + a^{n-1} \ , \quad n \geq 2 \ .$$

Since $0 \leq a < 1$, we have

$$a + \sum_{n=2}^{\infty} \left(a^n + a^{n-1}\right) = a + \frac{a^2 + a}{1 - a} < \infty \ .$$

Thus Weierstrass M-Test applies and we conclude that the series $\sum_{n=1}^{\infty} g_n$ and hence the sequence $\{f_n\}_{n\in\mathbb{N}}$ converges uniformly on $[0, a]$.

Finally, we have the following series version of Theorem 5.1.15.

Theorem 5.2.6. *Let $\sum_{n=1}^{\infty} f_n$ be a formal series of differentiable functions on an open set $S \subset \mathbb{R}$. Suppose that*

(i) *$\sum_{n=1}^{\infty} f_n'$ converges to some function g uniformly in S, and*
(ii) *there exists $x_0 \in S$ such that $\sum_{n=1}^{\infty} f_n(x_0)$ converges.*

Then

(a) *$\sum_{n=1}^{\infty} f_n$ converges to some function f uniformly in S, and*
(b) *f is differentiable, with $f' = \sum_{n=1}^{\infty} f_n'$ in S.*

Proof. It follows immediately from Theorem 5.1.15. In fact, consider the partial sums

$$F_n := \sum_{k=1}^{n} f_k, \quad n \in \mathbb{N},$$

of the formal series $\sum_{n=1}^{\infty} f_n$. Since each f_k is differentiable, each F_n is differentiable, with

$$F_n' = \left(\sum_{k=1}^{n} f_k \right)' = \sum_{k=1}^{n} f_k' \to \sum_{k=1}^{\infty} f_k' = g$$

uniformly in S. Furthermore,

$$\{F_n(x_0)\} = \left\{ \sum_{k=1}^{n} f_k(x_0) \right\} \to \sum_{k=1}^{\infty} f_k(x_0) \in \mathbb{R},$$

thus conditions (i) and (ii) of Theorem 5.1.15 are satisfied and so there exists a differentiable function $f : S \to \mathbb{R}$ such that

$$\sum_{n=1}^{\infty} f_n = \lim_{n\to\infty} \sum_{k=1}^{n} f_k = \lim_{n\to\infty} F_n = f$$

uniformly in S, with

$$f' = g = \sum_{k=1}^{\infty} f_k' \quad \text{in } S. \qquad \square$$

Exercise 5.2

Part A: True or False Questions

For each of the following statements, determine if it is true or false. If it is true, prove it. If it is false, give a counterexample or provide proper justification.

1. If $\sum_{n=1}^{\infty} f_n$ is uniformly convergent on $S \subset \mathbb{R}$, then $\{f_n\} \to 0$ uniformly on S.

   Answer: True.
   Proof. Since $\sum_{n=1}^{\infty} f_n$ is uniformly convergent on S, by Cauchy's Criterion for uniform convergence of infinite series, for any $\varepsilon > 0$, there exists $N \in \mathbb{N}$ such that

   $$\Big|\sum_{k=n+1}^{n+p} f_k(x)\Big| < \varepsilon \quad \text{for all } n \geq N,\ p \in \mathbb{N}, \text{ and all } x \in S\ .$$

   Taking $p = 1$, this reads

   $$|f_n(x)| < \varepsilon \quad \text{for all } n > N \text{ and all } x \in S\ ,$$

   which is equivalent to saying that $\{f_n\} \to 0$ uniformly on S.

2. $\sum_{n=1}^{\infty} \dfrac{1}{nx+1}$ is uniformly convergent in $(0, 1)$.
   Answer: False.
   Justification. Write $f_n(x) := \frac{1}{nx+1}$, $x \in (0, 1)$. It is clear that $\{f_n\} \to f \equiv 0$ pointwisely in $(0, 1)$. However, as

   $$\sup\Big\{\Big|f_n(x) - f(x)\Big| : x \in (0,1)\Big\} \geq \left|f_n\left(\frac{1}{n}\right)\right| = \frac{1}{2}$$

   for all $n \in \mathbb{N}$, the convergence is not uniform in $(0, 1)$. Hence by Exercise 5.2, Part A, Problem #1, $\sum_{n=1}^{\infty} f_n$ is not uniformly convergent in $(0, 1)$.

3. $\sum_{n=1}^{\infty} \dfrac{x}{nx+1}$ is uniformly convergent in $(0,1)$.

   Answer: False.

   Justification. Write $f_n(x) := \frac{x}{nx+1}$, $x \in (0,1)$. It is clear that $\{f_n\} \to f \equiv 0$ pointwisely in $(0,1)$. Furthermore, it is elementary to check that for each $n \in \mathbb{N}$, f_n is increasing in $(0,1)$ and so

$$\sup_{x\in(0,1)} |f_n(x) - f(x)| = \lim_{x\to 1^-} |f_n(x)| = \frac{1}{n+1} \to 0 \quad \text{as } n \to \infty\,.$$

   Thus the convergence $\{f_n\} \to f \equiv 0$ is actually uniform in $(0,1)$. However, this does not imply that $\sum_{n=1}^{\infty} f_n$ is uniformly convergent in $(0,1)$. In fact, as

$$\sum_{i=1}^{N} f_n\left(\frac{1}{2}\right) = \sum_{n=1}^{N} \frac{1}{n+2} \to \infty \quad \text{as } N \to \infty\,,$$

   $\sum_{n=1}^{\infty} f_n$ is not even pointwisely convergent in $(0,1)$, leave alone uniform convergence.

4. If $\{f_n\} \to 0$ uniformly on $S \subset \mathbb{R}$, then $\sum_{n=1}^{\infty} f_n$ is uniformly convergent on S.

   Answer: False.

   Example: Exercise 5.2, Part A, Problem #3 is a good example.

   Second Example: Consider the sequence of functions $\{f_n\}_{n\in\mathbb{N}}$ on $(0,1)$ defined by $f_n(x) := \frac{x^n}{n}$, $x \in (0,1)$. For any $\varepsilon > 0$, let $N := \left[\frac{1}{\varepsilon}\right] + 1 \in \mathbb{N}$. Then

$$|f_n(x) - 0| = \left|\frac{x^n}{n} - 0\right| < \frac{1}{n} \le \frac{1}{N} < \varepsilon$$

   for any $n \ge N$ and any $x \in (0,1)$. Thus $\{f_n\} \to 0$ uniformly in $(0,1)$. However, for $\varepsilon := \frac{1}{5}$, for any $N \in \mathbb{N}$, by taking $x_0 := 2^{-\frac{1}{2N}} \in (0,1)$, we have

$$\sum_{n=N+1}^{\infty} \frac{x_0^n}{n} \ge \sum_{n=N+1}^{2N} \frac{x_0^n}{n} \ge \sum_{n=N+1}^{2N} \frac{x_0^{2N}}{2N} = \frac{x_0^{2N}}{2} = \frac{1}{4} > \varepsilon\,.$$

   Thus $\sum_{n=1}^{\infty} \dfrac{x^n}{n}$ is not uniformly convergent in $(0,1)$.

5. Let $\phi : \mathbb{R} \to \mathbb{R}$ be a continuous even function with period 2 given by

$$\phi(t) := \begin{cases} 0 & \text{for } 0 \le t \le \frac{1}{3} \\ 3t - 1 & \text{for } \frac{1}{3} < t < \frac{2}{3} \\ 1 & \text{for } \frac{2}{3} \le t \le 1 \,, \end{cases}$$

and $f_n : [0,1] \to \mathbb{R}$ be defined by $f_n(t) := \frac{1}{2^n}\phi(3^{2(n-1)}t)$. Then $\sum_{n=1}^{\infty} f_n(t)$ converges uniformly on $[0,1]$.

Answer: True.

Proof. It is easy to check that $0 \le \phi \le 1$ and hence $|f_n| \le \frac{1}{2^n}$ on $\mathbb{R}$. Since $\sum_{n=1}^{\infty} \frac{1}{2^n}$ converges, by Weierstrass M-Test, $\sum_{n=1}^{\infty} f_n$ is uniformly convergent on $\mathbb{R}$.

6. The infinite series of functions

$$\sum_{n=1}^{\infty} \frac{\log(1+nx)}{nx^n}$$

is uniformly convergent on $[1+\alpha, \infty)$ for any $\alpha > 0$.

Answer: True.

Proof. It is elementary to check that

$$\begin{aligned} \left|\frac{\log(1+nx)}{nx^n}\right| &< \frac{\log(x+nx)}{nx^n} < \frac{x+nx}{nx^n} \le \frac{2nx}{nx^n} \\ &= \frac{2}{x^{n-1}} \le \frac{2}{(1+\alpha)^{n-1}} \end{aligned}$$

for any $x \in [1+\alpha, \infty)$ and any $n \in \mathbb{N}$. Since $\sum_{n=1}^{\infty} \frac{2}{(1+\alpha)^{n-1}}$ is convergent, by Weierstrass M-Test, the result follows.

7. The series $\sum_{n=1}^{\infty} \dfrac{\cos(nx)}{n^2}$ and $\sum_{n=1}^{\infty} \dfrac{\sin(nx)}{n^2}$ are uniformly convergent on $\mathbb{R}$.

Answer: True.

Proof. For any $x \in \mathbb{R}$ and $n \in \mathbb{N}$, we have

$$\left|\frac{\cos(nx)}{n^2}\right|, \quad \left|\frac{\sin(nx)}{n^2}\right| \leq \frac{1}{n^2} \, .$$

Since

$$\sum_{n=1}^{\infty} \frac{1}{n^2} < \infty \, ,$$

Weierstrass M-Test applies and so the two given series are uniformly convergent on $\mathbb{R}$.

8. $\sum_{n=1}^{\infty} \dfrac{x}{n^\alpha(1+nx^2)}$ converges uniformly on $\mathbb{R}$ for any $\alpha > \frac{1}{2}$.

Answer: True.

Proof. For $x \neq 0$, using A.M.-G.M. inequality, one has

$$n^{\frac{1}{2}} |x| \leq \frac{1}{2} \left(1 + nx^2\right) \, .$$

Thus for any $x \neq 0$ and any $n \in \mathbb{N}$,

$$\left|\frac{x}{n^\alpha(1+nx^2)}\right| = \frac{1}{n^\alpha} \frac{|x|}{1+nx^2} \leq \frac{1}{n^\alpha} \frac{1}{2n^{\frac{1}{2}}} = \frac{1}{2n^{\alpha+\frac{1}{2}}} \, .$$

For $x = 0$, we have trivially that $\left|\dfrac{x}{n^\alpha(1+nx^2)}\right| = 0 < \dfrac{1}{2n^{\alpha+\frac{1}{2}}}$ for all $n \in \mathbb{N}$. Hence we have

$$\left|\frac{x}{n^\alpha(1+nx^2)}\right| \leq \frac{1}{2n^{\alpha+\frac{1}{2}}} \quad \text{for all } x \in \mathbb{R} \text{ and all } n \in \mathbb{N} \, .$$

Since $\sum_{n=1}^{\infty} \dfrac{1}{2n^{\alpha+\frac{1}{2}}}$ converges for $\alpha > \frac{1}{2}$, by Weierstrass M-Test, $\sum_{n=1}^{\infty} \dfrac{x}{n^\alpha(1+nx^2)}$ converges uniformly on $\mathbb{R}$ whenever $\alpha > \frac{1}{2}$.

9. $\sum_{n=1}^{\infty} \dfrac{1}{1+n^2x}$ converges uniformly in $(0, \infty)$.

Answer: False.

Justification. For any $n \in \mathbb{N}$, write $f_n(x) := \dfrac{1}{1+n^2x}$, $x \in (0,\infty)$. As for every n, $p \in \mathbb{N}$,

$$\left|\sum_{k=n+1}^{n+p} f_k\left(\frac{1}{(n+1)^2}\right)\right| \geq \left|f_{n+1}\left(\frac{1}{(n+1)^2}\right)\right| = \frac{1}{2}\,,$$

$\sum_{n=1}^{\infty} f_n$ is not uniformly Cauchy and hence it is not uniformly convergent in $(0,\infty)$.

10. $\sum_{n=1}^{\infty}\left(\dfrac{x^{2n+1}}{2n+1} - \dfrac{x^{n+1}}{2n+2}\right)$ converges uniformly on $[0,1]$.

Answer: False.

Justification. It is elementary to verify that

$$\sum_{n=1}^{\infty}\left(\frac{x^{2n+1}}{2n+1} - \frac{x^{n+1}}{2n+2}\right)$$
$$= f(x) := \begin{cases} \frac{1}{2}\log(1+x) & \text{if } x \in [0,1) \\ \log 2 & \text{if } x = 1\ . \end{cases}$$

As each summand of the given infinite series is continuous but its limit f is not continuous at 1, the convergence is not uniform on $[0,1]$.

11. $\sum_{n=1}^{\infty}\dfrac{1}{n^2+x^2}$ is differentiable in $\mathbb{R}$.

Answer: True.

Proof. Write $f_n(x) := \dfrac{1}{n^2+x^2}$ and $f := \sum_{n=1}^{\infty} f_n$. We have

$$f_n'(x) = -\frac{2x}{(n^2+x^2)^2}\,, \quad f_n''(x) = \frac{2(3x^2-n^2)}{(n^2+x^2)^3}\ .$$

It is elementary to see that

$$|f_n'(x)| \leq \left|f_n'\left(\frac{n}{\sqrt{3}}\right)\right| = \frac{3^{3/2}}{8n^3} \quad \text{for all } n \in \mathbb{N} \text{ and all } x \in \mathbb{R}\ .$$

Since $\sum_{n=1}^{\infty} \dfrac{3^{3/2}}{8n^3} < \infty$, by Weierstrass M-Test, $\{\sum_{n=1}^{\infty} f_n'\}$ is uniformly convergent on $\mathbb{R}$. Furthermore, since $\sum_{n=1}^{\infty} f_n(0) = \sum_{n=1}^{\infty} \frac{1}{n^2} < \infty$, by Theorem 5.2.6, f is differentiable in $\mathbb{R}$ and $f' = \sum_{n=1}^{\infty} f_n'$ in $\mathbb{R}$.

12. Uniform convergence implies absolute convergence.
That is, if $\sum_{n=1}^{\infty} f_n$ converges uniformly on $S \subset \mathbb{R}$, then $\sum_{n=1}^{\infty} |f_n|$ converges pointwisely on S.

Answer: False.

Example: For any $n \in \mathbb{N}$, let $f_n(x) := (-1)^{n+1}\dfrac{x^n}{n}$, $x \in [0,1]$. Then for every fixed $x \in [0,1]$, $\sum_{n=1}^{\infty} f_n(x)$ is an alternating series. Since $|f_n(x)| \geq |f_{n+1}(x)|$ for all $n \in \mathbb{N}$, by Alternating Series Test, $\sum_{n=1}^{\infty} f_n(x)$ is convergent, and it is easy to see that $\sum_{n=1}^{\infty} f_n(x) = 0$. Since this is true for all $x \in [0,1]$, we see that $\sum_{n=1}^{\infty} f_n = f :\equiv 0$ pointwisely on $[0,1]$. Now since

$$\left|\sum_{k=1}^{n} f_n(x) - f(x)\right| \leq |f_{n+1}(x)| = \frac{x^{n+1}}{n+1} \leq \frac{1}{n+1}$$

for all $x \in [0,1]$ and all $n \in \mathbb{N}$, we see that $\sum_{n=1}^{\infty} f_n = f :\equiv 0$ uniformly on $[0,1]$. However, $\sum_{n=1}^{\infty} |f_n(x)| = \sum_{n=1}^{\infty} \dfrac{x^n}{n}$ does not converge at $x = 1$ and so $\sum_{n=1}^{\infty} f_n$ is not absolutely convergent on $[0,1]$.

13. Absolute convergence implies uniform convergence.
That is, if $\sum_{n=1}^{\infty} |f_n|$ converges pointwisely on $S \subset \mathbb{R}$, then $\sum_{n=1}^{\infty} f_n$ converges uniformly on S.

Answer: False.

Example: For any $n \in \mathbb{N}$, let $f_n(x) := \dfrac{x^2}{(1+x^2)^n}$, $x \in \mathbb{R}$. It is elementary to check that

$$\sum_{n=1}^{\infty} |f_n(x)| = \sum_{n=1}^{\infty} f_n(x) = f(x) := \begin{cases} 0 & \text{for } x = 0 \\ 1 & \text{for } x \neq 0 \end{cases}$$

pointwisely on $\mathbb{R}$. However, as f is not continuous, the convergence cannot be uniform.

Part B: Problems

1. For $n \in \mathbb{N}$, let $f_n(x) := \dfrac{x}{1+nx^2}$, $x \in \mathbb{R}$.
   (a) Compute $f := \lim_{n\to\infty} f_n$ and $g := \lim_{n\to\infty} f_n'$ on $\mathbb{R}$.
   (b) Determine whether $\{f_n\} \to f$ uniformly on $\mathbb{R}$.
   (c) Determine whether $\{f_n'\} \to g$ uniformly on $\mathbb{R}$. If not, would the convergence be uniform on any interval of the form $[a, b]$?

   Solution:
   (a) It is elementary to see that $f(x) := \lim_{n\to\infty} f(x) \equiv 0$, and

   $$g(x) := \lim_{n\to\infty} f_n'(x) = \lim_{n\to\infty} \frac{1-nx^2}{(1+nx^2)^2} = \begin{cases} 1 & \text{if } x = 0 \\ 0 & \text{if } x \neq 0\,. \end{cases}$$

   (b) By A.M.-G.M. Inequality, we have $1 + nx^2 \geq 2\sqrt{n}\,|x|$ for all $n \in \mathbb{N}$ and all $x \in \mathbb{R}$. Thus

   $$|f_n(x) - f(x)| = \left|\frac{x}{1+nx^2}\right| \leq \frac{|x|}{2\sqrt{n}\,|x|} = \frac{1}{2\sqrt{n}}$$

   for all $n \in \mathbb{N}$ and all $x \in \mathbb{R}$. Hence $\{f_n\} \to f \equiv 0$ uniformly on $\mathbb{R}$.
   (c) Since each f_n' is continuous on $\mathbb{R}$ but g is not continuous at 0, the convergence $\{f_n'\} \to g$ is not uniform on $\mathbb{R}$.

Next, let $[a, b]$ be any interval in $\mathbb{R}$. If $0 \in [a, b]$, by the same consideration as above, the convergence is not uniform on $[a, b]$. It thus remains to consider the case $0 \notin [a, b]$. Without loss of generality, assume $0 < a$. Then

$$|f_n'(x) - g(x)| = \left| \frac{1 - nx^2}{(1 + nx^2)^2} \right| \le \frac{1}{1 + nx^2} \le \frac{1}{na^2}$$

for all $n \in \mathbb{N}$ and all $x \in [a, b]$. Thus $\{f_n'\} \to g$ uniformly on $[a, b]$. Hence the convergence is uniform on every closed interval of $\mathbb{R}$ which does not contain the point 0.

2. For $n \in \mathbb{N}$, let $f_n(x) := \dfrac{1}{n}\, e^{-n^2x^2}$, $x \in \mathbb{R}$.
   (a) Compute $f := \lim_{n\to\infty} f_n$ and $g := \lim_{n\to\infty} f_n'$ on $\mathbb{R}$.
   (b) Determine whether $\{f_n\} \to f$ uniformly on $\mathbb{R}$.
   (c) Determine whether $\{f_n'\} \to g$ uniformly on $\mathbb{R}$. If not, would the convergence be uniform on any interval of the form $[a, b]$?

Solution.

(a) It is elementary to see that $f(x) := \lim_{n\to\infty} f(x) \equiv 0$, and $g(x) := \lim_{n\to\infty} f_n'(x) = \lim_{n\to\infty} \left(-2nx\, e^{-n^2x^2}\right) \equiv 0$.

(b) *Answer*: Yes, $\{f_n\} \to f$ uniformly on $\mathbb{R}$.
Proof. It is evident since

$$|f_n(x) - f(x)| = \left| \frac{1}{n}\, e^{-n^2x^2} \right| \le \frac{1}{n}$$

for all $n \in \mathbb{N}$ and all $x \in \mathbb{R}$.

(c) *Answer*: No, the convergence is not uniform.
Justification. Let $\varepsilon := \frac{1}{e}$. For any $n \in \mathbb{N}$, take $x := \frac{1}{n}$. Then

$$|f_n'(x) - g(x)| = \left| \frac{-2nx}{e^{n^2x^2}} \right| = \frac{2}{e} > \frac{1}{e} = \varepsilon\ .$$

Hence the convergence is not uniform on $\mathbb{R}$.

Finally, let $[a,b]$ be any interval in $\mathbb{R}$. If $0 \in [a,b]$, by the same consideration as above, the convergence is not uniform on $[a,b]$. It thus remains to consider the case $0 \notin [a,b]$. Without loss of generality, assume $0 < a$. Then

$$\begin{aligned}
&\sup\left\{ |f_n'(x) - g(x)| : x \in [a,b]\right\} \\
&= \sup\left\{ \left|\frac{2nx}{e^{n^2x^2}}\right| : x \in [a,b]\right\} \\
&\le \left|\frac{2nb}{e^{n^2a^2}}\right| \to 0 \quad \text{as } n \to \infty\,.
\end{aligned}$$

Thus $\{f_n'\} \to g$ uniformly on $[a,b]$. Hence the convergence is uniform on every closed interval of $\mathbb{R}$ which does not contain the point 0.

3. Let $\{a_n\}_{n\in\mathbb{N}}$ be a sequence of non-negative real numbers. Prove that the series $\sum_{n=1}^{\infty} a_n \cos(nx)$ converges uniformly on $\mathbb{R}$ if and only if $\sum_{n=1}^{\infty} a_n < \infty$.

   Proof. ($\Rightarrow$): Suppose that the series $\sum_{n=1}^{\infty} a_n \cos nx$ converges uniformly on $\mathbb{R}$. Then in particular, the series converges at $x = 0$. That is, $\sum_{n=1}^{\infty} a_n$ is convergent.
   ($\Leftarrow$): Suppose that $\sum_{n=1}^{\infty} a_n < \infty$. Since

   $$\left|a_n \cos nx\right| \le a_n \quad \text{for all } n \in \mathbb{N} \text{ and all } x \in \mathbb{R}\,,$$

   the uniform convergence of the series $\sum_{n=1}^{\infty} a_n \cos nx$ follows immediately from Weierstrass M-Test.

4. Let $\{f_n\}_{n\in\mathbb{N}}$ be the sequence of functions on $[0,1]$ defined by

   $$f_n(x) := \begin{cases} \frac{1}{n} & \text{if } \frac{1}{2^n} < x \le \frac{1}{2^{n-1}} \\ 0 & \text{otherwise}\,. \end{cases}$$

   Show that $\sum_{n=1}^{\infty} f_n$ converges uniformly on $[0,1]$.

Proof. An immediate attempt is to apply Weierstrass M-Test. By assumption, $|f_n| \le \frac{1}{n}$ on $[0,1]$. But then as $\sum_{n=1}^{\infty} \frac{1}{n}$ is not convergent, the condition for Weierstrass M-Test is not satisfied. However, as Weierstrass M-Test is only a sufficient condition for uniform convergence but not a necessary condition, the condition fails to hold does not necessarily imply that the series is not uniformly convergent. In fact, $\sum_{n=1}^{\infty} f_n$ indeed converges uniformly on $[0,1]$.

To see this, first observe that if we write $S_n := \left(\frac{1}{2^n}, \frac{1}{2^{n-1}}\right]$ for any $n \in \mathbb{N}$, then $f_n = \frac{1}{n}\chi_{S_n}$, where, as usual, χ_A stands for the characteristic function of set A. Observe also that the S_n's are pairwisely disjoint and cover the entire set $(0,1]$. In particular, for any fixed $x \in (0,1]$, there is a unique $k \in \mathbb{N}$ such that $x \in S_k$. Hence the formal series $\sum_{n=1}^{\infty} f_n(x)$ actually consists of one single summand, namely,

$$\sum_{n=1}^{\infty} f_n(x) = \sum_{n=1}^{\infty} \frac{1}{n}\chi_{S_n}(x) = \frac{1}{k}\chi_{S_k}(x) = \frac{1}{k},$$

where $k \in \mathbb{N}$ is the unique integer for which $x \in S_k$. Hence for any given $\varepsilon > 0$, let $N \in \mathbb{N}$ be such that $\frac{1}{N} < \varepsilon$. Then for any $n \ge N$, $p \in \mathbb{N}$, and any $x \in (0,1]$, we have

$$\begin{aligned}
\left|\sum_{m=n+1}^{n+p} f_m(x)\right| &= \sum_{m=n+1}^{n+p} \frac{1}{m}\chi_{S_m}(x) \\
&= \begin{cases} 0 & \text{if } x \notin \bigcup_{m=n+1}^{n+p} S_m \\ \frac{1}{k} & \text{if } x \in S_k \text{ for some } k = n+1, \dots, n+p\,. \end{cases} \\
&\le \frac{1}{n+1} \le \frac{1}{N+1} < \frac{1}{N} < \varepsilon\,.
\end{aligned}$$

Hence $\sum_{n=1}^{\infty} f_n$ is uniformly Cauchy in $(0,1]$. On the other hand, as $f_n(0) = 0$ for all $n \in \mathbb{N}$, it is evident that $\sum_{n=1}^{\infty} f_n$ is also uniformly Cauchy on $[0,1]$ and hence uniformly convergent there.

5. Let $\{f_n(x)\}_{n\in\mathbb{N}}$ be a sequence of functions on $[a,b]$ such that each f_n is continuous at b. Suppose that $\sum_{n=1}^{\infty} f_n(x)$ converges at every $x \in [a,b)$ but $\sum_{n=1}^{\infty} f_n(b)$ diverges. Determine whether $\sum_{n=1}^{\infty} f_n(x)$ is uniformly convergent on $[a,b)$.

   Answer: No.

   Justification. If $\sum_{n=1}^{\infty} f_n(x)$ were uniformly convergent on $[a,b)$, then for any $\varepsilon > 0$, there exists $N \in \mathbb{N}$ such that

$$\left|\sum_{k=n+1}^{n+p} f_k(x)\right| < \frac{\varepsilon}{2} \quad \text{for all } n \geq N, p \in \mathbb{N}, \text{ and all } x \in [a,b) .$$

   For any $k \in \mathbb{N}$, since f_k is continuous at b, there exists $\delta_k > 0$ such that whenever $x \in (b - \delta_k, b)$,

$$\left|f_k(x) - f_k(b)\right| < \frac{\varepsilon}{2^k} .$$

   For any fixed n, $p \in \mathbb{N}$, let $\delta_{n,p} := \min\{\delta_{n+1}, \dots, \delta_{n+p}\} > 0$. Pick $x_{n,p} \in (b - \delta_{n,p}, b)$. Then $x_{n,p} \in (b - \delta_k, b)$ for all $k = n+1, \dots, n+p$ and so

$$\left|f_k(x_{n,p}) - f_k(b)\right| < \frac{\varepsilon}{2^k} \quad \text{for all } k = n+1, \dots, n+p .$$

   Hence by triangle inequality, we have

$$\begin{aligned}
\left|\sum_{k=n+1}^{n+p} f_k(b)\right| &\leq \left|\sum_{k=n+1}^{n+p} f_k(x_{n,p})\right| + \left|\sum_{k=n+1}^{n+p} \Big[f_k(x_{n,p}) - f_k(b)\Big]\right| \\
&\leq \left|\sum_{k=n+1}^{n+p} f_k(x_{n,p})\right| + \sum_{k=n+1}^{n+p} \left|f_k(x_{n,p}) - f_k(b)\right| \\
&< \frac{\varepsilon}{2} + \sum_{k=n+1}^{n+p} \frac{\varepsilon}{2^k} < \varepsilon ,
\end{aligned}$$

which implies the Cauchyness and hence the convergence of $\sum_{n=1}^{\infty} f_n(b)$. This is a contradiction.

6. (*Dirichlet's Test for Uniform Convergence*) Let $I \subset \mathbb{R}$ be an interval. For any $n \in \mathbb{N}$, let f_n, $g_n : I \to \mathbb{R}$ be any functions satisfying
   (i) there exist a constant $B > 0$ such that for all $n \in \mathbb{N}$,
$$\left|\sum_{i=1}^{n} f_i\right| \leq B \quad \text{in } I\ ;$$
   (ii) $\{g_n\}_{n\in\mathbb{N}}$ is monotonically decreasing in I ;
   (iii) $\{g_n\} \to g \equiv 0$ uniformly on I.

   Prove that the series $\sum_{n=1}^{\infty} f_n g_n$ converges uniformly on I.

   Proof. For any $n \in \mathbb{N}$, write $F_n(x) := \sum_{i=1}^{n} f_i(x)$. Then we have $|F_n| \leq B$ on I for all $n \in \mathbb{N}$. Note that for any $m > n \in \mathbb{N}$, we can write
$$\begin{aligned}\sum_{i=n}^{m} f_i g_i &= \sum_{i=n}^{m}(F_i - F_{i-1})g_i \\ &= F_m g_m - F_{n-1}g_n + \sum_{i=n}^{m-1} F_i(g_i - g_{i+1})\end{aligned}$$
   and hence
$$\begin{aligned}\left|\sum_{i=n}^{m} f_i g_i\right| &\leq |F_m g_m| + |F_{n-1}g_n| + \sum_{i=n}^{m-1}\Big|F_i(g_i - g_{i+1})\Big| \\ &\leq B|g_m| + B|g_n| + B\sum_{i=n}^{m-1}|g_i - g_{i+1}|\ .\end{aligned}$$
   As $g_n \geq g_{n+1}$ for all $n \in \mathbb{N}$ and $\{g_n\} \to g \equiv 0$ uniformly on I, we have $g_n \geq 0$ and $g_n - g_{n+1} \geq 0$ in I for all $n \in \mathbb{N}$ and so
$$\left|\sum_{i=n}^{m} f_i g_i\right| \leq Bg_m + Bg_n + B\sum_{i=n}^{m-1}(g_i - g_{i+1}) = 2Bg_n\ .$$

As $\{g_n\} \to 0$ uniformly on I, for any $\varepsilon > 0$, there exists $N \in \mathbb{N}$ such that $g_n(x) < \dfrac{\varepsilon}{2B}$ for all $n \geq N$ and all $x \in I$. Consequently,

$$\left|\sum_{i=n}^{m} f_i g_i\right| < \varepsilon \quad \text{for all } m > n \geq N \text{ and all } x \in I\ .$$

Hence the series $\sum_{n=1}^{\infty} f_n g_n$ is uniformly Cauchy in I and so by Cauchy's Criterion, it is uniformly convergent there.

7. Show that $\sum_{n=1}^{\infty} \dfrac{\sin(nx)}{n}$ converges uniformly on $I := [\delta, 2\pi - \delta]$ for any $0 < \delta < \pi$.

Proof. For any $n \in \mathbb{N}$, write $f_n(x) := \sin(nx)$ and $g_n(x) := \frac{1}{n}$, $x \in I = [\delta, 2\pi - \delta]$. We have obviously that $g_n \geq g_{n+1}$ for all $n \in \mathbb{N}$ and $\{g_n\} \to g :\equiv 0$ uniformly on I. Hence

$$\begin{aligned} 2\sin\left(\frac{x}{2}\right)\sum_{n=1}^{N} f_n(x) &= \sum_{n=1}^{N} 2\sin\left(\frac{x}{2}\right)\sin(nx) \\ &= \sum_{n=1}^{N}\left[\cos\left(nx - \frac{x}{2}\right) - \cos\left(nx + \frac{x}{2}\right)\right] \\ &= \cos\left(\frac{x}{2}\right) - \cos\left[\frac{(2N+1)x}{2}\right] \end{aligned}$$

for any $N \in \mathbb{N}$ and any $x \in I$. Therefore,

$$\left|\sum_{n=1}^{N} f_n(x)\right| \leq \left|\frac{\cos\left(\frac{x}{2}\right) - \cos\left(\frac{(2N+1)x}{2}\right)}{2\sin\frac{x}{2}}\right| \leq \left|\frac{1}{\sin\frac{x}{2}}\right| \leq \left|\frac{1}{\sin\frac{\delta}{2}}\right|$$

for any $N \in \mathbb{N}$ and any $x \in I$. It follows from Exercise 5.2, Part B, Problem #6 (Dirichlet's test for uniform convergence) that

$$\sum_{n=1}^{\infty} \frac{\sin(nx)}{n} = \sum_{n=1}^{\infty} f_n g_n(x)$$

converges uniformly on I.

8. Let $\{g_n\}_{n\in\mathbb{N}}$ be a decreasing sequence of functions on an interval $I \subset \mathbb{R}$ such that $\{g_n\} \to 0$ uniformly on I, determine whether the series $\sum_{n=1}^{\infty}(-1)^n g_n$ converges uniformly on I.

   [Compare with Exercise 5.2, Part A, Problem #4.]

   Answer: Yes, $\sum_{n=1}^{\infty}(-1)^n g_n$ converges uniformly on I.

   Proof. For every $n \in \mathbb{N}$, write $f_n(x) := (-1)^n$ for all $x \in I$. Then we have

$$\left|\sum_{i=1}^{n} f_i\right| = \begin{cases} |-1| & \text{if } n \text{ is odd} \\ 0 & \text{if } n \text{ is even} \end{cases}$$
$$\leq 1$$

   for all $n \in \mathbb{N}$. Hence all conditions of Dirichlet's Test for Uniform Convergence (Exercise 5.2, Part B, Problem #6) are satisfied and so $\sum_{n=1}^{\infty}(-1)^n g_n = \sum_{n=1}^{\infty} f_n\, g_n$ converges uniformly on I.

9. (*Abel's Test for Uniform Convergence*) Let $\{f_n\}_{n\in\mathbb{N}}$ and $\{g_n\}_{n\in\mathbb{N}}$ be sequences of functions on $S \subset \mathbb{R}$ such that
   (i) $\sum_{n=1}^{\infty} f_n$ converges uniformly on S;
   (ii) $\{g_n\}_{n\in\mathbb{N}}$ is monotonically decreasing in S;
   (iii) $\{g_n\}_{n\in\mathbb{N}}$ is uniformly bounded on S.

   Prove that the series $\sum_{n=1}^{\infty} f_n g_n$ converges uniformly on S.

   Proof. Since $\{g_n\}_{n\in\mathbb{N}}$ is uniformly bounded, there exists $M > 0$ such that $|g_n| \leq M$ on S for all $n \in \mathbb{N}$. Let $\{F_n\}_{n\in\mathbb{N}}$ be the sequence of partial sums of $\sum_{n=1}^{\infty} f_n$, that is, for any $n \in \mathbb{N}$, $F_n := \sum_{k=1}^{n} f_k$. Since $\sum_{n=1}^{\infty} f_n$ is uniformly convergent on S, for any $\varepsilon > 0$, there exists $N \in \mathbb{N}$ such that

$$\left|F_n - F_m\right| < \frac{\varepsilon}{6M} \quad \text{on } S \text{ for any } n, m \geq N\,.$$

Hence for any $m > n \geq N$ and any $x \in S$, we have

$$\begin{aligned}
&\left|\sum_{k=m+1}^{n} f_k(x)g_k(x)\right| \\
&= \left|\sum_{k=m+1}^{n} \Big(F_k(x) - F_m(x) + F_m(x) - F_{k-1}(x)\Big)g_k(x)\right| \\
&= \Bigg|\sum_{k=m+1}^{n} \Big(F_k(x) - F_m(x)\Big)g_k(x) \\
&\quad - \sum_{k=m}^{n-1} \Big(F_k(x) - F_m(x)\Big)g_{k+1}(x)\Bigg| \\
&\leq \left|\sum_{k=m+1}^{n-1} \Big(F_k(x) - F_m(x)\Big)\Big(g_k(x) - g_{k+1}(x)\Big)\right| \\
&\quad + \Big|F_n(x) - F_m(x)\Big| \cdot \Big|g_n(x)\Big| \\
&\leq \frac{\varepsilon}{6M} \sum_{k=m+1}^{n-1} \Big(g_k(x) - g_{k+1}(x)\Big) + \frac{\varepsilon}{6M}\Big|g_n(x)\Big| \\
&\leq \frac{\varepsilon}{6M}\Big(g_{m+1}(x) - g_n(x)\Big) + \frac{\varepsilon}{6} \\
&\leq \frac{\varepsilon}{3} + \frac{\varepsilon}{6} < \varepsilon\ .
\end{aligned}$$

Therefore, $\sum_{n=1}^{\infty} f_n g_n$ is uniformly Cauchy on S and so by Cauchy's Criterion, it is uniformly convergent there.

10. Show that the series $\sum_{n=1}^{\infty} \dfrac{(-1)^n}{\sqrt{n}} \sin\left(1 + \dfrac{x}{n}\right)$ converges uniformly on every compact subset of $\mathbb{R}$.

Proof. As every compact subset of $\mathbb{R}$ is contained in $[0, a] \cup [-a, 0]$ for some $a > 0$, it suffices to show that the convergence is uniform on $[0, a]$ and $[-a, 0]$. As the cases for $[0, a]$ and $[-a, 0]$ can be dealt with by analogous arguments, we will only work on the case $[0, a]$ here. For any $n \in \mathbb{N}$, let $f_n(x) := \frac{(-1)^n}{\sqrt{n}}$ and $g_n(x) := \sin\left(1 + \frac{x}{n}\right)$, $x \in [0, a]$. Then for all $n \in \mathbb{N}$ large enough such that $\frac{a}{n} < \frac{1}{2}$,

we have

$$g_{n+1}(x) = \sin\left(1 + \frac{x}{n+1}\right) \le \sin\left(1 + \frac{x}{n}\right) = g_n(x)$$

for all $x \in [0, a]$. That is, $\{g_n\}_{n\in\mathbb{N}, n\gg}$ is monotonically decreasing in $[0, a]$. Furthermore, $\sum_{n=1}^{\infty} f_n(x) = \sum_{n=1}^{\infty} \frac{(-1)^n}{\sqrt{n}}$ is clearly uniformly convergent on $[0, a]$, and $|g_n(x)| = \left|\sin\left(1 + \frac{x}{n}\right)\right| \le 1$ for all $n \in \mathbb{N}$ and all $x \in [0, a]$. Hence all conditions of Abel's Test for Uniform Convergence (Exercise 5.2, Part B, Problem #9) are satisfied and so $\sum_{n=1}^{\infty} \frac{(-1)^n}{\sqrt{n}} \sin\left(1 + \frac{x}{n}\right) = \sum_{n=1}^{\infty} f_n\, g_n$ converges uniformly on $[0, a]$.

11. Let $\sum_{n=1}^{\infty} a_n$ be a convergent series of real numbers. Prove that $\sum_{n=1}^{\infty} \dfrac{a_n}{n^x}$ converges uniformly on $[0, \infty)$. Compute $\lim_{x\to 0+} \sum_{n=1}^{\infty} \dfrac{a_n}{n^x}$.

Proof. For every $n \in \mathbb{N}$, let $f_n(x) := a_n$ and $g_n(x) := \frac{1}{n^x}$ for all $x \in [0, \infty)$. Then $\sum_{n=1}^{\infty} f_n(x) = \sum_{n=1}^{\infty} a_n$ converges uniformly on $[0, \infty)$, $\{g_n\}_{n\in\mathbb{N}}$ is monotonically decreasing in $[0, \infty)$, and $|g_n(x)| = \frac{1}{n^x} \le 1$ for all $n \in \mathbb{N}$ and all $x \in [0, \infty)$. Hence all conditions of of Abel's Test for Uniform Convergence (Exercise 5.2, Part B, Problem #9) are satisfied and so $\sum_{n=1}^{\infty} \frac{a_n}{n^x} = \sum_{n=1}^{\infty} f_n\, g_n$ converges uniformly on $[0, \infty)$.

Finally, since $f_n\, g_n$ is continuous at 0 for every $n \in \mathbb{N}$, by Corollary 5.2.3, $\sum_{n=1}^{\infty} \frac{a_n}{n^x} = \sum_{n=1}^{\infty} f_n\, g_n$ is continuous at 0 and so

$$\lim_{x\to 0+} \sum_{n=1}^{\infty} \frac{a_n}{n^x} = \sum_{n=1}^{\infty} \left(\lim_{x\to 0+} \frac{a_n}{n^x}\right) = \sum_{n=1}^{\infty} a_n\,.$$

12. Let $\sum_{n=1}^{\infty} a_n$ be a convergent series of real numbers. Prove that $\sum_{n=1}^{\infty} a_n\, x^n$ converges uniformly on $[0,1]$, and hence show that $\lim_{x\to 1^-} \sum_{n=1}^{\infty} a_n\, x^n = \sum_{n=1}^{\infty} a_n$.

Proof. For every $n \in \mathbb{N}$, let $f_n(x) := a_n$ and $g_n(x) := x^n$ for all $x \in [0,1]$. Then $\sum_{n=1}^{\infty} f_n(x) = \sum_{n=1}^{\infty} a_n$ converges uniformly on $[0,1]$, $\{g_n\}_{n\in\mathbb{N}}$ is monotonically decreasing in $[0,1]$, and $|g_n(x)| = x^n \le 1$ for all $n \in \mathbb{N}$ and all $x \in [0,1]$. Hence all conditions of of Abel's Test for Uniform Convergence (Exercise 5.2, Part B, Problem #9) are satisfied and so $\sum_{n=1}^{\infty} a_n\, x^n = \sum_{n=1}^{\infty} f_n\, g_n$ converges uniformly on $[0,1]$.

Finally, since $f_n\, g_n$ is continuous at 1 for every $n \in \mathbb{N}$, by Corollary 5.2.3, $\sum_{n=1}^{\infty} a_n\, x^n = \sum_{n=1}^{\infty} f_n\, g_n$ is continuous at 1 and so

$$\lim_{x\to 1^-} \sum_{n=1}^{\infty} a_n\, x^n = \sum_{n=1}^{\infty} \left(\lim_{x\to 1^-} a_n\, x^n \right) = \sum_{n=1}^{\infty} a_n \ .$$

13. Let $\{f_n\}_{n\in\mathbb{N}}$ and $\{g_n\}_{n\in\mathbb{N}}$ be sequences of functions on $S \subset \mathbb{R}$ satisfying
    (i) f_1 is bounded on S;
    (ii) $\sum_{n=1}^{\infty} \big|f_{n+1}(x) - f_n(x)\big|$ converges pointwisely on S;
    (iii) $\sup_{x\in S} \sum_{n=1}^{\infty} |f_{n+1}(x) - f_n(x)| < \infty$; and
    (iv) $\sum_{n=1}^{\infty} g_n(x)$ converges uniformly on S.

    Show that $\sum_{n=1}^{\infty} f_n g_n$ converges uniformly on S.

Proof. Observe that

$$|f_n(x)| = \left| \sum_{k=1}^{n-1} \Big(f_{k+1}(x) - f_k(x) \Big) + f_1(x) \right|$$

$$\le \sum_{k=1}^{n-1} \Big| f_{k+1}(x) - f_k(x) \Big| + |f_1(x)|$$

$$\le \sum_{k=1}^{\infty} \Big| f_{k+1}(x) - f_k(x) \Big| + |f_1(x)| \ .$$

Since f_1 is bounded and $\sup_{x\in S}\sum_{k=1}^{\infty}|f_{k+1}(x)-f_k(x)|<\infty$, $\{f_n\}_{n\in\mathbb{N}}$ is uniformly bounded on S. Let K be the uniform bound of $\{f_n\}_{n\in\mathbb{N}}$ on S and

$$M := 1 + K + \sup_{x\in S}\sum_{k=1}^{\infty}\Big|f_{k+1}(x) - f_k(x)\Big| > 0\,.$$

Since $\sum_{k=1}^{\infty} g_k$ is uniformly convergent, for any $\varepsilon > 0$, there is $N \in \mathbb{N}$ such that

$$\sup_{x\in S}\left|\sum_{k=n+1}^{n+p} g_k(x)\right| < \frac{\varepsilon}{2M} \quad \text{for any } n \geq N \text{ and any } p \in \mathbb{N}\,.$$

Hence for any $n \geq N$, we have

$$\begin{aligned}
&\left|\sum_{k=n+1}^{n+p} f_k g_k\right| \\
=&\left|f_{n+1}\sum_{k=n+1}^{n+1} g_k + f_{n+2}\left(\sum_{k=n+1}^{n+2} g_k - \sum_{k=n+1}^{n+1} g_k\right)\right. \\
&\qquad\left. + \cdots + f_{n+p}\left(\sum_{k=n+1}^{n+p} g_k - \sum_{k=n+1}^{n+p-1} g_k\right)\right| \\
=&\left|\sum_{k=n+1}^{n+1} g_k\big(f_{n+1} - f_{n+2}\big) + \sum_{k=n+1}^{n+2} g_k\big(f_{n+2} - f_{n+3}\big)\right. \\
&\qquad\left. + \cdots + \sum_{k=n+1}^{n+p-1} g_k\big(f_{n+p-1} - f_{n+p}\big) + \sum_{k=n+1}^{n+p} g_k f_{n+p}\right| \\
\leq&\frac{\varepsilon}{2M}\Big(\big|f_{n+1} - f_{n+2}\big| + \cdots + \big|f_{n+p-1} - f_{n+p}\big| + \big|f_{n+p}\big|\Big) \\
\leq&\frac{\varepsilon}{2} < \varepsilon
\end{aligned}$$

on S. Therefore, $\sum_{n=1}^{\infty} f_n g_n$ is uniformly Cauchy on S and so by Cauchy's Criterion, it is uniformly convergent there.

14. Let $A := (a_{ij})_{n\times n}$ be an $n \times n$ real matrix. If

$$\sqrt{\sum_{i=1}^{n}\left(\sum_{j=1}^{n} a_{ij}\right)^2} \leq \alpha \quad \text{for some } 0 < \alpha < 1\ ,$$

show that $(I - A)^{-1}$ exists.

Proof. It suffices to show that for any $b \in \mathbb{R}^n$,

$$(I - A)x = b$$

has a unique solution. Equivalently, we need to show that for any $b \in \mathbb{R}^n$, the map $T : \mathbb{R}^n \to \mathbb{R}^n$ given by

$$T(x) := Ax + b$$

has a unique fixed point in $\mathbb{R}^n$. This suggests that we try and apply Banach's Fixed Point Theorem, and the whole proof now boils down to showing that T is a contraction, the process of which is routine: For any x, $y \in \mathbb{R}^n$,

$$\begin{aligned}
\|T(x) - T(y)\| &= \|A(x - y)\| \\
&= \sqrt{\sum_{i=1}^{n}\left(\sum_{j=1}^{n} a_{ij}\,(x_j - y_j)\right)^2} \\
&\leq \sqrt{\sum_{i=1}^{n}\left(\sum_{j=1}^{n} a_{ij}\,\|x - y\|\right)^2} \\
&= \sqrt{\sum_{i=1}^{n}\left(\sum_{j=1}^{n} a_{ij}\right)^2} \cdot \|x - y\| \\
&\leq \alpha\,\|x - y\|\ .
\end{aligned}$$

Hence T is a contraction with contraction constant $0 < \alpha < 1$ and so by Banach's Fixed Point Theorem, T has a unique fixed point.

15. Let $\{a_n\}_{n\in\mathbb{N}}$ be a decreasing sequence of positive numbers. Prove that the series $\sum_{n=1}^{\infty} a_n \sin nx$ converges uniformly on $\mathbb{R}$ if and only if $\{na_n\} \to 0$ as $n \to \infty$.

Proof. ($\Rightarrow$): Suppose the series $\sum_{n=1}^{\infty} a_n \sin nx$ converges uniformly on $\mathbb{R}$. Then for any $\varepsilon > 0$, there exists $N \in \mathbb{N}$ such that for every fixed $n \geq N$, we have

$$\left|\sum_{k=n}^{2n-1} a_k \sin kx\right| < \varepsilon \quad \text{for any } x \in \mathbb{R} .$$

Pick $x_0 = \frac{1}{2n}$. For $k = n, \ldots, 2n-1$, we have $\frac{1}{2} \leq kx_0 \leq \frac{2n-1}{2n}$ and so $\sin\frac{1}{2} \leq \sin kx_0 \leq \sin 1$. Hence

$$\begin{aligned} \varepsilon > \left|\sum_{k=n}^{2n-1} a_k \sin kx_0\right| &= \sum_{k=n}^{2n-1} a_k \sin kx_0 \\ &\geq \sum_{k=n}^{2n-1} a_{2n} \sin\frac{1}{2} = \left(\frac{1}{2}\sin\frac{1}{2}\right) 2na_{2n} . \end{aligned}$$

Since this is true for all $n \geq N$, this shows $\{2na_{2n}\} \to 0$ as $n \to \infty$. Similarly, we also have $\{(2n-1)a_{2n-1}\} \to 0$ as $n \to \infty$. Combining, we have $\{na_n\} \to 0$ as $n \to \infty$.

($\Leftarrow$): Suppose $na_n \to 0$ as $n \to \infty$. Observe first that in order to show the uniform convergence of $\sum_{n=1}^{\infty} a_n \sin nx$ on $\mathbb{R}$, it suffices to show the uniform convergence of the series on $[0, \pi]$, or in turn the uniform Cauchyness of the series in $[0, \pi]$.

So let $\varepsilon > 0$ be given. For any $n, p \in \mathbb{N}$, we need to show that

$$\left|\sum_{k=n+1}^{n+p} a_k \sin kx\right| < \varepsilon \quad \text{for any } x \in [0,\pi] \text{ whenever } n \gg .$$

Now for $x \in \left[0, \dfrac{\pi}{n+p}\right]$, we have

$$\begin{aligned} \left|\sum_{k=n+1}^{n+p} a_k \sin kx\right| &= \sum_{k=n+1}^{n+p} a_k \sin kx \leq \sum_{k=n+1}^{n+p} a_k kx \\ &\leq \left[\sum_{k=n+1}^{n+p} ka_k\right] \frac{\pi}{n+p} . \end{aligned}$$

On the other hand, for $x \in \left[\frac{\pi}{n+p}, \pi\right]$, by writing $m := \left[\frac{\pi}{x}\right]$ and using the convention that $\sum_{k=\alpha}^{\beta} z_k := 0$ whenever $\alpha > \beta$, we have

$$\begin{aligned}
&\left|\sum_{k=n+1}^{n+p} a_k \sin kx\right| \\
&\le \sum_{k=n+1}^{m} a_k \sin kx + \left|\sum_{k=m+1}^{n+p} a_k \sin kx\right| \\
&\le \sum_{k=n+1}^{m} a_k \sin kx + \left|\sum_{k=m+1}^{n+p} \frac{(\sin \frac{x}{2})\, a_k \sin kx}{\sin \frac{x}{2}}\right| \\
&\le \sum_{k=n+1}^{m} a_k \sin kx + \frac{1}{\sin \frac{x}{2}} \left|\sum_{k=m+1}^{n+p} a_k \sin \frac{x}{2} \sin kx\right| .
\end{aligned}$$

Note that

$$\begin{aligned}
&\left|\sum_{k=m+1}^{n+p} a_k \sin \frac{x}{2} \sin kx\right| \\
&= \frac{1}{2}\left|\sum_{k=m+1}^{n+p} \left\{a_k \cos\left[(k-\frac{1}{2})x\right] - a_k \cos\left[(k+\frac{1}{2})x\right]\right\}\right| \\
&= \frac{1}{2}\left|a_{m+1}\cos\left[(m+\frac{1}{2})x\right] + \sum_{k=m+1}^{n+p-1}(a_{k+1}-a_k)\cos\left[(k+\frac{1}{2})x\right]\right. \\
&\quad \left. -a_{n+p}\cos\left[(n+p+\frac{1}{2})x\right]\right| \\
&\le \frac{1}{2}\left\{a_{m+1}\left|\cos\left[(m+\frac{1}{2})x\right]\right|\right. \\
&\quad + \sum_{k=m+1}^{n+p-1}(a_k - a_{k+1})\left|\cos\left[(k+\frac{1}{2})x\right]\right| \\
&\quad \left. + a_{n+p}\left|\cos\left[(n+p+\frac{1}{2})x\right]\right|\right\} \\
&\le \frac{1}{2}\left\{a_{m+1} + \sum_{k=m+1}^{n+p-1}(a_k - a_{k+1}) + a_{n+p}\right\} = a_{m+1}
\end{aligned}$$

and so

$$\begin{aligned}\left|\sum_{k=n+1}^{n+p} a_k \sin kx\right| &\leq \sum_{k=n+1}^{m} a_k \sin kx + \frac{a_{m+1}}{\sin \frac{x}{2}} \\ &\leq \sum_{k=n+1}^{m} a_k\, kx + a_{m+1} \cdot \frac{x}{\pi} \\ &\leq \sum_{k=n+1}^{m} a_k\, kx + a_{m+1}(m+1)\,,\end{aligned}$$

where the last two inequalities follows from the facts that $\sin y \leq y$ for all $y \geq 0$, $\sin y \geq \frac{2y}{\pi}$ for all $y \in [0, \frac{\pi}{2}]$, and $m = \left[\frac{\pi}{x}\right]$.
This suggests that we take $N \in \mathbb{N}$ to be large enough such that

$$|na_n| = na_n < \frac{\varepsilon}{2(\pi+1)} \quad \text{for any } n \geq N\,.$$

Then for $x \in [0, \frac{\pi}{n+p}]$, we have

$$\begin{aligned}\left|\sum_{k=n+1}^{n+p} a_k \sin kx\right| &\leq \left(\sum_{k=n+1}^{n+p} ka_k\right) \frac{\pi}{n+p} \\ &\leq p \cdot \frac{\varepsilon}{2(\pi+1)} \cdot \frac{\pi}{n+p} < \varepsilon\end{aligned}$$

for any $n \geq N$. On the other hand, for $x \in [\frac{\pi}{n+p}, \pi]$, we have

$$\begin{aligned}\left|\sum_{k=n+1}^{n+p} a_k \sin kx\right| &\leq \sum_{k=n+1}^{m} a_k kx + a_{m+1}(m+1) \\ &\leq \begin{cases} (m-n)\left[\frac{\varepsilon}{2(\pi+1)}\, x\right] + \frac{\varepsilon}{2(\pi+1)} & \text{if } m > n \\ \frac{\varepsilon}{2(\pi+1)} & \text{if } m \leq n \end{cases} \\ &\leq mx\, \frac{\varepsilon}{2(\pi+1)} + \frac{\varepsilon}{2(\pi+1)} \\ &\leq \pi\, \frac{\varepsilon}{2(\pi+1)} + \frac{\varepsilon}{2(\pi+1)} < \varepsilon\end{aligned}$$

for any $n \geq N$. Combining, we see that $\sum_{n=1}^{\infty} a_n \sin nx$ is uniformly Cauchy in $[0, \pi]$ and hence uniformly convergent there.

16. For any $n \in \mathbb{N}$, let $f_n : (0, \infty) \to \mathbb{R}$ be defined by $f_n(x) = \dfrac{1}{n^x}$, $x \in (0, \infty)$. Let $a > 1$.
    (a) Show that $\sum_{n=1}^{\infty} f_n(x)$ converges uniformly to a function ζ on $[a, \infty)$, called the *Riemann-zeta function.*
    (b) Prove that $\sum_{n=1}^{\infty} f_n'(x)$ converges uniformly on $[a, \infty)$.
    (c) Prove that ζ is differentiable in $[a, \infty)$.
    (d) Does $\sum_{n=1}^{\infty} f_n(x)$ converge uniformly on $[1, \infty)$?

    Proof.

    (a) For any $x \in [a, \infty)$, consider the function $\varphi : (0, \infty) \to \mathbb{R}$ defined by $\varphi(t) := \dfrac{1}{t^x}$, $t \in (0, \infty)$. Clearly, φ is decreasing and so we have

    $$\int_{n-1}^{n} \frac{1}{t^x}\, dt \geq \frac{1}{n^x} \cdot \int_{n-1}^{n} 1\, dt = \frac{1}{n^x} \ .$$

    For any $\varepsilon > 0$, choose $N \in \mathbb{N}$ large enough such that $N^{a-1} > \dfrac{1}{(a-1)\varepsilon}$. Then for any $n \geq N$, $p \in \mathbb{N}$, and any $x \in [a, \infty)$, we have

    $$\begin{aligned} \left| \sum_{k=n+1}^{n+p} \frac{1}{k^x} \right| &\leq \sum_{k=n+1}^{n+p} \int_{k-1}^{k} \frac{1}{t^x}\, dt = \int_{n}^{n+p} \frac{1}{t^x}\, dt \\ &\leq \int_{n}^{n+p} \frac{1}{t^a}\, dt = \frac{n^{1-a} - (n+p)^{1-a}}{a-1} \\ &< \frac{n^{1-a}}{a-1} \leq \frac{N^{1-a}}{a-1} < \varepsilon \ . \end{aligned}$$

    Hence the series $\sum_{n=1}^{\infty} f_n(x)$ is uniformly Cauchy in $[a, \infty)$ and so by Cauchy's Criterion, it is uniformly convergent there.

    (b) It is elementary to see that $f_n'(x) = -n^{-x} \ln n$ for any $x \in (0, \infty)$. It is also easy to check that the function $\psi(t) :=$

$t^{-x}\ln t$ is decreasing for $t \in (e^{1/a}, \infty)$. Thus, for any $n \geq e^{1/a}+1$ and any $x \in (0,\infty)$, we have

$$\int_{n-1}^{n} t^{-x}\ln t dt \geq n^{-x}\ln n \cdot \int_{n-1}^{n} 1\,dt = \frac{\ln n}{n^x}\,.$$

Moreover, it is elementary to verify that the function $h(t) := t^{1-a}\ln t$ is decreasing for $t \in (e^{1/(a-1)}, \infty)$ and $\lim_{t\to\infty} h(t) = 0$. So for any $\varepsilon > 0$, we can choose $N \in \mathbb{N}$ large enough such that $N > e^{1/(a-1)}$ and $(a-1)h(N) + N^{1-a} < (a-1)^2\varepsilon/2$. Therefore, for any $n \geq N$, $p \in \mathbb{N}$, and any $x \in [a,\infty)$, we have

$$\begin{aligned}
\left|\sum_{k=n+1}^{n+p} \frac{\ln k}{k^x}\right| &\leq \sum_{k=n+1}^{n+p} \int_{k-1}^{k} t^{-x}\ln t\,dt = \int_{n}^{n+p} t^{-x}\ln t\,dt \\
&\leq \int_{n}^{n+p} t^{-a}\ln t dt \\
&= \left(\frac{t^{1-a}\ln t}{1-a} - \frac{t^{1-a}}{(1-a)^2}\right)\Bigg|_{t=n}^{t=n+p} \\
&< \frac{n^{1-a}\ln n}{a-1} + \frac{n^{1-a}}{(a-1)^2} \\
&\leq \frac{N^{1-a}\ln N}{a-1} + \frac{N^{1-a}}{(a-1)^2} \\
&= \frac{(a-1)h(N) + N^{1-a}}{(a-1)^2} < \varepsilon\,.
\end{aligned}$$

Hence $\sum_{n=1}^{\infty} f_n'(x)$ is uniformly Cauchy in $[a,\infty)$ and so by Cauchy's Criterion, it is uniformly convergent there.

(c) As (a) and (b) hold in $[a,\infty)$ for any $a > 1$, they also hold in $[1+\frac{a-1}{2},\infty)$. In particular, $\sum_{n=1}^{\infty} f_n'$ converges uniformly and also, $\zeta = \sum_{n=1}^{\infty} f_n$ uniformly in the open set $(1+\frac{a-1}{2},\infty)$. Hence, by Theorem 5.2.6, ζ is differentiable in $(1+\frac{a-1}{2},\infty)$ and so it is differentiable in $[a,\infty)$. Furthermore, by Theorem 5.2.6 again, we have $\zeta' = \sum_{n=1}^{\infty} f_n'$ uniformly on $[a,\infty)$.

(d) No, $\sum_{n=1}^{\infty} f_n(x)$ does not converge, even pointwisely, on $[1,\infty)$. In fact, the series is not convergent at $x = 1$:

$$\sum_{n=1}^{\infty} f_n(1) = \sum_{n=1}^{\infty} \frac{1}{n}$$

is divergent.

Bibliography

[1] Ansari, Q.H., *Metric Spaces*, Alpha Science, Oxford, 2010.

[2] Apostol, T., *Mathematical Analysis*, 2nd ed., Addison Wesley, Reading, Massachusetts, 1974.

[3] Kelley, J.L., *General Topology*, Springer-Verlag, New York, 1955.

[4] Kumaresan, S., *Topology of Metric Spaces*, Alpha Science, Harrow, Middlesex, 2005.

[5] Lipschutz, S., *General Topology*, Schaum's Outline Series, Int'l ed., McGraw-Hill, Singapore, 1965.

[6] Rudin, W., *Principles of Mathematical Analysis*, 3rd ed., McGraw-Hill, Singapore, 1976.

[7] Searcóid, M., *Metric Spaces*, Springer, London, 2007.

[8] Shirali, S., and Vasudeva, H.L., *Metric Spaces*, Springer, London, 2006.

[9] Simmons, G.F., *Introduction to Topology and Modern Analysis*, McGraw-Hill, Tokyo, 1963.

Index